Oxford Lecture Series in
Mathematics and its Applications 4

*Series editors*
John Ball Dominic Welsh

OXFORD LECTURE SERIES
In Mathematics and its Applications

1. J. C. Baez (ed.): *Knots and quantum gravity*
2. I. Fonseca and W. Gangbo: *Degree theory in analysis and applications*
3. P. L. Lions: *Mathematical topics in fluid mechanics, Vol. 1: Incompressible models*
4. J. E. Beasley (ed.): *Advances in linear and integer programming*

# Advances in Linear and Integer Programming

Edited by

J. E. Beasley

*The Management School*
*Imperial College, London*

CLARENDON PRESS • OXFORD

1996

*Oxford University Press, Walton Street, Oxford OX2 6DP*
*Oxford New York*
*Athens Auckland Bangkok Bogota Bombay*
*Buenos Aires Calcutta Cape Town Dar es Salaam*
*Delhi Florence Hong Kong Istanbul Karachi*
*Kuala Lumpur Madras Madrid Melbourne*
*Mexico City Nairobi Paris Singapore*
*Taipei Tokyo Toronto*
*and associated companies in*
*Berlin Ibadan*





*Published in the United States by*
*Oxford University Press Inc., New York*



*A catalogue record for this book is available from the British Library*

*Library of Congress Cataloging in Publication Data*
*(Data available)*

*ISBN 0 19 853856 1*

*Typeset by the author*
*Printed in Great Britain by*
*Biddles Ltd, Guildford and King's Lynn*

# Preface

Linear and integer programming are mathematical techniques that are concerned with optimization, that is with finding the best possible answer to a problem. They are often associated with the wider field of operations research. They have been studied and researched since the late 1940s and elements of them are now taught in undergraduate and graduate programmes in mathematics/operations research world-wide.

In recent years, after the advent of interior point methods, there has been an explosion of research into linear programming, as well as further steady advances in integer programming. This research has been reported, as one would expect, piece by piece in the research literature, i.e. at conferences and in journals. The reason for assembling this book was to bring together in a single text an accessible exposition of these advances.

With contributions from acknowledged experts in their field this book deals with:

- simplex algorithms for linear programming
- interior point methods for linear programming, from both a theoretical and a computational viewpoint
- interior point algorithms for network flow problems
- branch and cut algorithms for integer programming
- interior point algorithms for integer programming
- computational logic and integer programming.

Whilst some may read this book whole it is likely that the majority of readers will be most interested in particular chapters. For this reason each chapter has been written so that it can be read and studied separately.

## Acknowledgements

I would like to acknowledge the efforts of all the contributors to this book. Their patience and prompt replies to requests have been much appreciated. I would also like to thank Paul C. Chu for his help with assembling the LaTeXversion of this book. All errors that are found are, however, my responsibility.

Finally I would like to acknowledge the support of my wife and children who, as Christmas approaches, are probably hoping to see more of me than they have recently!

John Beasley
London, December 1995

# Contents

# Contributors

**John Beasley** has a BA in Mathematics from Cambridge University and an MSc and PhD in Management Science from Imperial College, London. His research interests are in linear optimization (principally integer programming/combinatorial optimization) and he is the author of some fifty papers.

**Sally Brailsford** obtained a BSc in Mathematics from King's College London and then qualified as a professional Russian translator. She has followed a varied career, including freelance translation of Russian mathematical journals and training as a nurse. Returning to academic study she obtained an MSc in Operational Research at the University of Southampton, followed by a PhD in modelling for HIV and AIDS. She then worked with Paul Williams on a research project concerned with the applications of logic to integer programming. She is currently working as a lecturer at the University of Southampton. Her research interests include integer programming, modelling for health care, simulation and goal programming.

**Jacek Gondzio** has an engineering degree and a PhD in Electronics, both from the Department of Electronics, Warsaw University of Technology. He is an Assistant Professor at the Systems Research Institute of the Polish Academy of Sciences, Warsaw, Poland. He is currently visiting the Department of Management Studies of the University of Geneva, Switzerland. His research interests are the theory and implementation of large-scale optimization methods. He has been involved in the development of the simplex, simplex-type and interior point methods for linear programming. He has published several articles in leading journals.

**Abilio Lucena** is currently an Associate Researcher at the National Laboratory for Scientific Computing (LNCC/CNPq) in Rio de Janeiro. He has BSc and MSc degrees from the Department of Electrical Engineering in the Catholic University of Rio de Janeiro and a PhD in Management Science from Imperial College, London. His research interests involve combinatorial optimization and mathematical programming and he has published in a number of leading journals.

**István Maros** has an MSc in Mathematics from Eötvös University, Budapest, Hungary and a PhD in Operations Research from the Hungarian Academy of Sciences, Budapest. He is currently a Visiting Professor at Brunel University. His research interests are large-scale computational optimization with his primary interest being in continuous LP, structure-exploiting specialized algorithms, performance analysis and algorithm animation. He has published over seventy papers.

**John Mitchell** has a BA in Mathematics from Cambridge University and an MSc and PhD in Operations Research and Industrial Engineering from Cornell University. He is currently an Associate Professor in the Department of Mathematical Sciences at Rensselaer Polytechnic Institute. His research interests lie in optimization. In particular he is interested in interior point methods for linear programming and in linear programming approaches for solving integer programming problems, such as cutting planes and branch and bound algorithms. He is also interested in the use of integer programming methods to solve problems in logic. He has published over twenty papers.

**Gautam Mitra** is a Professor of Computational Optimization and Modelling in the Department of Mathematics and Statistics at Brunel University. He has a PhD in Computational Methods in Operational Research from the Institute of Computer Science, University of London. His research interests cover the solution of linear programming problems using the sparse simplex method and the interior point method, and implementing algorithms in software systems using serial and parallel computers. He is also interested in the interaction of logic and integer programming and mathematical programming modelling systems. He has authored and edited five books and over sixty papers.

**Panos Pardalos** has a BS degree in Mathematics from the University of Athens, Greece and a PhD degree in Computer Sciences from the University of Minnesota, USA. He is currently co-director of the Center for Applied Optimization and Professor of Industrial and Systems Engineering at the University of Florida. He is editor-in-chief of the *Journal of Global Optimization*, has researched in global and combinatorial optimization and published over one hundred papers and fifteen books in these areas.

**Mauricio Resende** is a Member of Technical Staff at the Mathematical Sciences Research Center at AT&T Bell Laboratories, in Murray Hill, New Jersey. He has an Electrical Engineering degree from the Catholic University of Rio de Janeiro, an MSc in Operations Research from Georgia Institute of Technology, and a PhD in Operations Research from the University of California at Berkeley. His research interests include combinatorial optimization, design and analysis of computer algorithms, graph theory, interior point methods, mathematical programming, network flows, network design, operations research modelling, parallel computing in mathematical programming, scientific computing, and software design and development. He has published over thirty papers.

**Cornelis Roos** has an MSc and a PhD from Delft University of Technology, Delft, The Netherlands. He is an Associate Professor in the Department of Applied Mathematics and Computer Science, Delft University of Technology. His current research interests are complexity analysis of interior point algorithms for linear and non-linear programming, mainly in cooperation with Tamás Terlaky and Jean-Philippe Vial. Recent research has been concerned with target-following methods for linear and non-linear (convex) problems. He is the author of more than eighty papers.

**Tamás Terlaky** has an MSc and a PhD from Eötvös University, Budapest, Hungary. He is an Assistant Professor in the Department of Applied Mathematics and Computer Science, Delft University of Technology, Delft, The Netherlands, and is on leave from the Department of Operations Research, Eötvös University, Budapest, Hungary. His current research mainly concerns interior point algorithms for linear and non-linear programming. He also has a long-standing interest in pivoting methods for linear and quadratic programming, linear complementarity problems, and oriented matroid programming. He has contributed to the duality theory of $l_p$ programming, proposed algorithms for the modular design problem, the channel capacity and the colour matching problem. He has published more than eighty papers.

**Jean-Philippe Vial** has an engineering degree from Ecole Centrale, Paris, France, a PhD in Operations Research from University of Louvain, Belgium and a Doctorat es-Sciences (France). He is currently Professor of Operations Management at the University of Geneva. His research interests are mainly in the field of optimization, both theory and applications. He has been very much involved in the early developments of interior point methods and in the application of these methods to non-differentiable optimization. He has published fifty papers in leading journals.

**Paul Williams** has a BA and MA from Cambridge University in Mathematics and a PhD in Mathematical Logic from Leicester University. He is currently Professor of Operational Research at Southampton University. His interest in mathematical programming began when he was at IBM working on the development of software and helping clients build models. His major interest is in integer programming. He was one of the first people to recognize the advantages of improved modelling techniques and the connections with mathematical logic. He has published over fifty papers and books.

# 1
# Simplex algorithms

**István Maros**

*Brunel University, Uxbridge UB8 3PH, England*
*Computer and Automation Institute, Hungarian Academy of Sciences, Budapest, Hungary*

**Gautam Mitra**

*Brunel University, Uxbridge UB8 3PH, England*

## 1 Introduction

Linear inequality systems were studied by Fourier, Motzkin and Farkas (see [14]) but only in the late 1940s did the work of Dantzig in the USA and Kantorovich in the former USSR make linear programming a leading research topic with wide ranging applications to problems of planning and scheduling as found in business, industry and government [14]. Dantzig developed the celebrated *simplex method* (in the tableau form) for finding the maximum or minimum of a linear function subject to linear restrictions. The resulting optimization problem because of its many planning applications was called linear programming (LP). Since that time there has been considerable curiosity driven research, as well as applicable research, on the topic of LP.

In Table 1 we have highlighted the major landmarks in theoretical developments. In Table 2 we have summarized some of the major computational developments.

We would, however, like to outline how a number of important research topics sprung out of LP and the simplex method and then became mature in their own right. All through the 1960s and 1970s the industrial application of equation solving technologies of large sparse systems continued to gain importance and applicability. The basis factorization (inversion) strategies of the simplex method were studied with considerable interest. A series of special conferences on the topic of sparse matrices and sparse equation solving methods was started during this period (Willoughby *et al.* [63, 69], Reid [60], Bunch and Rose [10], Duff and Stewart [19]). Today there is a wide body of literature which covers these developments. This has made a major contribution to the now established field of computational methods for large sparse systems where data structures, software implementation and exploitation of machine architecture continue to be leading research issues.

It can be claimed that the study of computational complexity grew out of

**Table 1** Landmark contributions in theoretical developments of the simplex method

| | Topic Area | Main Researchers | Date |
|---|---|---|---|
| 1. | Primal simplex method | Dantzig | 1947 |
| 2. | Duality theory and dual simplex algorithm | Lemke | 1954 |
| 3. | Degeneracy and cycling in LP | Beale | 1955 |
| | Lexicographic rules for anti-degeneracy | Dantzig, Orden, Wolfe | 1955 |
| | | Wolfe | 1963 |
| | Other approaches | Bland | 1977 |
| 4. | Complexity of simplex algorithms | | |
| | Worst case behaviour | Klee, Minty | 1972 |
| | Probabilistic/average behaviour | Borgwardt, Smale | 1982 |
| 5. | Crossover: recovering an optimal basis from a pair of primal and dual optimal solutions | Megiddo | 1991 |

the intractability of integer linear programs in contrast to linear programs. The algorithmic behaviour of the simplex has been well understood. Its average behaviour and worst case behaviour have been studied and explained by Borgwardt [6] and Klee and Minty [44], respectively. Karp in his Turing award lecture in [42] explained the central role occupied by LP and its solution methods in the area of computational complexity or in more practical terms the scalability of algorithms.

Indeed Khachiyan [43] was motivated to find a (worst case) polynomial algorithm for LP, however, computationally it performed very poorly. Karmarkar's work [41] and the subsequent development of the interior point method (IPM) resulted from his motivation to find a method which can dominate the simplex method computationally both in its average behaviour and in the worst case. The overwhelming computational success of IPM in turn strongly challenged the simplex algorithm designers who responded with a whole series of extensions and improvements. As a result the simplex algorithm continues to be the chosen method within software systems designed to solve large LP problems of widely varying structures. We believe simplex will continue to be widely used for the following reasons:

(1) There are many industrial applications in which "warm start" from an earlier solution of a slightly perturbed neighbouring LP problem provides considerable computational advantage. In these cases IPM performs poorly in contrast to simplex. Successive linear programming to solve non-linear programming problems in the oil industry is an example of such an application. Solution of subproblems in integer programming and post optimality analysis of LPs provide other examples of repeated use of warm start.

**Table 2** Landmark contributions to the simplex-based computational solution methods

| | Topic Area | Main Researchers | Date |
|---|---|---|---|
| 1. | Tableau simplex method | Dantzig | 1947 |
| 2. | Revised simplex method | Dantzig, | |
| | | Orchard-Hays, Wolfe | 1953/54 |
| 3. | Simple upper bound algorithm | Dantzig | 1954 |
| 4. | Generalized upper bound algorithm | Dantzig, Van Slyke | 1967 |
| 5. | Basis factorization and the elimination | Markowitz | 1954 |
| | form of the inverse (EFI) | Beale | 1971 |
| | | Hellerman, Rarick | 1971/72 |
| 6. | Sparse update procedures | | |
| | Bartels, Golub | Bartels, Golub | 1969 |
| | Forrest, Tomlin | Forrest, Tomlin | 1972 |
| | Reid | Reid | 1976 |
| 7. | Presolve procedures | Brearley, | |
| | | Mitra, Williams | 1975 |
| 8. | Composite Phase-I procedures | Wolfe | 1965 |
| | | Maros | 1986 |
| 9. | Combined price and pivot choice | | |
| | DEVEX procedure | Harris | 1973 |
| | Steepest edge pricing | Goldfarb, Reid | 1977 |
| | | Forrest, Goldfarb | 1992 |

(2) Shadow prices (costs) play a key role in the descriptive analysis of many economic planning and business applications of LP. In these situations the established theory relating to optimal basis simplex multipliers and dual solution values have wide acceptance. This in turn requires that the simplex method is used to compute these items of (economically) meaningful descriptive information.

The simplex method has undergone very substantial improvement over the past two decades. These improvements are mainly computational and have the sole purpose of making the simplex solution method efficient as well as reliable in respect of solving a wide range of practical problems. This chapter sets out a summary account of these developments.

The rest of the chapter is organized in the following way. In Section 2 we state the linear programming problem and introduce different types of variables. In Section 3 the revised simplex method is set out in a summary form. Section 4 introduces the sparse simplex (SSX) method and the algorithmic developments relating to preprocessing, starting basis, pricing, degeneracy, inverse represen-

**Table 3** Types of variables

| Feasibility range | Type | Reference |
|---|---|---|
| $x_j = 0$ | 0 | Fixed variable |
| $0 \leq x_j \leq u_j < +\infty$ | 1 | Bounded variable |
| $0 \leq x_j \leq +\infty$ | 2 | Non-negative variable |
| $-\infty \leq x_j \leq +\infty$ | 3 | Free variable |

tation and sparse updates. In short, we consider the major developments in the algorithmic components of SSX. In Section 5 we cover the computational issues relating to SSX. Thus we outline the information flow and computational use of SSX, we present an analysis of benchmark problems and provide profiling information for benchmark problems in respect of the major algorithmic components. We also set out comparative performance summaries of the two leading solvers, FortMP and MINOS, offered by academic institutions and two full strength commercial solvers, OSL and CPLEX. In Section 6 we summarize the recent extensions of SSX for parallel computing architecture and we also discuss the related difficulties and explain why there is very limited development in this field. We consider the integration of IPM and SSX in Section 7 and outline our view of future research directions in Section 8.

## 2 Problem statement

### 2.1 The primal problem

We consider the following primal linear programming (LP) problem:

$$\begin{aligned} \text{minimize} \quad & c^T x, \\ \text{subject to} \quad & Ax = b, && (2.1) \\ & l \leq x \leq u, && (2.2) \end{aligned}$$

where $A$ is an $m \times n$ matrix, $c$, $x$, $l$, and $u$ are $n$ vectors, and $b$ is an $m$ vector. Some or all components of $l$ and $u$ can be $-\infty$ or $+\infty$, respectively. For computational convenience, it is assumed that all finite lower bounds in (2.2) are translated to zero. $A$ itself contains a unit matrix $I$, that is, $A = [I, \bar{A}]$, so it is of full row rank. Variables which multiply columns of $I$ transform every constraint to an equation and are often referred to as *logical variables.* Variables which multiply columns of $\bar{A}$ are called *structural variables.*

Based on the relation in (2.2), the variables (whether logical or structural) can be categorized as shown in Table 3 (for further details, see [58]).

An $x$ vector that satisfies (2.1) is called a *solution.* If, additionally, it satisfies (2.2) it is called a *feasible solution.*

LP systems allow for other types of variables and constraints. Namely, a variable can have infinite lower bound and finite upper bound (*minus type variable*). Such variables are transformed during the setup phase into type-2, that is, non-negative variables with zero lower bound. Similarly, a constraint can be of "greater-than-or-equal" ($\geq$) type to which a non-positive logical variable should be added to make it an equality. This is avoided by converting the constraint into "less-than-or-equal" ($\leq$) type and adding a non-negative logical variable to it to obtain the form of (2.1) and (2.2).

Any $m \times m$ non-singular submatrix $B$ selected from the columns of $A$ is called a *basis* (at least one exists because $A$ is of full row rank). Variables corresponding to the columns of $B$ are called *basic variables,* while the remaining variables are called *non-basic variables.* For the purpose of convenience we use the term basis to refer both to the set of basic variables and to the set of indices of those variables. Without loss of generality, we can assume that the columns of the given $A$ are permuted such that the columns of the basis matrix are in the first $m$ positions of the matrix. Thus the matrix $A$ is partitioned as $[A_B, A_N]$, or equivalently, $[B, N]$ (with respective interpretation of $N$); the components of the solution vector are then partitioned accordingly and $x$ is expressed as $x = [x_B, x_N]$. With this notation, (2.1) can be written in the following form:

$$\begin{aligned} A_B x_B + A_N x_N &= b \\ \text{or} \quad B x_B + N x_N &= b. \end{aligned} \tag{2.3}$$

Clearly, the values of the non-basic variables uniquely determine the values of the basic variables:

$$x_B = B^{-1}(b - N x_N) \tag{2.4}$$

This $x_B$ is often referred to as $\beta$.

In the upper bounded simplex method the non-basic variables can take a value equal to their lower or (finite) upper bound. (It means that the non-basic variables always satisfy (2.2), that is, they are always feasible.) We call the corresponding $x_B$ a *basic solution.*

The simplex method is an iterative procedure that starts with a basic (not necessarily feasible) solution and moves over to a new, neighbouring, basis that is different from the previous one in one column only and has an objective value not worse than the objective value at the previous basis. Such a step implies that it is necessary to select a column (and the associated variable) to enter the basis and to select another column (and the associated variable) to leave the basis. Usually there is a large number of such neighbouring bases indicating that there is large freedom in determining the next iteration. The total number of iterations and also the total computational effort greatly depend on the way the basis changes are determined. Unfortunately, there are no known rules that specify these exchanges in such a way that would guarantee an overall best performance.

## 2.2 The dual problem

Duality plays a basic role in the theory and practice of linear programming. A slightly simplified (but mathematically equivalent) form of the primal problem of equations (2.1) and (2.2) is the case when all the variables are of type-2:

$$\begin{array}{lll} \text{minimize} & c^T x, & \\ \text{subject to} & Ax = b, & (2.5) \\ & x \geq 0. & (2.6) \end{array}$$

This is called the standard form. Its dual pair is:

$$\begin{array}{lll} \text{maximize} & b^T y, & \\ \text{subject to} & A^T y \leq c, & (2.7) \end{array}$$

where $y$ is an $m$-vector of type-3 dual variables which are unrestricted in their values $(-\infty < y < +\infty)$. It is often practical to use a slightly modified version of (2.7), where all inequalities are converted into equations. This form can be written as

$$\begin{array}{lll} \text{maximize} & b^T y, & \\ \text{subject to} & A^T y + z = c, & (2.8) \\ & z \geq 0, & (2.9) \end{array}$$

where $z$ is an $n$-vector of dual slack variables.

Though this chapter is not devoted to the detailed discussion of the theoretical aspects of duality, it is still important to state the basic theorem that establishes the link between the solutions of the primal and the dual (for proof, c.f. [14]).

**Theorem 2.1. (Duality theorem)** *(i) If any of the primal or dual problems have an optimal solution, then so does the other, and they have the same optimal solution, that is,* $\min c^T x = \max b^T y$. *(ii) If any of the problems has an unbounded solution, then the other has no feasible solution.*

Additionally, it is also true that if they both have an optimal solution then an optimal basis $B$ of the primal can be used to determine an optimal solution for the dual, namely in such a case

$$c_B^T x_B = b^T y = y^T b. \tag{2.10}$$

The left-hand side of the first equality can be further rewritten as $c_B^T x_B = c_B^T B^{-1} b$. Compared with the second equality of (2.10), this yields $y^T = c_B^T B^{-1}$.

Because of the symmetry between the primal and the dual, if we solve any of them, the solution for the other can easily be created as an outcome of the above observation. This fact has very important practical consequences, namely, in many situations the dual of a problem can be solved more efficiently than the primal and the corresponding dual simplex algorithm is the preferred solution method.

## 3 The revised simplex method

This section describes the main computational steps of the revised simplex method (RSM). It also sets out a framework for the discussion of the sparse simplex method. A theoretical discussion of the revised simplex method can be found in any of the LP textbooks. The interested reader is referred to, for example, [14]. The main steps of the RSM are summarized below.

(1) **Initialize**
In this step, a starting basis $B$ for problem (2.1)–(2.2) is determined. This can either be the unit matrix present in $A$ or an advanced basis obtained through the use of a heuristic method.

(2) **Factorize**
In a number of RSM steps operations have to be performed with the inverse $B^{-1}$ of the basis. To make these steps economical, the inverse or rather an equivalent form of $B^{-1}$ is determined. Using Gaussian elimination, we get $B = LU$, where $L$ is a lower, $U$ is an upper triangular matrix and the inverse is expressed as $B^{-1} = U^{-1}L^{-1}$. The inverses of the triangular matrices can be obtained at very little computational expense. It is useful to represent and store $U^{-1}$ and $L^{-1}$ as products of elementary elimination matrices that differ from the unit matrix in one column or in one row only. Additionally, the non-zeros of the non-trivial column (row) are located on one side of the diagonal.

(3) **Form (price) vector**
Depending on which phase (Phase-I: solution is infeasible, or Phase-II: solution is feasible) we are in, the (price) form vector $v$ is computed by two alternative rules. If the current basic solution is infeasible (Phase-I) then a distinction is made between categories of infeasibility. The index set of basic variables below their lower bound (i.e. negative, since all finite lower bounds have been translated to 0) is denoted by $M$ and the index set of variables above their upper bound (applicable only to type-0 and type-1 variables) is denoted by $P$. In Phase-I, $v_i$ is determined by the feasibility status of the $i$-th basic variable in the following way:

$$v_i = \begin{cases} 1 & \text{if } i \in M, \\ -1 & \text{if } i \in P, \\ 0 & \text{otherwise.} \end{cases}$$

If the current solution is feasible then $v_i = c_{Bi}, \; i = 1, \ldots, m$. Computationally, this step involves checking feasibility of the basis in order to establish the phase and to identify sets $M$ and $P$ when in Phase-I.

(4) **BTRAN**
The name BTRAN refers to *backward transformation*. The elementary elimination matrices are used in the reverse order of their creation to compute the $\pi$ simplex multipliers: thus $\pi^T = v^T B^{-1}$.

(5) **Price**
Theoretically this step is simple, however, for computational efficiency this can be one of the most intricate parts of RSM. Here, we check whether there is an improving vector or not which is determined by examining the sign of the computed reduced costs ($d_j$) of the non-basic variables. This is computed as $d_j = \pi^T a_{.j}$ in Phase-I and $d_j = c_j - \pi^T a_{.j}$ in Phase-II with $a_{.j}$ being the $j$-th column of $A$. To be an improving variable, if $x_j$ is at lower bound, then $d_j < 0$ is needed, if $x_j$ is at upper bound, then $d_j > 0$ is needed, while if $x_j$ is of type-3 $d_j \neq 0$ is sufficient. Type-0 variables can have $d_j$ of any sign, they never have to be priced. Any improving variable can be selected to enter the basis. The Dantzig criterion suggests selecting the one with largest $|d_j|$. If no improving variable can be found by the above criterion, the solution is optimal if it is detected in Phase-II, or no feasible solution exists if this occurs in Phase-I.

(6) **FTRAN**
Let $k$ be the index of the selected non-basic variable. In this step its coordinates in the current basis are computed by $\alpha_{.k} = B^{-1}a_{.k} = U^{-1}L^{-1}a_{.k}$. FTRAN refers to *forward transformation*, the sequence in which the elementary elimination matrices are used (namely, in the order of their creation) in computing $\alpha_{.k}$.

(7) **Choose pivot**
Sometimes this step is referred to as CHUZR (choose row) to express that a pivot row is chosen that determines the outgoing variable, that is, the one that leaves the basis. Choice of pivot row involves a set of logical steps and different procedures are employed in Phase-I and Phase-II. The selection in Phase-II is done in the following way.
The incoming variable $k$ can enter positively (type-1-2-3 variables with $d_j < 0$) or negatively ($d_j > 0$ for a type-1 variable at upper bound or a type-3 variable). Let us denote the corresponding direction by $\gamma = 1$ and $\gamma = -1$, respectively.

**For** $i = 1, \ldots, m$
  **If** the $i$-th basic variable is of type-3 **then** do nothing;
  **else** let $u_{Bi}$ denote the (finite or infinite) upper bound of the $i$-th basic variable and $d = \gamma\alpha_{ik}$. Compute

$$\theta_i = \begin{cases} +\infty & \text{if } d = 0, \\ \beta_i d & \text{if } d > 0, \\ (\beta_i - u_{Bi})/d & \text{if } d < 0. \end{cases}$$

  **endif**
**endfor**

If we compute

$$\theta = \min\{\min_i \theta_i,\ u_k\} \tag{3.1}$$

then the following cases may occur:

(a) $\theta = +\infty$. In this case the solution is unbounded.

(b) $\theta = u_k$. In this case no outgoing variable is determined, $x_k$ changes bound (goes from lower to upper bound or vice versa).

(c) $0 \leq \theta < u_k$. This situation results in a basis change. Namely, variable $k$ enters the basis and the basic variable for which the (finite) minimum in (3.1) was attained leaves the basis.

In Phase-I the situation is a bit more complicated since there are several different ideas that may guide the selection of the pivot row. The traditional method works in such a way that (i) the incoming variable takes a feasible value, (ii) the outgoing variable leaves the basis at a feasible level, and (iii) the feasible basic variables remain feasible (that is, the number of infeasibilities does not increase during an iteration). Any of these requirements can be relaxed resulting in more sophisticated but usually more efficient Phase-I algorithms (see [49, 71]). $\theta$ is often referred to as the *steplength* since this is the length the incoming variable moves.

(8) **Update**
This step completes the iteration cycle. If there was a change in the basis then the new basis inverse is determined. Depending on the representation this may mean a product form update (known as creation of an $\eta$ vector) or an $LU$ update. In both cases, the basic solution is updated. If there was no change in the basis then only the solution update is made. The algorithm returns to step (3) or step (2) depending on the need for refactorization.

# 4 The sparse simplex (SSX) algorithm and its developments

Before defining the sparse simplex algorithm (or rather the family of sparse simplex algorithms), a few words have to be said about the real-life problems. First of all, formally they can be categorized by size as small, medium and large problems. The real meaning of these categories has been changing all the time. The size is usually determined by three factors: (i) the number of rows (which is $m$ for $A$ and $\bar{A}$), (ii) the number of columns in $\bar{A}$ (that will be denoted by $n_{\bar{A}}$), and (iii) the number of non-zero elements (denoted by $K$) in $\bar{A}$. Presently (1996), it is accepted that if $n_{\bar{A}}$ is not much greater than $m$, problems with less than 1,000 rows are small, problems with $1{,}000 \leq m \leq 5{,}000$ are medium, and if the number of rows exceeds 5,000, problems are said to be large. However, as $m$ increases, the number of columns and non-zeros start playing an important role in categorizing the problem. Therefore, problems are best characterized by these three parameters: $(m, n_{\bar{A}}, K)$.

The purpose of this categorization is to get an initial estimate of the effort required to solve a problem. Based on the size, it is usually (but not always) possible to estimate if a problem can be solved with the memory and the processing power of a given computer.

Medium to large-scale LP problems tend to have very few non-zero entries in $\bar{A}$ in comparison with the total number of entries. The *density* of $\bar{A}$ is defined by $\rho = K/(m \times n_{\bar{A}})$. It is typical that medium to large-scale problems have a density between 0.05% and 1%. Problems with this low density are often referred to as *sparse* problems. There is an observation that for a given problem structure the average number of non-zeros per column is a small constant (typically between 5 and 10), therefore, larger instances of the same structure tend to be sparser.

It is a natural idea to design and implement solution procedures that exploit sparsity as much as possible. The revised simplex method seems ideal for these purposes since the $A$ matrix does not change during iterations. However, if the $B^{-1}$ matrix is kept in explicit form it usually fills up considerably during transformations. The fill-in of $B^{-1}$ easily leads to a fully dense matrix and there is no scope of sparse manipulation since the explicit representation of $B^{-1}$ is unique. In real terms it means that a problem with 5,000 rows would require 25 million non-zeros to compute and store for $B^{-1}$. This is hardly possible and definitely not efficient even with the present-day high-performance computers. This is why simplex implementations use exclusively the product form of the *LU* decomposition of the basis. It has the favourable feature that the number of non-zeros needed to represent the *LU* form can be kept under control during factorization. Additionally, triangularity can be maintained during simplex iterations using matrix algebra based algorithmic procedures and the number of newly generated non-zeros can be controlled.

Variants of the simplex method that explicitly take advantage of the sparsity of the problems are called *sparse simplex algorithms.* They are characterized by carefully designed data structures and such algorithmic components that are best suited for sparsity.

Though in the 1970s the simplex implementations were considered quite mature and "not much more can be done", with the emergence of efficient interior point methods (IPM) for linear programming a fierce competition started between SSX and IPM algorithm designers. Interestingly enough, it resulted in a vast improvement in all aspects of SSX. It is commonplace to say that SSX was a big beneficiary of the appearance of IPM. Nearly all the main steps of SSX have undergone a considerable improvement.

### 4.1 Data structures

As a result of the inherent sparsity of large-scale LP problems, $A$ is usually kept in a column-wise sparse form. This form can be as simple as in Fig. 1. There is a slightly modified version of RSM where the objective function is included in the $A$ matrix making some computational forms more unified in the case of the primal simplex algorithm. As algorithmic richness of the LP solvers increased it became clear that it is better to keep the objective row separate. At the same time, storage of $A$ has also become a bit more elaborate. Namely, columns of $A$ are stored in the linear array by a pointer to the starting element and the length (number of non-zeros) of the column. This scheme does not assume that columns are stored consecutively and enables more flexibility in deleting and adding elements to a column.

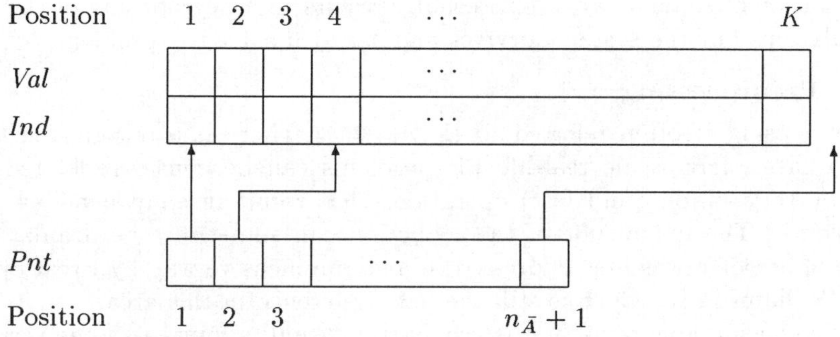


FIG. 1. A column-wise storage structure of $\bar{A}$

In most of the cases column-wise access of $A$ is sufficient. However, there are operations where row-wise access is more desirable. If there is sufficient memory available $A$ can be kept in both row and column-wise representation. The idea of row-wise representation is the same as that of column-wise.

The active *rim vectors* that are needed during RSM (objective row, right-hand side, and upper bounds, $c$, $b$, and $u$, respectively) are usually kept in explicit (dense) form.

Data structures for the *LU* decomposition are more complicated than those for $A$. This is because the corresponding representation needs to be changed (expanded) dynamically; the expansion can happen in two places: (i) fill-in during factorization, (ii) addition of new factors during simplex iterations. Additionally, factorization requires the storage of many pointers and lists (most of them are integer $m$ arrays), and occasionally row and/or column-wise representation of the basis. It is important that this data structure is carefully designed because the memory requirement of this step is the determining factor in the total memory required.

There is a playground for tradeoff between speed and memory and implementations can differ considerably in the way their data structures are organized. It is generally true that the more additional arrays that are used the more efficient the resulting solver is. However, this idea cannot be followed to the extreme. Though the memory of present-day computers (even 486 processor based PCs) is vastly expandable, this does not come for free, if possible at all. In a single user environment this is usually just a question of money because memory is still a serious cost factor. In a multiuser environment (workstations and upwards) memory is even more costly and installing more memory usually requires a corporate decision that may not be easy to obtain. As a result, it is still important to be reasonably memory-cautious with data structures. The best implementations work in such a way that they recognize the size of the available memory and as a function of it they dynamically use as many auxiliary arrays as possible. "Dynamically" means that during solution the rate of fill-in may require the elimination of some arrays. This usually results in a degradation of the further calculations but the system survives and can still solve the problem.

### 4.2 Preprocessing

Preprocessing is often referred to as *presolve*. There is a recognition that, in many cases, large-scale real-life LP problems can be considerably reduced in size by very simple (and fast) operations that result in equivalent but smaller problems. The magnitude of the reduction can sometimes be dramatic. The issue of presolve was first addressed in a comprehensive way by Brearley, Mitra and Williams in [7] which is still the basic reference in this area.

If a considerable reduction is achieved, it usually points to some weaknesses in formulating the model and it is therefore justifiable to claim that presolve, though it is not part of the simplex algorithm, can have an impact both on modelling and problem solving.

Basically, presolve applies simple primal and dual type tests on the rows and columns (logical and structural variables) of the problem and identifies (i) joint constraints that can be replaced by individual bounds on variables, (ii) constraints that are redundant and can be removed from the problem, and (iii) variables that can be fixed at a certain level and thus eliminated.

The main idea of the primal tests is to determine the lower and upper values a constraint can take and use this information to define new bounds for row and column variables. As such, a *lower row sum* for constraint $i$ is computed by

$$\underline{R}_i = \sum_{a_{ij}>0} a_{ij} l_j + \sum_{a_{ij}<0} a_{ij} u_j,$$

if none of the $l_j$ and $u_j$ values in the summations is infinite. $\underline{R}_i$ is the smallest value the row can take with the current bounds. The *upper row sum* is defined in a similar sense:

$$\overline{R}_i = \sum_{a_{ij}>0} a_{ij} u_j + \sum_{a_{ij}<0} a_{ij} l_j.$$

$\overline{R}_i$ is the largest value the row can take with the current bounds. If any of these sums is finite it can be used to define new bounds on the variables with $a_{ij} \neq 0$. If a bound is tighter than the one used in the calculation of the row sums then it replaces the old one and for all rows that intersect this column row sums are recomputed (or more precisely, updated). This procedure is repeated until no tighter bound is found. Similar tests can be defined for the dual of the problem with consequences on the shadow prices of the variables which again can lead to new bounds on constraints and variables.

The above description provides a preliminary conceptual outline of the procedure. There are many refinements that lead to deeper levels of logical analysis and in turn achieve further reduction of the model.

An interesting consequence of these tests is that if bounds on a variable were tightened during the tests then this variable can be freed simply because its value is limited by other constraints more strictly than by its original bounds.

In some cases, presolve can "solve" the problem in as much as it detects the infeasibility or unboundedness of the problem. It is important that presolve should be fast since it is just the setup of the solution procedure. The (simple) tests described can be carried out with little computational effort.

Recently, presolve has been extended by the technique of aggregation. Aggregation means that if we have an equality constraint of type-2 variables of the form $\sum_{j,j\neq k} x_j - x_k = 0$, then $x_k$ can be expressed and substituted back into the remaining constraints thus decreasing the number of constraints and variables by one. If there are many constraints of this type then the reduction in the number of variables can be considerable. Unfortunately, this procedure generates fill-in in the matrix that may slow down the solution process even on the resulting smaller problem.

At the expense of additional effort, *dominated variables* and *dominated constraints* can also be identified and eliminated. Furthermore, more complicated linear dependency of rows/columns can also be identified. However, these procedures may involve elimination steps that slow down presolve, though they can be very effective in certain cases.

The steps of presolve are such that they create a smaller but mathematically equivalent problem. After this reduced problem is solved, the solution has to be expressed in terms of the original problem. Depending on the steps in presolve, this may or may not require *postsolve.* In other words in some cases it is possible to reinstate the eliminated/fixed constraints and variables to their value determined at presolve stage resulting in a primal-dual optimal basis and solution for the original problem. In other cases this procedure will not produce a primal-dual optimal basic solution. The objective value is usually optimal and the solution is primal feasible but "clean-up" steps are needed that are usually of dual type to recover dual feasibility of the original problem. From the primal point of view, these steps are degenerate because they do not change the value of the objective function.

### 4.3 Starting basis

Since the simplex method iterates on basic solutions a basis must be found so that the algorithm can start.

The textbook LP algorithms do not pay much attention to determining the starting basis. They usually create an equivalent problem to which a feasible starting basis is readily available. The simplest way to achieve this is to multiply all rows in (2.1) by $-1$ for which the corresponding $b_i$ element is negative and augment the problem by a unit matrix of *artificial variables* $x_{a1}, \ldots, x_{am}$. These variables are defined to be of type-2. In this way the all-artificial basis is a feasible starting basis to the modified problem. The objective at this stage is to drive all artificial variables to zero and make them non-basic in order to obtain feasibility in terms of the original variables of the problem. This can always be achieved if the problem has a feasible solution. The important thing is that in this situation the algorithm can work entirely with the Phase-II pivot rule since the basis is always feasible for the modified problem. There are some variants and refinements of this simple idea (see [3]) but the main point is that they all start with a basis that is rather unrelated to the variables of the original problem and therefore usually a large number of iterations (and more time) are needed to solve the problem.

The situation changes when we allow infeasible bases to start with. Of course, it requires the availability of a complete Phase-I algorithm. In such a case we can get started with the originally present all-logical basis which is usually not feasible. The advantage of this basis selection is that it does not require the creation of an enlarged problem and has some meaning in terms of the LP problem being solved.

However, it has long been recognized that nearly any starting basis containing some (or more) structural variables is better than the all-logical basis (see, e.g. [58]). Such a basis is often called an *advanced basis*. To measure the quality of the starting basis, we can consider the total solution time. A slightly simpler but equally accepted measure is the number of iterations in Phase-I. The importance of this measure lies in the fact that iterations in Phase-I are computationally more expensive than in Phase-II.

Interestingly enough, not much has been done to find algorithms that can determine good starting bases though there have been some attempts. Even the need for an advanced basis was not widely recognized. The argument was that a start from the all-logical basis is computationally very favourable since the first few iterations are fast and accurate. Recently, the solution of real large-scale problems has become routine. Experience revealed that for such problems the simple all-logical starting basis is not good enough.

Presently two main alternatives appear to be the most efficient replacements for the trivial all-logical basis. One of them, the *crash*, is actually a family of algorithms to find a starting basis. Its main idea is to include as many structural variables in the basis as possible in such a way that they form a triangular matrix. If it does not result in a matrix of full row rank then it is augmented by logical variables as necessary to form a non-singular triangular matrix that can be used as a starting basis for the simplex algorithm. Originally, this idea was described by Carstens in [11]. The other alternative, introduced by Gould and Reid in [31], can be considered as a generalization of the idea of triangular crash, namely, this is a sort of a *block triangular crash*. We discuss one symbolic triangular (see definition later) and the block triangular (which is a numerical) crash in some detail.

### 4.3.1 *Triangular crash*

Crash algorithms are mainly characterized by the fact that they create a triangular basis. As such, "Triangular" could be omitted from the title of this subsection. We still keep it to distinguish it from the block triangular crash which is discussed in the next subsection.

It is easy to see that two types of triangular $B$ bases can be defined in the LP context. In the first, the *lower triangular* case, the non-zeros of $B$ are located in and below the main diagonal. In the second, the *upper triangular* case, the non-zeros of $B$ are located in and above the main diagonal. In both cases the main diagonal is zero-free.

The simplest way to trace the possible triangular structure is to introduce row and column counts, $R_i$ and $C_j$, which are defined as the number of non-zeros in row $i$ of $\bar{A}$, and the number of non-zeros in column $j$ of $\bar{A}$, respectively.

The diagonal elements of a lower triangular basis have only zero elements on their right. Therefore, for the first diagonal element (upper left corner) one can select a pivot row $i$ such that $R_i = \min_k\{R_k\}$. The pivot column is uniquely determined if $R_i = 1$. However, if $R_i > 1$, a column with non-zero in row $i$ having the smallest column count has to be found. The selections are usually based on some additional criteria if there is a tie. The selected element becomes the current pivot and the pivot position is logically permuted (by row and column permutations) to the top left corner. All columns having a non-zero element in row $i$ are marked unavailable for further consideration. In this way, the remaining columns do not have to be transformed at later stages. Row and column counts are recomputed (updated) for the remaining rows and columns (the *active part*) of $\bar{A}$. These steps are repeated for the active part of $\bar{A}$. This simple procedure is the basis of the lower triangular crash procedures. The pivot order will be the natural order as the pivot positions have been identified.

An upper triangular (sub) matrix can be found in a similar way. The difference is that the role of rows and columns is interchanged.

In each step of the lower triangular procedure one row (the pivot row) and all the columns that intersect with this row are marked unavailable. In the upper triangular procedure one column (the pivot column) and all rows that intersect with this column become unavailable. The two procedures may find triangular parts of different size.

The *quality of the basis* is determined by several factors, such as density, condition number, feasibility, and the corresponding objective value. They, or their combination, play an important role in breaking row or column ties.

If a procedure does not take into account the actual numerical values of the coefficients but only utilizes the location of non-zeros, then we speak about a *symbolic crash* procedure. If the numerical values (or even if the signs) of the coefficients are taken into account then this is called a *numerical crash* procedure.

If in the above described framework we do not use any numerical values of the matrix at any stage of breaking the tie we are actually doing a symbolic crash. One may oppose this idea because there is no safeguard to prevent too small pivot elements appearing. However, it should be noted that in most of the cases the matrix is scaled and therefore the range of elements is reduced. The scaled elements are obtained by multiplicative operations carrying very small relative error (in the magnitude of the relative accuracy of floating point numbers). Thus, values after scaling can be considered free from generated arithmetic error. Therefore, a choice of a relatively small pivot does not necessarily lead to immediate numerical troubles, since it is more or less well understood that not necessarily the small but rather the inaccurate values are the main sources of numerical problems.

To illustrate how symbolic crash algorithms can be created, we describe a lower triangular symbolic crash procedure designed to find an advance basis that is as feasible as possible under the circumstances. This algorithm is identified as *CRASH(LTSF)*.

We proceed with the general procedure of the lower triangular symbolic crash as described above. If at every stage the selection is unique (i.e. there is a unique minimum row count and a unique minimum column count for the positions in the selected row) then this procedure is CRASH(LTSF). Evidently, in this case we are unable to give consideration to issues of feasibility.

In practice, this happens very rarely and there are multiple choices (ties) that have to be broken. The main idea is that if column selection for triangularity is not unique then we give preference to

- removal of logical variables with small feasibility range, and
- inclusion of structural variables with large feasibility range.

To investigate the usefulness of this algorithm, the method used in [3] can be followed, namely, observe the number of iterations in Phase-I as a measure of the power of a starting procedure. The results of a limited comparison on some of the more interesting problems in NETLIB (the standard library of real-life LP problems assembled for algorithm testing, see [27]) are shown in Table 4. Here, the CRASH(LTSF) column was obtained by LP systems FortMP (see [21]) and MILP (see [50]). We have implemented CRASH(LTSF) in these two systems. For comparison, relevant figures with the all-logical and the proprietary advanced starting procedure of CPLEX, as described in [3], are quoted. It is important to mention that the staring procedure of CPLEX falls into the category of numerical crash. Since the solution algorithms of FortMP and MILP are probably rather different from that of CPLEX even the number of Phase-I iterations must be viewed with caution. It is likely that the Phase-I algorithms in these three solvers are similar, at least in determining the outgoing variable. The same may not be true for Phase-II, therefore the total number of iterations would not demonstrate the real effect of the starting procedure.

The figures of Table 4 underline well that it is difficult to expect that one single algorithm dominates all the others on all problems. By additional refinements, more sophisticated and more efficient algorithms can be created in the category of symbolic crash procedures.

#### 4.3.2 *Block triangular crash*

The crash algorithm of Gould and Reid [31] generalizes the idea of triangularization in a straightforward way and tries to find a basis which is close to feasibility. It is a numerical crash procedure involving even the solution of a series of small dense LP problems. They introduce non-negative "undershoot" and "overshoot" variables $v_1, \ldots, v_m$ and $w_1, \ldots, w_m$, respectively, and consider the following problem:

$$\text{minimize} \qquad e^T(v + w),$$

$$\text{subject to} \qquad Ax + v - w = b,$$
$$0 \le x \le u, \; v \ge 0, \; w \ge 0,$$

where $e$ is a vector of 1's.

In the first step the $A$ matrix is permuted into a lower block triangular form:

$$A = \begin{bmatrix} A_{11} & & & & \\ A_{21} & A_{22} & & & \\ \vdots & \vdots & \ddots & & \\ A_{r1} & A_{r2} & \dots & A_{rr} & \\ A_{r+1,1} & A_{r+1,2} & \dots & A_{r+1,r} & 0 \end{bmatrix} \tag{4.1}$$

**Table 4** Number of Phase-I iterations on some NETLIB problems with different starting bases (a number in italic denotes the best in each case. The FortMP column stands also for MILP)

| | Number of | | Iterations in Phase-I | | |
|---|---|---|---|---|---|
| Problem | Constraints | Equalities | Logical | CPLEX | FortMP |
| 80bau3b | 2262 | 0 | 4056 | 2227 | *1124* |
| bnl1 | 643 | 232 | 2241 | 2401 | *2034* |
| bnl2 | 2324 | 1327 | 4640 | 2858 | *1802* |
| degen2 | 444 | 221 | 674 | *547* | 553 |
| degen3 | 1503 | 717 | 3188 | *2873* | 3146 |
| finnis | 497 | 47 | 670 | 229 | *139* |
| fit2p | 3000 | 3000 | 5271 | 3076 | *0* |
| gfrd-pnc | 616 | 548 | 910 | 293 | *0* |
| greenbea | 2392 | 2199 | 3688 | *2966* | 3098 |
| greenbeb | 2392 | 2199 | 3695 | 3171 | *2643* |
| israel | 174 | 0 | 298 | *7* | 76 |
| maros | 846 | 323 | 1215 | 760 | *442* |
| pilot87 | 2030 | 233 | 5213 | *4053* | 4288 |
| pilotnov | 975 | 701 | 2864 | 2262 | *752* |
| pilot-we | 722 | 583 | 1618 | 675 | *439* |
| sc105 | 105 | 45 | 102 | *0* | 5 |
| scagr7 | 129 | 87 | 146 | 51 | *23* |
| scorpion | 388 | 280 | 365 | 131 | *97* |
| sctap3 | 1480 | 620 | 1703 | 337 | *0* |
| Total | | | 46171 | 30287 | 22221 |
| Winner | | | 0 times | 6 times | 14 times |

where the $A_{ii}$ diagonal blocks are dense and have the property that there are at least as many columns as there are rows, that is, $A_{ii} \in \mathbb{R}^{m_i \times n_i}$, $m_i \leq n_i$, $i = 1, \ldots, r$. Vectors $x$, $u$, $b$, $v$ and $w$ are partitioned accordingly.

The algorithm moves along the block diagonal and in step $k$ solves the following problem:

$$\begin{array}{ll} \text{minimize} & e^T(v_k + w_k), \\ \text{subject to} & A_{kk}x_k + v_k - w_k = b_k - \sum_{i=1}^{k-1} A_{ki}\bar{x}_i, \\ & 0 \leq x_k \leq u_k, v_k \geq 0, w_k \geq 0, \end{array}$$

where $\bar{x}_i$ is the solution vector (of dimension $n_i$) of the $i$-th subproblem, $x_k$ and $u_k$ are $n_k$ vectors, and $v_k$, $w_k$, and $b_k$ are $m_k$ vectors. This pattern is used for $k = 1, \ldots, r$ with the obvious interpretation of a sum being zero in case of $k = 1$. A subproblem is said to have a satisfactory solution if the optimal value of (4.2) is zero. It is hoped that this will be the case for most of the subproblems even if some artificial (undershoot or overshoot) variables remain basic. The $(r+1)$th block contains the empty and the structurally dependent rows. It has to be treated in a different but simple way.

Solving each subproblem we get a solution and a set of basic vectors. Jointly, they form a starting basis to the original problem. Some artificial variables may be present even at positive level which means an infeasible solution in terms of the original problem.

The main point in this procedure is the permutation of the original problem into the form of (4.1). It is hoped that with a modified version of the $P^5$ algorithm of Erisman *et al.* [22] the number of subproblems ($r$) will be large enough so that their dimension is small in terms of the number of rows. Gould and Reid noticed that in case of real-life problems where sparsity is present it is very often true that $\max_i\{m_i\} \leq 5$. In such cases an efficient dense LP solver can be used to solve the small subproblems.

Though this description was a little simplified (for further details the reader is referred to the original paper [31]) one may feel its power. The computational study supports this feeling and in the category of numerical crash algorithms this is substantial progress. Computationally this procedure is more demanding than a symbolic crash but is still within a tolerable level.

### 4.4 Pricing

Perhaps this is the algorithmic unit of SSX that has undergone the most substantial progress in recent times. Pricing (the selection of an incoming variable) has a large freedom in picking up one variable, usually from many candidates. The candidates are characterized by having a favourable reduced cost. The theory of the simplex method says that any of the candidates can be chosen and under proper treatment of degeneracy (see Section 4.5) reaching an optimal solution is guaranteed (if one exists). In practice, however, the choice of candidate makes a big difference to the number of iterations. In other words, the strategy of selecting the incoming variable is of vital importance.

As early as 1968, this importance was recognized and efforts were made to make this choice economical and effective (see [58]). By "economical" we mean that the selection is computationally not too expensive and by "effective" we mean that the choice contributes to reducing the number of iterations. These two factors usually work against one another in the sense that there are some effective methods that require a large amount of work per selection but considerably reduce the iteration count, while there are cheap selection rules that usually lead to a large number of iterations. To say which is better we have to evaluate the total computational effort needed to solve the problem.

Interestingly enough, the simplest possible selection rule has a crucial theoretical importance. According to this rule, start scanning the non-basic variables starting at the variable with lowest index, move ahead and select the first candidate (the one with favourable $d_j$). This is the first stage of the theoretically finite (anti-degeneracy) pivot selection rule of Bland [5]. However, in practice this rule is very poor. It usually results in a very large number of iterations and requires much computational effort.

The most widely known selection rule is the Dantzig rule [14] that scans all the non-basic variables, called *full pricing*, and selects the one with the largest favourable $|d_j|$. It is easy to see that this rule is sensitive to the initial scaling of the matrix.

Computing the $d_j$ for each non-basic variable can be quite expensive and, in general, is not necessary. By *partial pricing* only a part of the matrix is scanned and the best candidate is selected. There is a great variety of partial pricing strategies depending on the way the parts (sections) are defined. They can be preassigned and static (for example to take advantage of some known structure of the matrix) or dynamically determined. It also makes a difference how the scanning is performed. For instance, it can be a strategy to scan the next section (in a cyclic order) fully. Another possibility is to scan several sections partially. The complexity can be increased by the introduction of some learning capability, that is, to study the success of past pricing and adjust if necessary. If to adjust then how? This again multiplies the possibilities.

Though all pricing strategies for the simplex method are local, it has long been understood that the best strategy would be to select that variable in each step which results in the largest improvement in the objective function. The improvement is determined by $\theta$ and $d_j$ (in Phase-II it is $\theta d_j$). The computational requirement of this idea is prohibitive. Since at pricing in the SSX the steplength $\theta$ is unknown we cannot be sure about the quality of the selection. This becomes known when the pivot row is also determined after a FTRAN and a "choose row" operation.

Another variation of column selection is *multiple pricing* and the subsequent *suboptimization* first described in [58]. In this idea several candidates are selected during the scanning pass and a subproblem with the basic and the selected variables is solved, sometimes to its optimality. This technique can reduce the work per iteration and was especially popular when the main memory of computers was relatively small and pricing usually required access to background memory. With the appearance of large memory the importance of multiple pricing faded a little but it is still extensively used today. One remarkable feature of the suboptimization is that it is carried out on the full tableau of the subproblem which makes it possible to follow the principle of largest improvement resulting in efficient steps. At the same time, usually it is not a good idea to make too many iterations in suboptimization since later steps become very local to the subproblem and the total number of iterations increases unnecessarily.

The above pricing strategies belong to the category of cheap pricing because none of them requires more effort per iteration than the Dantzig rule. At the same time, some ingeniously designed strategies (mostly combinations of partial and multiple pricing) proved to be very effective. As such, the Dantzig rule is never directly used in the present-day advanced LP systems.

Until quite recently, full pricing was considered computationally too expensive and therefore was discarded as a realistic alternative. However, recent developments in pricing are all based in some way on full pricing. To this end, the idea of keeping $d_N$ updated was introduced.

Let us assume that the incoming variable is $x_q$ and row $p$ is pivotal in the current iteration. The elements of $B^{-1}$ (the inverse of the previous iteration) are denoted by $\beta_{ij}$ and the updated $q$-th column by $\alpha_{\cdot q}$, that is $\alpha_{\cdot q} = B^{-1}a_{\cdot q}$, with $\alpha_{pq}$ as its $p$-th component. The $p$-th row of $B^{-1}$, $\beta_{p\cdot}^T$ can be obtained by $\beta_{p\cdot}^T = e_p^T B^{-1}$. Using the notation

$$h^T = \beta_{p\cdot}^T, \lambda = -d_q/\alpha_{pq}, \tag{4.2}$$

the values of $d_j$, $j \in \bar{N}$ ($\bar{N}$ is the set of new non-basic variables) are obtained by

$$\bar{d}_{\bar{N}}^T = d_{\bar{N}}^T + \lambda h^T A_{\bar{N}}. \tag{4.3}$$

If there is enough memory to store $A$ row-wise as well then computation in (4.3) can be very fast.

One problem with non-basic $d_j$'s is that only their sign is really meaningful. As indicated before, their magnitude depends on initial scaling. Therefore, they are not proper indicators of how much progress in the objective function can be anticipated. Experience shows that in columns with large $d_j$ the other elements are also large. Since these elements are potential denominators in the ratio test, the resulting $\theta$ steplength and progress are often very small.

It is easy to see that if $d_j$ is replaced by some normalized value, $\hat{d}_j = d_j/||\alpha_{\cdot j}||$, it will be independent of initial scaling. The big difficulty lies in computing and/or updating the norm as the basis changes.

As early as in 1977, Goldfarb and Reid [30] gave formulas for the exact update of the following column norm:

$$\gamma_j = ||\alpha_{\cdot j}|| = (\alpha_{\cdot j}^T \alpha_{\cdot j} + 1)^{1/2}. \tag{4.4}$$

While $d_j$ represents the unit change of the objective function along the $j$-th axis, if $d_j$ is normalized by (4.4) then $\hat{d}_j$ gives the unit change of the objective function along the edge from the vertex defined by $B$ to the vertex of a new basis $\bar{B}$ with the $j$-th variable. Clearly, this latter has more relevance to the LP problem. If column selection is based on $\hat{d}_j$ then we talk about *steepest edge (SE) pricing*.

In 1977 it was still uneconomical to use SE. With the appearance of large-memory computers the situation has changed. In 1992 Forrest and Goldfarb [23] introduced additional SE pricing algorithms for both the primal and dual simplex method and reported favourable practical performance. Their solution times were usually 2 to 5 times less than with Dantzig pricing on a series of large-scale real-life problems. In some extreme cases the speedup factor was as much as about 50. The rate of improvement in the number of iterations was even larger. From this point on it was evident that while SE pricing is expensive it is more than worth considering.

There is a significant computational difference between their primal and dual type SE. Part of the computational effort needed for the exact update of the norms is carried out to update $d_N$ for full pricing anyhow. The extra computational effort in the primal case corresponds to one BTRAN and a dot product for the non-basic columns. To be more precise, if $x_q$ enters the basis, then $v$ has to be obtained from $B^T v = \alpha_{\cdot q}$ followed by the computation of $v^T A_N$. In the dual case, only the solution of $B^T v = \alpha_{\cdot q}$ for $v$ is needed. If $n \gg m$ then the difference is very substantial. Therefore, in the primal simplex SE is usually reserved as a spare engine to be activated if necessary. In the dual simplex SE is a good choice for the default pricing strategy.

The extra cost of normalized SE pricing can be reduced if we are satisfied with an approximation to the norm. One such scheme was developed in 1973 by Harris [36], actually well before the SE pricing was introduced. She calls her pricing DEVEX. Using a reference framework $R$ of non-basic variables at a certain stage of the algorithm, the approximation $T_j$ of $\gamma_j$ is updated in the following way (see [4]). First, compute

$$\rho = \left( \delta(q) + \sum_{B_i \in R} \alpha_{iq}^2 \right)^{1/2}, \tag{4.5}$$

which is the exact norm of the entering variable, where $B_i$ is the original index of the $i$-th basic variable and

$$\delta(q) = \begin{cases} 1 & \text{if } q \in R, \\ 0 & \text{otherwise.} \end{cases}$$

Let

$$r_{\bar{N}}^T = \frac{\rho}{\alpha_{pq}} h^T A_{\bar{N}}$$

with $h$ being the $p$-th row of the basis inverse before the current iteration as defined in (4.2). The new approximation of $\gamma_j$ for each non-basic variable is

$$\bar{T}_j = \max\{T_j, r_j\}, \tag{4.6}$$

assuming that the $T_j$ value of basic variables is 1. From (4.6) it is clear that the $T_j$ values keep growing. It necessitates the monitoring of the accuracy of the approximation. A vehicle for that is the incoming variable at each iteration since its exact norm is computed in (4.5) anyhow. When the approximation becomes poor a new reference framework is set up and the norms are reinitialized to 1. The extra work of DEVEX per iteration is the computation of $h^T A_{\bar{N}}$ and $\rho$.

DEVEX is meant to be a very efficient approximate SE pricing scheme though in some cases the extra work does not pay off. See also Greenberg and Kalan who provided formulas for the exact update of the $T_j$'s at the expense of additional computations [32].

The most advanced LP systems are equipped with some flexible cheap and some expensive pricing techniques that can be activated whenever necessary. Here, the question of "necessary" requires some more specification. Presently, it is answered by the taste and best knowledge of each individual LP designer.

### 4.5 Degeneracy

Degeneracy has long been known to be a major obstacle to the numerical progress of the simplex method. A basis is called *degenerate* if at least one of the basic variables is at their lower or upper bound. The problem is that in such a case the requirement of maintaining feasibility of the basic variables may force a zero steplength for the incoming variable. This, in turn, keeps the objective value unchanged. Theoretically, this can lead to cycling unless proper provision is made to prevent it.

The problem is not simply theoretical. Many real-life problems tend to be rather degenerate. Though the sophisticated pivot procedures of the advanced LP systems prevent cycling long sequences of degenerate iterations (*stalling*) are still possible.

Several (though not too many) authors have addressed the issue of efficient anti-degeneracy and anti-stalling pivot rules in the framework of the simplex method. Nowadays, these rules, combined with SE-type pricing, are believed to be adequate for the proper practical treatment of degeneracy. Of course, it does not mean that the problem is completely resolved.

Actions against stalling/cycling can be taken at two points, namely at column selection (pricing) and at row selection. The idea of anti-degeneracy pricing is that the chance of a non-degenerate iteration can be increased if proper candidates are selected. The ideal situation would be if the selected and transformed column had zeros or elements with the proper sign in the degenerate basic positions. During pricing, it is practically impossible to check this exactly. Approximate methods can be used at the expense of an extra BTRAN and a dot product with the promising candidates. Interestingly enough, the opposite can be checked exactly, that is, with the same amount of extra work those columns that will necessarily lead to degenerate iterations can be identified. If nothing else is available, this idea can be used to avoid these columns as long as others are available. Unfortunately, it is not true that the remaining columns will lead to non-degenerate steps. The only thing which is true is that if there is any such column it will be among the non-rejected ones.

Based on this discussion, the anti-cycling procedures can be categorized as column action and row action methods. The column action methods have to be aware of degeneracy while the row action methods do not necessarily.

The first, and long forgotten, contribution to this area was Wolfe's work [70] in 1963. This is a row action method and requires degeneracy awareness that has to be checked at each step. The challenging feature of his algorithm is that it is theoretically sound and practically efficient.

The Bland rule [5] has never been tested seriously in the case of degeneracy though it would not require explicit knowledge of the situation. Gal and Geue [26] have developed a theory of degeneracy graphs and have given a new pivot rule. Their procedure is a degeneracy conscious row action method. Again, it requires a careful computational investigation.

Presently, the two most frequently used anti-cycling heuristics are *bound perturbation* and *bound shifting*.

In the case of bound perturbation the bounds of the variables are perturbed by a (usually randomly selected) small value and the modified problem is solved to optimality. At optimum, the original bounds are restored and a reoptimization takes place starting from the perturbed optimal basis. The monitoring mechanism of this method is very cheap (comparing the objective value between iterations) and the extra computational effort is minimal. During iterations the algorithm does not use any degeneracy information.

The basic idea of bound shifting introduced by Gill *et al.* [29] is the following. If we are able to improve the value of the objective function at each iteration then no basis can be repeated. However, to achieve this and maintain feasibility at the same time is not always possible within the traditional framework of the simplex algorithm. If we allow some variable(s) to go slightly infeasible then the objective can decrease but the emerging infeasibility must be kept under control. The authors call their procedure EXPAND since it expands the feasibility tolerance at each step by a small amount. If in an iteration the steplength is smaller than a predefined minimum then a forced step is made with this predefined minimum steplength. This will drive some variables beyond the current feasibility tolerance, but they will stay inside the newly expanded feasibility bounds for the next iteration. If the number of iterations is large, the changing feasibility bounds can go beyond a reasonable limit. Therefore, the bounds are periodically reset to their initial value. The same happens when an optimum is reached. Subtleties of the algorithm are omitted here. A remarkable feature of this procedure is that it is unaware of degeneracy and the amount of extra computations is negligible. At the same time, it has proved its practical efficiency in a large number of cases. Clearly, it is a row action method. It has an additional remarkable feature, namely in the selection of the pivot element it has some flexibility (similar to DEVEX) that can be used to control numerical accuracy.

Sometimes other procedures have favourable anti-cycling properties as a side effect. A typical representative is the FEWPHI Phase-I procedure of Maros [49] where the largest possible step towards feasibility more frequently results in a non-degenerate step than a traditional Phase-I method. Additionally, FEWPHI also enables the selection of proper sized pivots.

If the degeneracy situation is serious then a SE-type column selection combined with an efficient row selection method can be used. As a result of this, very resistant degenerate problems can be solved efficiently. However, it is to be noted that interior point methods are not sensitive to degeneracy and in very degenerate situations they are usually more efficient than the simplex method.

### 4.6 Basis factorization and sparse updates

The methods of computing and updating the inverse basis matrices of large sparse linear programming problems are dominated by the following major considerations: (a) the non-zero structure of the matrix needs to be analysed and the rows and columns of the matrix reordered: the ANALYSE phase, (b) pivots have to be chosen for (Gaussian) elimination and alternative methods of factorization of the matrix have to be considered: the FACTOR phase, (c) a method has to be derived whereby the inverse can be used in the solution process implicit in the backward and forward transformations (BTRAN and FTRAN, respectively): SOLVE procedures, (d) suitable data structures have to be designed to represent the original sparse matrix and the sparse basis inverse, (e) a complete lower block triangular reordering must be determined. In addition to the compact inverse representation (f) an efficient update procedure has to be designed taking into account (b) and (d) above. Finally, (g) basic factorization which takes advantage of static structures can be exploited in many practical problems. We refer the reader to papers by Hellerman and Rarick [37, 38], Bartels and Golub [1], Forrest and Tomlin [24], Gill *et al.* [28], Suhl and Suhl [66], and Reid [61, 62] which have addressed, expanded and dealt with most of these issues.

(a) The ANALYSE phase

Given a non-singular matrix $\bar{B}$ of order $m$ the reordering algorithms make use of the corresponding graph of the adjacency matrix derived from the non-zero elements. In the symmetric case the aim is to derive specifications for symmetric row and column interchanges given by permutation matrices $P$ and $P^T$ such that we obtain a reordered matrix $\bar{B}$, $B = P\bar{B}P^T$ which is presented as compactly as possible in band diagonal form. The heuristic methods exploit the underlying undirected graph and make use of the degrees of the vertices. In the case of non-symmetric matrices, the general situation with LP basis matrices, it is well established (see Mitra [54], Duff *et al.* [18]) that the Lower Block Triangular (LBT) is the most desirable form. The LBT reordering is achieved by considering the underlying directed graph and applying the following algorithms.

- **Maximal matching**

  We can apply the maximum matching algorithm of Hall [35] whereby the matrix $\bar{B}$ is reordered to $B^*$ which has a zero-free diagonal. As a result $B^* = \bar{B}Q$; the permutation matrix $Q$ specifies the reordering as derived by this algorithm.

- **Finding the strong components**
  To the matrix $B^*$ which has a zero-free diagonal the well-known algorithm of Tarjan [67] can be applied to obtain the strong components (the diagonal blocks of the matrix) of the directed graph. This leads to the derivation of orderings given by the permutation matrices $P$ and $P^T$ whereby the original matrix can be reordered into the lower block triangular (LBT) form $B$, $B = PB^*P^T$ or $B = P(\bar{B}Q)P^T$. These algorithms have the complexity of low-order polynomials and also execute very fast in practice. Beale [2] conjectured that a band form for the non-symmetric case may be equally attractive as the LBT form. Although we have done some preliminary work on this topic this has never been properly followed up by any other researchers. It is interesting to note that Hellerman and Rarick did not make use of Tarjan's work as these results were not known at that time. They also combined the ANALYSE and at least a part of the FACTOR phase as they preassigned pivot positions: the implications of this are discussed in Section 5.1.

(b) The FACTOR phase

The main motivation of this phase is to compute a sparse yet accurate $LU$ decomposition of the matrix B. If the $LU$ decomposition is computed in the product form of the $L$ and $U$ transformation matrices then this reduces non-zero growth. If pivot choice is made dynamically during factorization with a view to control accuracy then this can lead to a stable solution. Hellerman and Rarick's P4 algorithm finds spikes which are postponed for pivoting and constructs pivot agenda based entirely on a static analysis. The main aim is to minimize fill-in. Unfortunately, however, it is necessary to modify the algorithm [22] to deal with all possible contingencies otherwise the method can lead to structural singularity as well as unstable solutions. Erisman and his coworkers have studied the comparative performance of a modified and improved P5 (precautionary partitioned preassigned pivot procedure) with the modified Markowitz procedure with threshold tolerance due to Reid. The authors conclude that not only is the latter superior in non-LP applications, even in LP applications, the latter is equally good if not superior, taking into account accuracy and robustness. It is well-known that the choice of the largest admissible non-zero entry at each step of Gaussian elimination can provide accurate $LU$ decomposition but this strategy will result in considerable fill-in. We have found Reid's recommendation [18] of a threshold tolerance with restricted Markowitz merit criteria very effective indeed. In this scheme, at the $k$-th elimination step, pivots are chosen by the following criteria:

$$(r_p^k - 1)(c_q^k - 1) = \min_{i,j} (r_i^k - 1)(c_j^k - 1), \tag{4.7}$$

$$|b_{ij}^k| \geq u\left(\max_l |b_{lj}^k|\right), \quad 0 < u \leq 1, \tag{4.8}$$

where $r_i^k$, $c_j^k$ are row and column non-zero counts of row $i$ and column $j$ respectively, $b_{ij}^k$ is the $ij$-th entry in the remaining submatrix at the $k$-th elimination step, and $u$ is a *threshold pivot tolerance.*

Thus the pivot chosen is $b_{pq}^k$ such that $|b_{pq}^k/u|$ is greater than the largest absolute entry of the non-zero elements available for pivot in the column $q$ of the matrix at the $k$-th step. Duff, Erisman and Reid suggest using a near minimum as opposed to exact minimum merit count and a threshold value of $u = 0.1$ which leads to fairly good results both for sparsity of the inverse representation and accuracy of the solution.

(c) The SOLVE procedures

Let $e$ denote an $m$ component vector with $\pm 1$ as one or more of its components and zeros in the remaining positions ($e$ is known as the form vector). Let $a_j$ denote a column of the $A$ matrix, $\bar{a}_j$ its update and $\pi$ the vector of shadow prices. There are two well-known steps of the revised simplex, namely backward transformation (BTRAN) and forward transformation (FTRAN), which make use of the $B^{-1}$ representation. These transformation steps can be expressed as two equivalent equation solving steps given by the relations

$$e^T B^{-1} = \pi^T, \quad \pi = (B^{-1})^T e, \tag{4.9}$$

and

$$\bar{a}_j = B^{-1} a_j. \tag{4.10}$$

Having computed the inverse in the factored form it becomes necessary to compute the solutions shown in (4.9) and (4.10) making use of the data structures and the factored inverse representation.

(d) Choice of data structure

In the 1960s and 1970s in mainframe as well as mini computers the main memory modules (also known as core store) were very expensive and for an LP system to work over a range of machines with varying memory sizes both the $A$ and the $B^{-1}$ matrices were stored in backing (magnetic) stores in the form of sequential files. Today main memory is no longer at a premium even for desk top machines. Yet it is still necessary to design highly compact data structures to represent the $A$ and $B^{-1}$ matrices. There are well-known storage schemes for sparse and supersparse matrices [29, 30, 42]. Since we wish to have both $A$ and $B^{-1}$ memory resident it is to our advantage to use as much of $A$ as possible to represent $B^{-1}$, as a result it is possible to accommodate larger models in memory and considerably enhance the upward limit of the system.

(e) Lower block triangular (LBT) factorization

After reordering the matrix $\bar{B}$ into the LBT form $B$ we can factorize it in the $LU$ form and derive the elimination form of the inverse (EFI) as it is derived by the

**Table 5** Summary information for the test bases

| Problem name | Matrix order | Iteration number | Basis non-zeros |
|---|---|---|---|
| ATLAS | 315 | 900 | 1406 |
| ATLAS | 315 | 1200 | 1472 |
| BPTEST | 822 | 2500 | 4765 |
| BPTEST | 822 | 5500 | 4777 |
| GANGES | 1310 | 303 | 3453 |
| GANGES | 1310 | 603 | 5374 |

**Table 6** Comparison of non-zeros in the (basis) factors

| | MINOS non-zero | | FortLP non-zero | |
|---|---|---|---|---|
| Problem name | Total non-zeros | EFI | EFI | PEFI |
| ATLAS | 1406 | 1793 | 1836 | 1694 |
| ATLAS | 1472 | 1857 | 1833 | 1736 |
| BPTEST | 4765 | 6550 | 6537 | 6191 |
| BPTEST | 4777 | 6707 | 6521 | 6402 |
| GANGES | 3453 | 3424 | 3424 | 3424 |
| GANGES | 5374 | 5367 | 5369 | 5369 |

Gaussian elimination operations. Alternatively we can first factorize the matrix and then apply the elimination operation only partially to the diagonal blocks of the matrix $B$ and call this the partial elimination form of the inverse (PEFI). Mitra and Tamiz [55] provide a detailed discussion of the factorization and use of PEFI. Duff *et al.* [18] call the same representation the reducible factors of the original (reducible) matrix. If we do not take into account the possible cancellation of non-zero elements to zero values, then it is easy to show that PEFI will always produce a sparser inverse than EFI. Our computational experience bears this out and Erisman *et al.*, as above, make the same statement.

In Table 5 we have summarized statistics of bases extracted during LP iterations; we used MINOS for these experiments. The corresponding basis specification was then introduced into FortLP (earlier version of FortMP) to carry out the factorization of the bases and the results are summarized in Table 6. We can easily deduce from this table that PEFI produces a sparser representation of the inverse basis.

(f) Update procedures

If there was a basis change in a simplex iteration then the representation of the basis inverse needs updating. This update can be done in several different ways. The product form update is the simplest that adds one elementary transformation matrix to the product form. This usually suffers from heavy fill-in since the only non-trivial column of such a matrix is directly generated from the transformed entering column. Better update schemes have been developed that maintain triangularity obtained at the last factorization and can heavily reduce the growth of fill-in. The two most successful triangular updating schemes are due to Bartels and Golub [1] and Forrest and Tomlin [24].

The basis inverse update procedure of Bartels and Golub requires the original basis to be factored in the $LU$ form. Their work and subsequent studies (e.g. Reid [62]) demonstrated that this was superior to product form update in all considerations: sparsity, stability and speed. Forrest and Tomlin introduced a clever variation of this update procedure which is better than the latter in speed and sparsity but is poorer on a stability consideration. This issue, and its close relationship to the original problem of computing a sparse yet stable inverse, was well addressed by Reid who produced the celebrated LA05 subroutines with modified Bartels–Golub update. These subroutines have been widely quoted in many studies and have been adopted as the main compute engine in successful LP systems. Gill *et al.* [28] have also considered this issue in considerable depth.

- *Bartels–Golub update*
  In the Bartels–Golub scheme at the $k$-th iteration after inversion we can express the factorization of the basis matrix $B_k$ and its inverse as

$$\begin{aligned} B_k &= LT_1T_2\ldots T_{k-1}U_k, && (4.11)\\ \text{or}\quad (B_k)^{-1} &= U_k^{-1}T_{k-1}^{-1}\ldots T_2^{-1}T_1^{-1}L^{-1}. && (4.12)\end{aligned}$$

  In (4.11) and (4.12) we make use of the original $L$ and derive $T_1,\ldots,T_{k-1}$ and $U_k$ from the original $U$. The method of computing these is well established [28] and is not repeated here. The Forrest–Tomlin update can be considered to be similar to Bartels and Golub with an additional row interchange. The representation of $(B_k)^{-1}$ can be expressed as

$$(B_k)^{-1} = Q_k\ldots Q_1V_kR_{k-1}\ldots R_1L^{-1}, \tag{4.13}$$

  where $V_k$ is obtained from $U_k$ with row deletion and row interchange which in each step is specified by the permutation matrices $Q_1, Q_2, \ldots, Q_k$.
- *Forrest–Tomlin LU update*
  The Forrest–Tomlin update procedure to represent the "modified" basis inverse is well established. Here we briefly outline algebraic representations underpinning this procedure. Consider the inverse of the basis matrix $B$ after the reinversion in $LU$ form:

$$B^{-1} = U^{-1}L^{-1} = U_1^{-1}U_2^{-1}\ldots U_m^{-1}L_m^{-1}\ldots L_1^{-1}. \tag{4.14}$$

  The transformation matrices $U_k^{-1}$, $L_k^{-1}$ in (4.14) are naturally known as $U$-etas and $L$-etas. Let the column $q$ ($a_q$) pivot into the $p$-th row in the next simplex iteration. Let $\bar{a}_q$ and $\alpha_q$ be the fully updated and partially

**Table 7** Non-zero entries in the factored form of $B^{-1}$

| Problem | FortLP statistics | | | MINOS statistics | |
|---|---|---|---|---|---|
| | EFI at invert | 50 iterations: Product form | 50 iterations: F - T update | EFI at invert | 50 itns: B - G update |
| ATLAS | 1836 | 14118 | 3791 | 1793 | 2781 |
| ATLAS | 1833 | 14233 | 3721 | 1857 | 2817 |
| BPTEST | 6537 | 32432 | 9560 | 6550 | 8143 |
| BPTEST | 6521 | 34204 | 9223 | 6707 | 7905 |
| GANGES | 3424 | 5884 | 4245 | 3424 | 3908 |
| GANGES | 5369 | 12050 | 6255 | 5369 | 5824 |

updated vectors of $a_q$, respectively. That is

$$\alpha_q = L^{-1}a_q \quad \text{and} \quad \bar{a}_q = U^{-1}\alpha_q. \tag{4.15}$$

The updated vectors $\bar{a}_q$ and $\alpha_q$ are computed within one complete FTRAN operation. The column $a_q$ is unpacked to a work area and processed using the L-etas. The partially updated vector $(\alpha_q)$ is copied to the second work area. The FTRAN is completed by processing $\alpha_q$ using the U-etas to obtain the $\bar{a}_q$. The *LU* structure of the new basis matrix $\bar{B}$ is preserved by the following update steps:

(1) Delete the $p$-th U-eta.
(2) Eliminate the non-zero entries in the $p$-th row of the U-etas $(u_{pk})$ and place these in a row transformation vector $R_p$ (also called row-eta).
(3) Update the row-eta $R_p$ by using the modified U-etas, $U^{-1}$.
(4) Add a new U-eta $R_p^{-1}\alpha_q$ to the left of the $\bar{U}^{-1}$ and add a new row-eta $(R_p^{-1})$ to the right of the $L^{-1}$.

The transformation of $U^{-1}$ to $\bar{U}^{-1}$ and computation of the $\pi$ and $R_p$ vectors are carried out at the same time during the processing of the U-etas in the next BTRAN operation. The transformation matrix $U_p^{-1}$ is deleted implicitly by negating the pivot row pointer in the packed eta. In Table 7 we have summarized some experimental results which indicate the computational advantage of Bartels–Golub and Forrest–Tomlin in respect of non-zero build-up during the simplex iterations.

(g) Structured models and dynamic basis factorization

Although the use of a generalized upper bound (GUB) has been discontinued in simplex algorithms exploitation of structure has been recently advocated by Brown and Olson [9]. The early work in the area of dynamic basis factorization is due to Zoutendijk [72], Powell [59] and Brown and McBride [8]. Brearley, Mitra and Williams [7] extended the GUB concept to the exclusive row structure (ERS) concept. Brown [9] has considered dynamic basis factorization and describes its implementation in respect of GUB, pure network (PN) and generalized network (GN) substructures. He reports computational results which appear extremely promising when compared with OSL and CPLEX for a number of substantial industrial test problems.

A computationally oriented presentation of how factorization can be performed efficiently is well described in [65]. The same authors give details of an implementation of the Forrest–Tomlin ([24]) triangular updating procedure in [66]. Both papers address the issue of data structure which is essential for efficiency.

## 5 Design and computational issues of SSX

Since computational LP has become a matter of research as well as a highly relevant industrial software tool, benchmarking and performance evaluation have been introduced. For instance, the mathematical programming community has gathered together to collect a wide range of practical yet demanding test problems which are known as the NETLIB set of LP test problems. In this section we first consider the design objectives of SSX systems. We then consider the profiling information, that is, the relative computational effort made in different components of the SSX. Finally, we analyse the computational performance of a few well established SSX solvers.

### 5.1 Design objectives

The numerical processing of sparse matrices relies upon an established strategy that only the non-zeros of the problem data and the intermediate results are stored. The non-zero values are stored together with their positions. For matrix elements it is necessary that a position is identified by two coordinates. If the elements are arranged in some order then one of the coordinates can be specified implicitly. In practice, matrices are usually stored column-wise and the non-zero elements and their corresponding row indices are stored in two parallel linear arrays. Let us denote the array of values by *Val* and the array of row indices by *Ind.* Both of them are $K$ long (the number of non-zeros in the matrix). An additional pointer array *Pnt* is also needed. Its $k$-th element points to the first element of column $k$ in the linear array. In the case of the LP matrix $\bar{A}$ the size of *Pnt* is $n_{\bar{A}} + 1$. The last element of *Pnt* points to the first free position after the last non-zero element of the matrix. This storage scheme assumes that there is no free space left between the elements of two consecutive columns. The structure is schematically represented in Fig. 1. If the elements are stored as double precision numbers in 8 bytes and the indices and pointers as 4 byte integers then this simplest sparse storage scheme requires $12K + 4n_{\bar{A}}$ bytes for storing just the matrix of structural variables.

Implementing the sparse simplex algorithm means translating it into a computer program that works according to the logic of SSX and is able to solve LP problems numerically. There are many different ways SSX can be implemented and this gives rise to many simplex-based LP systems around the world.

Common in all the present-day systems is that they all work in the main random access memory (RAM) of the host computer. It means that all data that are needed during the iterations reside in RAM. Some computers extend the available memory by an intelligent memory paging method in a way transparent to the user. However, if this happens a performance degradation will result.

The main purpose of designing new variants of the revised simplex method is to be able to create ever better simplex-based LP systems. The criteria, on the basis of which systems are evaluated, can be summarized as follows.

**Reliability:** The system does not break down and gives a correct answer to all the problems attempted.

**Accuracy:** There is a sufficient number of accurate significant digits in the solution vector, including the objective value. This "sufficient" number usually means 6 or more significant digits.

**Efficiency:** The solution is achieved with a relatively small computational effort. This is typically a relative measure of performance. What one can say is that one system is more/less efficient on a given problem (or set of problems) than another system. A similar statement can be that this version of a system is more/less efficient than the other version of the same system (important information for the developers). A good basis for comparison is the number of operations made until solution. This measure requires a careful analysis of the code so that the internal counters can properly be incremented in case of evaluation. Another, and widely accepted, measure is central processor unit (CPU) time of the solution provided the system works in RAM. It is easy to measure but not very accurate in heavy multiprogramming environments. In this respect, single user environments (like PCs where CPU and wall clock times are nearly the same) are more suitable for testing efficiency. The different special features of present-day advanced hardware also influence efficiency. In this respect, the two most important features are *cache memory* for single processor computers and *parallel computer architecture*. Codes can be designed to exploit them as much as possible.

**Algorithmic richness:** Refers to the algorithmic capabilities of a system. In the case of SSX, the main solver is the primal simplex algorithm controlled by many numeric and algorithmic parameters. In addition to the primal, the availability of a dual algorithm, a variety of starting procedures (including start from a saved solution), problem analysis and sensitivity analysis are desirable features.

**Capacity:** What is the largest problem that a given system can reasonably solve? This requires a little explanation. Namely, some LP systems can increase the problem size at the expense of declining efficiency. As long as it is done by resorting to simpler (less memory hungry) algorithmic procedures, we call it reasonable. If it is achieved by storing part of the essential data (that is required in every iteration) on disk then we do not call it reasonable because it will result in serious deterioration in solution time.

**Portability:** LP solvers are usually (but not necessarily) written in some high level programming language. *Source level portability* means that the source code can be compiled on different computers and/or different compilers practically without any modification and the resulting program will run the same way. Portability is a very desirable feature and source level portability ensures that an LP system can be made available on a wide variety of computing platforms.

**User friendliness:** Often also referred to as *ease of use*, it means different requirements for different levels of users. It is useful if a system (i) can be used interactively and also in batch mode, (ii) supports both straightforward and sophisticated usage, (iii) provides different levels of clear and structured output, (iv) is available in stand-alone and callable library version, and (v) has good documentation, a help system and program maintenance.

Presently, no system is known to be the best by all criteria in all cases. However, there exist some systems that fulfil the above criteria to a large extent. They are viable alternatives to the interior point algorithms even in the case of large-scale problems.

### 5.2 Analysis of benchmark problems

In the context of linear programming, as in any other algorithm and software testing situation, benchmarking usually means a comparison of different solvers. The main issues in benchmarking are reliability and computational efficiency. As we already pointed out, presently there is no system that can be considered the best in all aspects and for all problems. This is quite evident if we simply consider the fact that the set of "all problems" is never well defined. Some authors reported remarkable computational results in solving some very large problems. However, it is unclear whether these problems can be solved more efficiently by other solvers because, due to the proprietary nature of these problems, they are not made available to the authors of other solvers. It is therefore important to work with a set of widely available standard benchmark problems to compare computational results and the capabilities of the solvers.

This chapter is not meant to be a detailed computational study of different solvers. At the same time, it is instructive to study comparative test results. We have selected seven problems from the NETLIB set and one other (`bgprob`) from an industrial application. These problems represent different classes and therefore form a suitable set to demonstrate some important aspects of benchmarking and the conclusions that can be drawn from it. Some of these problems are used to carry out profiling, as described in Section 5.3.

The main characteristics of these medium sized problems included in the investigations are summarized in Table 8. In real-life applications it is typical that problems fall into this category. In the summary we have included only those constraint types and variable types which are important from a computational point of view.

**Table 8** Characteristics of the selected benchmark problems

| | Number of | | | Equ. | Variable types | | |
|---|---|---|---|---|---|---|---|
| Problem | Constrs. | Vars. | Non-zs. | Constrs. | Non-neg. | Bounded | Other[a] |
| 25fv47 | 822 | 1571 | 11127 | 516 | 1571 | | |
| bgprob | 1651 | 1425 | 8722 | 1175 | | 1300 | 125 |
| bnl2 | 2325 | 3489 | 16124 | 1327 | 3489 | | |
| d2q06c | 2172 | 5167 | 35674 | 1507 | 5167 | | |
| degen3 | 1504 | 1818 | 26230 | 717 | 1818 | | |
| greenbea | 2393 | 5405 | 31499 | 2199 | 4996 | 290 | 119 |
| maros-r7 | 3137 | 9408 | 151120 | 3136 | 9408 | | |
| pilotnov | 976 | 2172 | 13129 | 701 | 1628 | 340 | 204 |

[a]This includes type-0 (fix) and type-3 (free) variables

From a numerical point of view, the benchmark problems are moderately difficult, except `maros-r7` which is relatively difficult.

### 5.3 CPU processing profile of SSX components

The different algorithmic components of SSX require different computational effort. For algorithm and software designers therefore it is informative to employ software tools to extract this information as a proportion of the total CPU processing time (computational effort). This information can guide the designer toward the direction and scope of improvement in the computational algorithm. Also for parallel algorithm designers it provides pointers as to how best to redesign the algorithm structure. To demonstrate the usefulness of profiling, we used the eight problems introduced in Section 5.2 and carried out the investigations on them.

Profiling tools require that during compilation of the program some extra code invoking probe programs is included. During run, this probe carries out a “sampling” that collects information about the time spent in each subroutine. Though the process itself is rather time consuming, the percentages provide relative execution time of subsystems and are unaffected by the profiling procedure.

For profiling, we selected FortMP. Since DEVEX pricing modifies the profiling pattern, two sets of profiling are presented here. DEVEX was disabled for problems `25fv47`, `bgprob` and `bnl2`, while problems `greenbea`, `maros-r7` and `pilotnov` were run without DEVEX. All parameters were left at their default setting. In the non-DEVEX case it means, among other things, that dynamic partial pricing was used. When DEVEX was enabled, full single pricing was used.

The main functional units of interest are: `Initialize, Factorize, Form price, Btran, Price, Ftran, Pivot,` and `Update`. The category `Others` is also included to add up to 100%. It is to be noted that `Form price` includes

**Table 9** Profiling: percentages of the total time spent in the different functional units of SSX on the selected problems

| | With DEVEX | | | Without DEVEX | | |
|---|---|---|---|---|---|---|
| Unit | 25fv47 | bgprob | bnl2 | greenbea | maros-r7 | pilotnov |
| Initialize | 0.4 | 2.9 | 1.2 | 1.1 | 3.0 | 1.9 |
| Factorize | 6.3 | 21.2 | 2.7 | 9.5 | 18.9 | 10.7 |
| Form price | 5.4 | 5.1 | 8.9 | 8.5 | 5.2 | 5.9 |
| Btran | 16.8 | 9.7 | 10.8 | 13.8 | 12.6 | 8.3 |
| Price | 17.5 | 3.6 | 19.0 | 12.5 | 12.2 | 9.8 |
| Ftran | 13.1 | 19.7 | 13.5 | 24.9 | 21.8 | 23.7 |
| Pivot | 8.9 | 15.4 | 12.2 | 18.9 | 11.9 | 17.4 |
| Update | 30.5 | 16.9 | 29.9 | 8.5 | 5.6 | 18.9 |
| Others | 1.1 | 4.9 | 1.2 | 0.8 | 1.5 | 3.2 |

checking feasibility.

The results of profiling are summarized in Table 9. It clearly highlights some interesting aspects. First of all, the different problems use the main functional units at different rates and the variation can be quite large. It can be seen also that the largest part of the time on all problems is spent on operations involving the representation of the basis inverse (`Factorize, Btran, Ftran`).

The relatively high percentage of `Ftran` (in the case of non-DEVEX runs) is a result of the suboptimization strategy of FortMP. It selects several (in this case four) candidates and updates them all to be able to use the principle of greatest improvement of the objective function. Since not all of them will finally enter the basis, some updated columns will not be used and their updating effort is lost. This is an expensive but usually powerful procedure.

The percentages look quite different if DEVEX or other normalized pricing is used. In such cases updating the approximate or exact scale factors requires a great deal of extra work as described in Section 4.4.

### 5.4 Performance summaries

Two academic and two commercial solvers were included in this limited benchmark study. The academic solvers are FortMP 1.03F and MINOS 5.3, while the commercial ones are CPLEX 2.1 and OSL 1.2. These solvers were available to us on the selected computer.

Carrying out benchmarking requires great care for many reasons. First of all, the influence of the very different performance characteristics of different computing platforms has to be eliminated. To this end, we used the same computer for all runs, namely an Elonex PC with 90 MHz Pentium processor with 16 MB of main memory (RAM) and 256 KB secondary cache. We have observed that performance is quite heavily dependent on the presence or absence of a fast secondary cache memory.

**Table 10** Number of iterations and solution times in seconds on the benchmark LP problems, without presolve, main algorithm: primal. Computer: Elonex PC with Pentium 90 MHz processor, 16 MB RAM and 256 KB cache. " — " denotes "unable to solve"

| | MINOS 5.3 | | FortMP 1.03F | | CPLEX 2.1 | | OSL 1.2 | |
|---|---|---|---|---|---|---|---|---|
| Problem | Itns | Secs | Itns | Secs | Itns | Secs | Itns | Secs |
| 25fv47 | 6265 | 69.75 | 2555 | 55.26 | 2868 | 27.36 | 1692 | 26.69 |
| bgprob | 2840 | 55.78 | 587 | 24.28 | 1063 | 26.64 | 1004 | 23.73 |
| bnl2 | 5830 | 127.11 | 3597 | 146.87 | 4534 | 54.93 | 2427 | 56.74 |
| d2q06c | 45555 | 1330.01 | 9545 | 621.32 | 9883 | 261.28 | 5418 | 218.44 |
| degen3 | 7724 | 169.36 | 4018 | 143.30 | 4544 | 73.27 | 1985 | 51.08 |
| greenbea | 24588 | 655.61 | 10123 | 343.07 | 7346 | 105.29 | 4687 | 174.55 |
| maros-r7 | — | — | 3025 | 315.44 | — | — | 4040 | 438.85 |
| pilotnov | 2726 | 29.97 | 1115 | 22.52 | 2602 | 25.26 | 2552 | 49.05 |

The next important influencing factors are the programming language, the compiler and the way of compilation (e.g. level of code optimization) used to create the executable program. Most solvers are written in some high level language (Fortran or C). Some of them use assembly routines in some critical parts of the program that results in substantial speedup. The academic solvers are available in source format (both FortMP and MINOS are written in Fortran) while the commercial solvers are provided in library or executable form.

Finally, the most important factors are the algorithmic capabilities of the solvers. Usually, there are a number of parameters that control the actual solution process. This number can be anything between 10 and 100. Their detailed discussion is beyond the scope of this chapter. They can be classified as *numeric* (e.g. the value of a zero tolerance), *strategic* (e.g. the number of variables in multiple pricing), and *algorithmic* (e.g. primal or dual) parameters. Practically all the parameters have default values that reflect the best choice for most of the problems. On the other hand, the advanced solvers are equipped with an adaptive feature which enables them to modify the parameters "on the fly" to achieve some goal, like increased accuracy, or iteration speed.

To preserve integrity of the solvers, we used them in the following way. It has been noted that presolving problems prior to solution is usually very useful in terms of reducing the computational effort to solve problems. Therefore, we solved the problems in two different ways: first without presolve and then with presolve. All other parameters were left at default giving the chance that the algorithm freely decides about the actual values. The two most interesting performance figures are the number of iterations and solution time.

The results of the testing are presented in Tables 10 and 11. They deserve some interpretation. As a preliminary remark, it has to be emphasized that all the solution times in the tables are extremely good. This is partly due to the fast

**Table 11** Number of iterations and solution times in seconds on benchmark LP problems, with presolve, except MINOS which does not have this feature. Main algorithm: primal. Computer: Elonex PC with Pentium 90 MHz processor, 16 MB RAM and 256 KB cache. " — "denotes "unable to solve"

| | MINOS 5.3 | | FortMP 1.03F | | CPLEX 2.1 | | OSL 1.2 | |
|---|---|---|---|---|---|---|---|---|
| Problem | Itns | Secs | Itns | Secs | Itns | Secs | Itns | Secs |
| 25fv47 | 6265 | 69.75 | 2765 | 57.84 | 2402 | 23.35 | 1760 | 27.46 |
| bgprob | 2840 | 55.78 | 551 | 22.14 | 918 | 20.82 | 1037 | 23.62 |
| bnl2 | 5830 | 127.11 | 2864 | 116.38 | 4532 | 34.99 | 2129 | 45.92 |
| d2q06c | 45555 | 1330.01 | 9698 | 628.04 | 10033 | 250.96 | 5644 | 226.73 |
| degen3 | 7724 | 169.36 | 4018 | 144.63 | 4613 | 82.06 | 1863 | 45.16 |
| greenbea | 24588 | 655.61 | 5792 | 181.70 | 5324 | 54.27 | 6362 | 229.37 |
| maros-r7 | — | — | 3005 | 190.32 | — | — | 4040 | 436.22 |
| pilotnov | 2726 | 29.97 | 540 | 9.67 | 3397 | 33.23 | 2551 | 44.10 |

computer but very importantly due to the substantial algorithmic developments in the sparse simplex method achieved in the last decade. It is widely accepted that under identical computing conditions these solvers are 10 to 100 times faster than the best ones available around the mid 1980s.

Looking at the results of the tables, the first striking observation is that there is a rather wide deviation in the performance figures. The second remarkable feature is that academic solvers are not necessarily worse than the commercial ones, though the iteration speed of the latter is usually, but not always, better. The third thing is that there can be problems that may turn out to be too hard even for very good solvers. If this happens then one can try manual tuning of the algorithm by "playing" with the parameters. Next, we have to note that application of presolve helps improve efficiency in many cases, but, again, not always. The final conclusion that can be drawn from these results, and in general, from results of much bigger sets of real-life problems, is that no single system can be said to be the best in all cases.

## 6 SSX on parallel machines

Essentially there are two inherent difficulties which have inhibited the wide spread adoption and extension of the simplex algorithm to parallel computers.

(1) The use of parallel computers for solving symmetric and non-symmetric systems of linear equations has been well investigated and the main research results may be summarized as follows. Parallel machines are best suited to deliver high speed-up when processing dense matrices. The impact of this issue is appreciated from results which show that for densities as low as 20%, dense processing is faster than sparse processing. Parallel machines are also exploited by algorithms which take advantage of

well-known matrix structures, such as tridiagonal and pentadiagonal matrices. In contrast to this, the efficient processing of unstructured sparse problems by parallel machines is still an open question and no substantial performance speed-up has been reported. Parallel machine architectures are generally well suited for dense vector and matrix processing mainly because hardware scatter and gather instructions do not fully cope with the problem of representing sparse data on these machines. The SSX algorithm is unfortunately affected by this as it is based on unstructured, non-symmetric, linear sparse equation solving.

(2) The profiling information presented in Section 5.3 shows that the main computational work is spread over a number of algorithmic subcomponents such as PRICE, BTRAN, FTRAN, INVERT, etc. The relative computational effort in these modules varies from test problem to test problem. Whereas on serial machines sparsity preservation techniques like *LU* update [24] can provide computational efficiency, these cannot be effectively applied in parallel machines.
Through some ingenuity and data reorganization Forrest and Tomlin [25] speeded up the PRICE step of SSX on vector computers. Although a number of other implementations have been reported (Chen *et al.* [12], Stunkel and Reed [64], Hall and McKinnon [34]) that due to Forrest and Tomlin is still the only credible and robust implementation. If we take into account Amdahl's law and the logic of the serial parts of the SSX algorithm spread over a number of subcomponents, it is easy to see why there is a fairly tight limit on the scope of speed-up of the SSX. Essentially, we cannot easily adapt the SSX for parallel computation since it is difficult to represent list structures on SIMD machines and to balance the processing load on MIMD machines. Whereas in serial machines the use of sparse vector structures leads to speed-up by reducing the number of operations, the same representation due to indirect addressing becomes a hindrance on parallel machines and may slow down processing.

# 7 Sparse simplex within an integrated solver

## 7.1 Integration of interior point methods and sparse simplex

SSX is not only well established, its progress through continued research has made it an indispensable optimization tool for a wide range of commercial and industrial applications. The processing speed advantage of IPM over SSX is marginal for the small to medium size model range which covers most practical LP applications. Applications of IPM to integer programming and postoptimal analysis are currently at a research stage. Nonetheless IPM can still provide a number of benefits, which are summarized below. For really large LPs (10,000 rows and 20,000 columns or more) IPM is undoubtedly the faster of the two solvers provided sufficient RAM is available for the IPM processing. Whereas occasionally SSX may require adaptation and control parameter tuning from one model class to another, default settings of IPM are sufficient to process a wide class of LPs. IPM is not only robust in this way, its progress is not hindered by degeneracy or the stalling problem of the SSX; indeed it reaches the "near-optimal" solution very quickly. This analysis of the relative advantages and disadvantages of the two methods led us to conclude that SSX and IPM are likely to coexist and future developments will harness these two techniques within a common framework. Our observation is vindicated by the recent upsurge of interest in the topic of crossover from IPM solution to an optimum basic SSX solution (Tomlin and Forrest [68], Megiddo [53]). We also note that in OSL the IPM and SSX procedures work from a common underlying model (data) definition; in FortMP's SSX and IPM systems a comparable strategy has been adopted and the designers of CPLEX and OB1 (later developed to CPLEX barrier) also combine their respective SSX and IPM solvers.

Recently a number of researchers have tried to separate the theory underlying IPM algorithms from the simplex terminology. Several researchers [39, 33] have developed alternative proofs to duality theorems (hitherto derived using basis representation) which use the existence or otherwise of the primal and dual feasible solutions. In fact, it is claimed with some justification that an interior solution provides a better starting point for investigating the sensitivity of the solution to changes in the right-hand side values or the objective coefficients than the extreme point solutions.

All the same there are several reasons why it is necessary to compute the extreme point optimum solution and the associated optimum basis. To start with the interior point algorithm does not give a real optimum solution; instead the solution lies in the interior of the polytope in close proximity to the optimal face. In other words all the variables are active in the solution. Levkovitz and Mitra [46], El-Bakry *et al.* [20] have shown that in many cases a very high solution precision is required to identify the values which can be set (rounded) to their bound values. The numerical values of the coefficients of the LP problem may sometimes make it difficult to achieve this solution precision within the framework of IPM. In such cases it is better to retrieve the extreme point solution by completing the search using SSX.

The second situation arises when the LP problem is slightly perturbed as in the case of successive LP or integer programming tree search. In these cases the reapplication of SSX is more efficient than reapplying IPM. This is because the current solution is no longer optimal but lies in a close neighbourhood of a non-optimal face. Therefore by using the optimal basis of the original problem and applying SSX the optimum solution to the new problem may be computed in relatively few iterations.

Given the wide applicability of the SSX algorithmic procedures most of the research effort towards integration has been directed towards identifying an optimum basis starting from a non-extreme point optimal IPM solution. This procedure is often referred to as *basis crossover* or *basis recovery.*

For the purpose of clarity in the following exposition we have considered an LP with only non-negative variables. There are essentially three related problems that lie at the heart of the basis crossover procedure. These are stated as:

(1) Given a set of primal feasible interior point solution values $x^1$, find a superior basic feasible solution $x_B$ such that $c_B^T x_B \leq c^T x^1$.

(2) Given a set of primal optimal solution values, find $B_{opt}$, such that

$$x_B^* = B_{opt}^{-1} b \geq 0 \qquad \text{and} \qquad [c_{B_{opt}^T} B_{opt}^{-1} A]^T \geq c^T.$$

(3) Given a set of primal and dual optimal solution values, find $B_{opt}^{-1}$ as in (2).

We note that if IPM stops at a primal feasible but suboptimal interior point we have to apply procedure (1); on the other hand if the IPM terminates on an optimal face, then solution procedures for (2) or (3) may be applied.

The recovery of a superior or an optimum basic solution from a primal feasible non-extreme point solution is well understood in the context of SSX. For instance, the BASIC procedure within MPSX finds a superior extreme point solution and the corresponding basis. Mitra *et al.* [56] and also Marsten and Shanno [52] set out comparable pivotal schemes for a primal only approach to this problem.

The difficulty experienced in the basis recovery (simplex) steps can be ascribed to the cases where the optimal solution is primal or dual degenerate. Taking into account these cases Megiddo [53] stated and proved the following two theorems:

**Theorem 7.1** *If there exists a strongly polynomial time algorithm that finds an optimal basis, given as optimal solution for either the primal or the dual, then there exists a strongly polynomial algorithm for the general LP problem.*

**Theorem 7.2** *There exists a strongly polynomial time algorithm that finds an optimal basis, given an optimal solution for both the primal and the dual.*

Megiddo gives a constructive proof of Theorem 7.2 and shows that no strongly polynomial algorithm exists if we take into consideration only the primal or the dual optimal solution. This explains the importance of problem (3) stated above.

Though the proof is constructive it is not a computational description of the procedure. Therefore, its implementation requires further considerations. First of all, an optimal solution obtained by an interior point solver is just $\epsilon$-optimal and primal/dual feasibility and complementarity are true within a tolerance. During crossover this or a chosen level of optimality, feasibility and complementarity have to be maintained.

Let us denote the primal solution (including the components for the logical variables) by $x$ and the dual solution with dual logicals by $d$. By complementarity we have $x^T d = 0$. Since $x \geq 0$ and $d \geq 0$, it follows that $x_j d_j = 0, \ \forall j$. This defines three mutually exclusive and collectively exhaustive index sets for all $j$:

$$\begin{aligned} J_1 &= \{j \mid x_j > 0, \ d_j = 0\}, \\ J_2 &= \{j \mid x_j = 0, \ d_j = 0\}, \\ J_3 &= \{j \mid x_j = 0, \ d_j > 0\}. \end{aligned}$$

To begin the crossover, an initial basis $B$ has to be constructed. This can be, by default, the ever-present identity matrix. Since this is usually not a good choice it is advisable to use a basis with as many variables from $J_1$ and $J_2$ as possible. For this purpose, crash algorithms can be used with success. In the resulting primal representation the components of $x$ which are positive but not in the basis are called *superbasic variables.* No value has to be recomputed because $x$ is a solution to $Ax = b$ which is now partitioned as $Bx_B + Nx_N = b$. There can be two problems with this representation: (a) one or more components of $x_N$ are non-zero (primal superbasics), (b) one or more components of $d_B$ are non-zero (dual superbasics).

In the first part of the crossover the non-zero components of $x_N$ are driven to zero. Formally, if $x_j > 0$ for $j \in N$ (and, therefore, $j \in J_1$) then an attempt is made to push it to zero. This, of course, causes changes in the values of the basic variables. If during the push a basic variable reaches its feasibility bound then a pivot step has to be made with $x_j$ as incoming and the blocking variable as outgoing variable. Otherwise, $x_j$ can be pushed to zero. This latter step is similar to the bound change of the bounded variable simplex algorithm. Whatever step is made the number of primal superbasic variables decreases by one. At termination, problem (a) is completely resolved. The procedure can be considered as a series of cheap simplex iterations where no BTRAN and pricing are needed, since the candidates are taken from a predetermined list: $J_1 \setminus B$. Primal feasibility ($x$) as well as optimality is maintained ($d_j = 0,\ j \in J_1 \cup J_2$) throughout the steps.

In the second part of the crossover the non-zero components of $d_B$ are driven to zero. (This $B$ may be different from the one used at the beginning of part one.) Formally, if $d_j > 0$ for $j \in B$ (and, therefore, $j \in J_3$) then the corresponding basic variable is removed from the basis. The entering variable is determined in such a way that the feasibility of the dual solution is preserved. Practically this requires a dual pivot step. Subtleties are omitted here. The important thing is that at each step the number of dual superbasics is reduced by one. Computationally, the elimination of one dual superbasic variable is usually more expensive than the elimination of a primal one.

It is easy to see that the above procedure makes degenerate steps in the sense that the primal/dual objective values remain unchanged. Since the values were optimal at the beginning, at the end we have a primal/dual optimal basic solution. The beauty and the great practical value of this crossover procedure is that the number of steps needed is never more than $n$.

### 7.2 Use of simplex within mixed-integer programming

In solving a family of slightly modified problems it is natural to reuse SSX and the so-called "warm start" from the previous optimal basis can lead to a substantial reduction in solution time. This is in contrast to IPM which cannot be exploited in the same way for a family of related LP problems. Mixed-integer programming (MIP) methods which employ branch and bound or branch and cut techniques therefore use SSX as a reoptimization procedure in a routine fashion. Since MIP procedures by and large require investigation of a series of slightly modified problems stored in a hierarchical (tree) structure, reuse of stored optimal bases and applying dual simplex leads to substantial efficiency gains. Even for those models for which IPM solves the relaxed LP much faster than SSX, if MIP tree search is required then it becomes necessary to carry out basis crossover at the end of the continuous LP relaxation phase.

### 7.3 The outline of information flow

SSX occupies a central position within a mathematical programming system. It is a solver in its own right, it is also an important algorithmic component for mixed-integer programming, postoptimality analysis, and for basis crossover from an interior point solution. Whether a primal algorithm or a dual algorithm is used in SSX the main data structure remains the same.

The information and control flow diagram (Fig. 2) illustrates how the different algorithmic components are related to one together. Figure 2 also highlights the rather paramount and central role taken by the SSX algorithm and the related basis (inverse) representation. It can be deduced from Fig. 2 that if presolve is carried out at the beginning then it is necessary to carry out postsolve at the end, and hence recover an optimum basis for the original problem. Similarly, an IPM solution requires postprocessing, that is, basis recovery should MIP or postoptimal analysis be needed. On the other hand, there are many situations where it is not necessary to recover an optimal basis and a postsolve followed by output provides the required information. Basis save and restart procedures are invoked where appropriate, to achieve computational efficiency.

## 8 Future directions

The continued interest in many topics surrounding computational aspects of the sparse simplex algorithm and the central position occupied by the SSX in all the industrial-strength LP solvers guarantee that this topic is going to remain high on the agenda of the mathematical programming and numerical analysis specialists.

In our opinion there remain at least two important challenges for researchers. First, it is necessary to find out the best ways of combining applied research results which have proven to be successful in restricted classes of models. Typical examples of these are preprocessing, advanced basis, normalized pricing, recognition and exploitation of network/GUB structure, alternative basis factorization and update strategies. Adaptive algorithms which can incorporate these algorithm components and algorithm control parameters intelligently are likely to provide more robust as well as better performing solution engines. Second, mapping the SSX algorithm components to multiprocessor architectures remains a genuinely difficult and thorny task; parallel extensions of SSX demonstrating speedup and scale up in respect of benchmark models will definitely be considered major contributions in the field of computational linear programming.

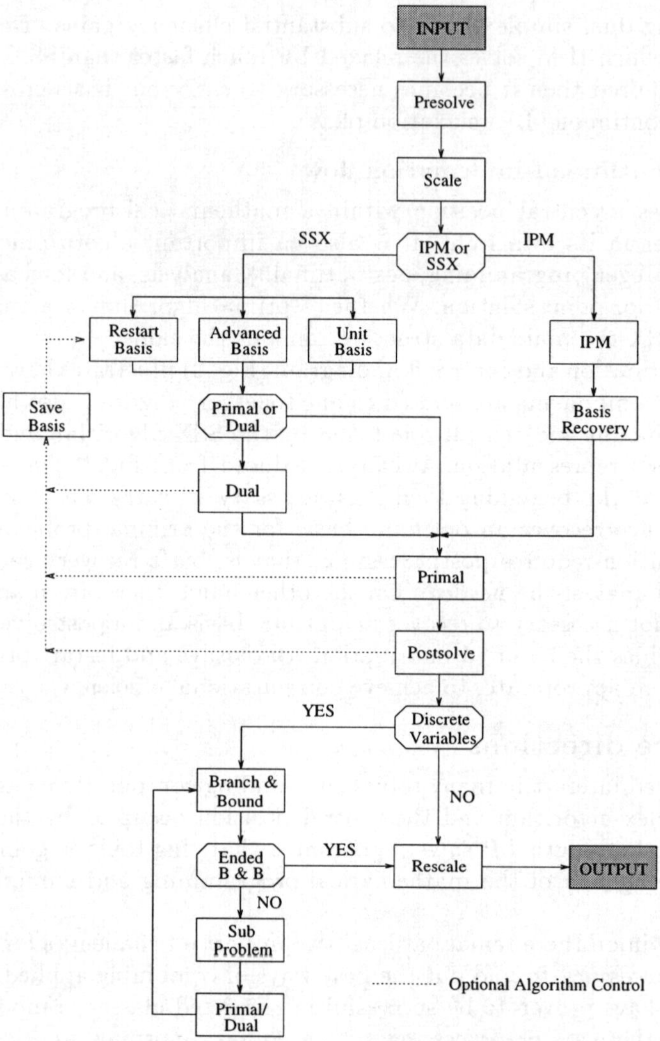


FIG. 2. Outline of information and control flow in an integrated optimization system

## Bibliography


1. R.H. Bartels and G.H. Golub. The simplex method of linear programming using LU decomposition. *Comm. ACM*, pages 266–268, 275–278, 1969.
2. E.M.L. Beale, 1984. Private Communication.
3. R.E. Bixby. Implementing the simplex method: the initial basis. *ORSA Journal on Computing*, 4:267–284, 1992.
4. R.E. Bixby. Progress in linear programming. *ORSA Journal on Computing*, 6:15–22, 1994.

5. R.G. Bland. New finite pivoting rules for the simplex method. *Mathematics of Operations Research*, 2:103–107, 1977.
6. K.H. Borgwardt. The average number of pivot steps required by the simplex method is polynomial. *Zeitschrift für Operational Research*, 26:157–177, 1982.
7. A.L. Brearley, G. Mitra and H.P. Williams. Analysis of mathematical programming problems prior to applying the simplex method. *Mathematical Programming*, 8:54–83, 1975.
8. G.G. Brown and R.D. McBride. Solving generalized networks. *Management Science*, 30:1497–1523, 1984.
9. G.G. Brown and M.P. Olson. Dynamic factorization in large scale optimization. Technical Report NPSOR-93-008, Naval Postgraduate School, Monterey, California, 1993.
10. J.R. Bunch and D.J. Rose, (eds.). *Sparse Matrix Computation.* Academic Press, 1976.
11. D.M. Carstens. Crashing techniques. In W. Orchard-Hays *Advanced Linear-Programming Computing Techniques*, 131–139. McGraw-Hill, 1968.
12. G.H. Chen, H.F Lina and J.P. Shen. Data mapping of linear programming on fixed sized hypercubes. *Parallel Computing*, 13:235–243, 1990.
13. G.B. Dantzig. Maximization of a linear function of variables subject to linear inequalities. In T.C. Koopmans, (ed.), *Activity analysis of production and allocation*, Chapter XXI, Wiley, New York, 1951.
14. G.B. Dantzig. *Linear Programming and Extensions.* Princeton University Press, Princeton, N.J., 1963.
15. G.B. Dantzig and W. Orchard-Hays. Notes on linear programming: Part V. — Alternative algorithm for the revised simplex method using product form of the inverse. The RAND Corporation, *Research Memorandum* RM-1268, November 1953.
16. G.B. Dantzig, A. Orden and P. Wolfe. A generalized simplex method for minimizing a linear form under linear inequality constraints. *Pacific Journal of Mathematics*, 5:183–195, 1955.
17. G.B. Dantzig and R.M. Van Slyke. Generalized upper bounding techniques. *Journal of Computer and System Sciences*, 1:213–226, 1967.
18. I.S. Duff, A.M. Erisman and J.K. Reid. *Direct Methods for Sparse Matrices.* Clarendon Press, Oxford Science Publications, Oxford, 1986.
19. I.S. Duff and G.W. Stewart, (eds.). *Sparse Matrix Proceedings 1978.* Society for Industrial and Applied Mathematics, Philadelphia, 1978.
20. A.S. El-Bakry, R.A. Tapia and Y. Zhang. A study of indicators for identifying zero variables in interior point methods. *SIAM Review*, 36:45–72, 1994.
21. E.F.D. Ellison, M. Hajian, R. Levkovitz, I. Maros, G. Mitra and D. Sayers. *FortMP Manual.* Brunel, The University of West London, May 1994.

22. A.M. Erisman, R.G. Grimes, J.G. Lewis and W.G. Poole, Jr. A structurally stable modification of Hellerman-Rarick's $P^4$ algorithm for reordering unsymmetric sparse matrices. *SIAM Journal on Numerical Analysis*, 22:369–385, 1985.
23. J.J. Forrest and D. Goldfarb. Steepest edge simplex algorithms for linear programming. *Mathematical Programming*, 57:341–374, 1992.
24. J.J. Forrest and J.A. Tomlin. Updating triangular factors of the basis to maintain sparsity in the product-form simplex method. *Mathematical Programming*, 2:263–278, 1972.
25. J.J. Forrest and J.A. Tomlin. Vector processing in simplex and interior point methods. *Annals of Operations Research*, 22:71–100, 1990.
26. T. Gal and F. Geue. A new pivoting rule for solving various degeneracy problems. *Operations Research Letters*, 11:23–32, 1992.
27. D.M. Gay. Electronic mail distribution of linear programming test problems. *Mathematical Programming Society COAL Newsletter*, 13:10–12, 1985.
28. P.E. Gill, W. Murray, M.A. Saunders and M.H Wright. Sparse matrix methods in optimization. *SIAM Journal of Scientific and Statistical Computing*, 5:562–589, 1984.
29. P.E. Gill, W. Murray, M.A. Saunders and M.H. Wright. A practical anti–cycling procedure for linearly constrained optimization. *Mathematical Programming*, 45:437–474, 1989.
30. D. Goldfarb and J. Reid. A practicable steepest-edge simplex algorithm. *Mathematical Programming*, 12:361–371, 1977.
31. N.I.M. Gould and J.K. Reid. New crash procedures for large systems of linear constraints. *Mathematical Programming*, 45:475–501, 1989.
32. H.J. Greenberg and J.E. Kalan. An exact update for Harris' tread. *Mathematical Programming Study*, 4:26–29, 1975.
33. O. Guler, C. Roos, T. Terlaky and J.P. Vial. Interior point approach to the theory of linear programming. Research Report 1992-3, Faculté des Sciences Economiques et Sociales, Université de Genéve, February 1992.
34. J.A.J. Hall and K.I.M. McKinnon. An asynchronous parallel algorithm to the revised simplex method. Presented to APMOD95, London, 1995.
35. M. Hall. An algorithm for distinct representatives. *American Mathematical Monthly*, 63:616–617, 1956.
36. P.M.J. Harris. Pivot selection method of the DEVEX LP code. *Mathematical Programming*, 5:1–28, 1973.
37. E. Hellerman and D. Rarick. Reinversion with the preassigned pivot procedure. *Mathematical Programming*, 1:195–216, 1971.
38. E. Hellerman and D. Rarick. The partitioned preassigned pivot procedure (P4). In D.J. Rose and R.A. Willoughby, (eds.), *Sparse Matrices and Their Applications*. Plenum Press, 1972.
39. B. Jansen, C. Roos, T. Terlaky and J.P. Vial. Interior point methodology

for linear programming: Duality, sensitivity analysis and computational aspects. Technical Report 93-28, Faculty of Mathematics and Informatics, Delft University, Delft, The Netherlands, 1993.

40. J.E. Kalan and J.F. Kalan. KRASH, an adaptive relaxation method for solving linear programs. Department of Computer Science, University of Texas at Austin, July 1986.
41. N. Karmarkar. A new polynomial-time algorithm for linear programming. *Combinatorica*, 4:373–395, 1984.
42. R. Karp. Combinatorics, complexity and randomness. *Comm. ACM*, 29:98–109, 1986.
43. L.G. Khachiyan. A polynomial algorithm in linear programming. *Doklady Akademii Nauk SSSR*, 244:1093–1096, 1979. Translated into English in *Soviet Mathematics Doklady*, 20:191–194, 1979.
44. V. Klee and G. Minty. How good is the simplex algorithm? In O. Sisha, (ed.), *Inequalities III*, 159–179. Academic Press, New York, 1972.
45. C.E. Lemke. The dual method of solving the linear programming problem. *Naval Research Logistics Quarterly*, 1:36–47, 1954.
46. R. Levkovitz and G. Mitra. Experimental investigations in combining primal-dual interior point method and simplex based LP solvers. *Annals of Operations Research*, 58:19–38, 1995.
47. H.H. Markowitz. The elimination form of the inverse and its applications to linear programming. *Management Science*, 3:255–269, 1954.
48. I. Maros. Adaptivity in linear programming, II. (In Hungarian) *Alkalmazott Matematikai Lapok* (Journal of Applied Mathematics), 7:1–71, 1981.
49. I. Maros. A general Phase-I method in linear programming. *European Journal of Operational Research*, 23:64–77, 1986.
50. I. Maros. *MILP Linear Programming System, User's Guide for Version V3.40.* Computer and Automation Institute (MTA SZTAKI), Budapest, Hungary, January 1990.
51. I. Maros and C. Mészaros. An exact implementation of the simplex method. *Annals of Operations Research*, 58:3–17, 1995.
52. R. Marsten and D. Shanno. *Interior point methods for LP; Ready for production use.* Presented at an ORSA/TIMS Workshop, New York, October 1989.
53. N. Megiddo. On finding primal and dual optimal bases. *ORSA Journal on Computing*, 3:63–65, 1991.
54. G. Mitra. Sparse inverse in the factored form and maintaining sparsity during simplex iterations. In D.J. Evans, (ed.), *Software for Numerical Mathematics*, 85–97. Academic Press, 1976.
55. G. Mitra and M. Tamiz. Alternative methods for representing inverse of LP basis matrices. In S. Kumar, (ed.), *Recent Developments in Mathematical Programming*, 273–302. Gordon and Breach, 1991.

56. G. Mitra, M. Tamiz and J. Yadegar. Experimental investigation of an interior search method within a simplex framework. *Communications of the ACM*, 31:1474–1482, 1988.
57. B.A. Murtagh and M. Saunders. MINOS 5.1 user's guide. Technical Report, SOL 83-20R, Stanford University, Stanford, California, 1987.
58. W. Orchard-Hays. *Advanced Linear-Programming Computing Techniques.* McGraw-Hill, 1968.
59. S. Powell. A development of the product algorithm for the simplex method using reduced transformation vectors. *Mathematical Programming*, 9:93–107, 1975.
60. J.K. Reid, (ed.). *Large Sparse sets of Linear Equations.* Academic Press, 1971.
61. J.K. Reid. Fortran subroutines for handling sparse linear programming bases. Report AERE-R8269, Harwell, 1976.
62. J.K. Reid. A sparsity-exploiting variant of the Bartels–Golub decomposition for linear programming bases. *Mathematical Programming*, 24:55–69, 1982.
63. D.J. Rose and R.A. Willoughby, (eds.). *Sparse Matrices and Their Applications.* Plenum Press, 1972.
64. C.B. Stunkel and D.A. Reed. Hypercube implementation of the simplex algorithm. In *Proceedings of the Third Conference on Hypercube Concurrent Computers and Applications*, 1473–1482, 1988.
65. U.H. Suhl and L.M. Suhl. Computing sparse LU factorizations for large-scale linear programming bases. *ORSA Journal on Computing*, 2:325–335, 1990.
66. U.H. Suhl and L.M. Suhl. A fast LU update for linear programming. *Annals of Operations Research*, 43:33–47, 1993.
67. R.E. Tarjan. Depth-first search and linear graph algorithms. *SIAM J. Computing*, 1:146–160, 1972.
68. J.A. Tomlin and J.J.H. Forrest. Switching from interior to vertex solution in OSL, 1991. Presented to IBM Europe Institute, Oberlech, Austria.
69. R.A Willoughby, (ed.). *Proceedings of the Symposium on Sparse Matrices and their Applications*, Yorktown Heights, New York, 1969. IBM report RA1.
70. P. Wolfe. A technique for resolving degeneracy in linear programming. *SIAM Journal of Applied Mathematics*, 11:205–211, 1963.
71. P. Wolfe. The composite simplex algorithm. *SIAM Review*, 7:42–54, 1965.
72. G. Zoutendijk. A product form algorithm using contracted transformation vectors. In J. Abadie, (ed.), *Integer and Non-linear Programming.* North Holland, Amsterdam, 1970.

# 2
# Interior point methods

**Cornelis Roos**

*Delft University of Technology,*
*Faculty of Technical Mathematics and Computer Science,*
*P.O. Box 5031, 2600 GA Delft, The Netherlands*

**Jean-Philippe Vial**

*Department of Management Studies, University of Geneva,*
*102 Bd Carl-Vogt, CH-1211 Geneva 4, Switzerland*

## 1 Introduction

The interest in interior point methods for linear programming emerged from Karmarkar's contribution in 1984. This field has soon become one of the most active in the area of mathematical programming. It introduced new ideas and techniques that now have received their own place among the basic tools in optimization. In this chapter we intend to provide an introduction to the theory of interior point methods.

To get a good understanding of interior point methods, one must keep in mind that at the heart of this theory is the analysis of complexity of algorithms. For a long time the simplex algorithm was the only practical algorithm for linear programming. Many people tried to justify the remarkable efficiency of the method by providing a theoretical bound on the number of simplex iterations. To date, no one has yet given a polynomial bound for general problems. The more basic question of the mere existence of an algorithm that would solve any instance of linear programming in polynomial time remained open until Khachiyan [27] settled the issue in 1979. Khachiyan's contribution was fundamental for the theory of linear programming and related topics. Unfortunately his ellipsoid method did not fulfil its promise as a practical solution method for linear programming. Shortly after Khachiyan's contribution, Karmarkar, in a seminal paper [26], patched the gap between theory and practice. He proposed a totally new method that enjoyed both polynomial complexity and practical efficiency. Since then, it has been the rule for each new algorithmic contribution that the author provides a complexity analysis and at least a bound on the total number of iterations. Consequently, a good deal of the developments in this chapter will be devoted to establishing bounds on the number of iterations.

In presenting interior point methods we are faced with a difficult choice. Due to the intense research activity in the field, literally hundreds of algorithms have been devised and analysed. The bibliography of Kranich [31, 32], in its most recent update of mid-1993, mentions approximately 1,300 papers, almost all of them dealing with some "new" interior point method. Although those methods obviously share similarities, they still look pretty much different. This very diversity discourages the non-specialist, and in the meantime challenges the specialist in search of a unifying framework. In this chapter we hope to give at least a partial answer to that query.

We organize this chapter around four main themes. After a short review of the definition relative to complexity we shall first revisit, in Section 3, the theory of linear programming itself, in the light of the ideas of interior point methods. We will put special emphasis on the concept of the central path, a privileged continuous curve that is interior to the feasible set and converges to an optimal point. Most interior point methods follow, or are related to, the central path. To describe those methods we shall use the notion of target sequence, an idea that was informally introduced by Mizuno [36, 37] and systematized in [25]. The concept captures the basic ideas that underlie some of the most prominent interior point methods that appeared in the literature. It provides a unified framework for a variety of methods, by-passing the need for separate analyses. This will be dealt with in Section 4. The last two sections will address two important issues and discuss algorithms that give answers to them. The first issue, dealt with in Section 5, is the asymptotic behaviour of interior point methods. It is often observed that shortly before convergence interior point methods drastically accelerate. This phenomenon can be explained relatively easily for the method named predictor–corrector. Section 6 will be devoted to the issue of infeasible starts. Indeed the vast majority of interior point algorithms suppose that an initial interior feasible point is at hand. Some special procedure must be set to produce this initial feasible point. However interior point methods have the nice property of enforcing feasibility quite naturally. This has recently been an area of very active research. See, e.g. [29, 38, 41, 44, 53, 57]. Interestingly enough a variant of Karmarkar's algorithm provides a simple solution to the problem. This gives us the opportunity to discuss Karmarkar's contribution and to conclude the chapter at the point where the field of interior point methods for linear programming started.

## 2 Complexity and convergence

Much of the theoretical analysis of algorithms has to do with the convergence properties. In the analysis of interior point methods, essentially two types of convergence are used. The first one is global convergence: does the algorithm converge to a solution and after how many iterations? The second one has to do with the asymptotic rate of convergence: is the progress towards a solution steady or can we expect an acceleration after some stage?

### 2.1 Complexity and global convergence

The main point of a complexity analysis is to estimate the total computational effort to solve any given linear programming problem. (The memory space used by the algorithm is another criterion for measuring complexity. We shall not discuss it here.) This estimation depends on the total work per iteration (mainly made up from linear algebra operations) and on a bound on the total number of iterations. Usually there is not so much room within a given method to reduce the computational complexity per iteration. This legitimizes focusing the analysis on the number of iterations.

To give more specific definitions, we must clarify two fundamental notions. What exactly do we mean by "solution" and by "problem size"? Due to finite precision calculations on all computing machines, we have to accept approximations in optimality as well as in feasibility. Therefore a complexity result has to be related to a user-specified precision level $\varepsilon$. Consider for example the problem

$$\min\{c^T x \,:\, Ax = b,\, x \geq 0\},$$

with an (unknown) optimal value $z^*$. We declare that any $x \geq 0$, such that $||Ax - b|| < \varepsilon$ and $c^T x - z^* < \varepsilon$, solves the problem at the $\varepsilon$ precision level.

The complexity theory provides a precise definition of the size of a problem: it is the length of a binary encoding of the data. Quantities such as the number of constraints, the number of variables, the initial gap, the norm of the initial infeasibilities and so on, do not fully capture the theoretical notion of problem size. However they are much easier to handle. Therefore it has become customary to use them as substitutes for the problem size in the computation of the complexity bound.

The last point in measuring the computational effort is the importance of the number of iterations. The reason is that the computational effort in one single iteration is very much the same from iteration to iteration. It essentially involves well-studied operations, such as vector–vector or matrix–vector products, matrix factorizations, etc. Their complexity is given in the literature.

A time complexity result consists of a function of the problem size that forms a bound on the total computational effort to solve the problem. The algorithm is polynomial if this function is polynomial. In view of our discussion, we will substitute for this formal definition a more operational one. Firstly the bound will be on the number of iterations and not on the computational effort. Secondly the function will be in terms of the number of constraints, the number of variables, the initial gap, the norm of the initial infeasibilities and so on. However we must distinguish between these quantities. If we think of the true definition of the problem size, we see that quantities related to problem data (e.g. numerical values of the entries of the matrix $A$) are related to the problem size through their logarithms. This applies to quantities like the initial infeasibility and duality gap and the precision level. On the contrary quantities like the number of variables and/or constraints can appear as such.

To be more concrete, let us present a typical complexity result in the terms invoked earlier. Assume that a pair of primal-dual feasible solutions $(x^0, s^0)$ is given at the start. Convergence will be met if the current primal-dual pair $(x, s)$ is feasible and the associated duality gap $x^T s$ is below the precision level $\varepsilon$. A typical complexity statement will be of the form: the algorithm stops after at most

$$K\sqrt{n}\ln\frac{(x^0)^T s^0}{\varepsilon}$$

iterations, where $n$ is the number of variables and $K$ is some constant.

The bound is polynomial if it meets several conditions. First, $K$ must be independent of the problem. Second, the bound should be polynomial in the number of variables and/or the number of constraints. Lastly the quantities pertaining to numerical values of the data (initial gap, initial infeasibilities, ..., etc.) and the precision level should appear under a logarithm.

Not all interior point methods meet the last requirement. We shall give examples in this chapter of a complexity result where a quantity pertaining to the numerical values of the data, namely the ratio between the mean and the max of the coordinates of a vector of data, appears directly as a factor in $K$. In that case, we cannot formally consider the bound to be polynomial.

## 2.2 Quadratic convergence

A polynomial bound on the number of iterations does not account for a phenomenon commonly observed in practice: a sudden increase in the progress towards a solution. While in a complexity analysis one focuses on steady guaranteed progress towards a solution, in the present analysis one concentrates on the rate of this progress. This rate of convergence has to be evaluated through some kind of proximity measure between the current iterate $x$ and a target point (solution) $\hat{x}$. The standard proximity measure is the distance $\|x - \hat{x}\|$ but in our analyses we will often use a different concept. The proximity measure can be any non-negative continuous function $\delta$ of $x$ and $\hat{x}$ with the property that $\delta(x, \hat{x}) = 0$ implies $x = \hat{x}$. Clearly $\delta(x^k, \hat{x})$ tends to zero as $x^k$ tends to $\hat{x}$. The rate of convergence between successive iterates $x^k$ and $x^{k+1}$ is measured by the ratio

$$r_\nu^k = \frac{\delta(x^k, \hat{x})}{\delta(x^{k-1}, \hat{x})^\nu}$$

where $\nu \geq 1$ is a given constant. In this chapter we shall be concerned only with the case $\nu = 2$. We say that an algorithm converges quadratically (with respect to the proximity measure $\delta$) in a neighbourhood of $\hat{x}$, if, in that neighbourhood, $\delta < 1$ and the ratio $r_2$ is uniformly bounded by some constant $c < 1$. (If the constant $c$ is larger than 1, the convergence may be analysed with respect to the proximity measure $\tilde{\delta} = \delta/\sqrt{c}$.) Quadratic convergence is very fast; entering a convergence quadratic neighbourhood is nearly equivalent to obtaining an exact solution in a very few iterations. (For instance, if $c = 1$ and $\delta \leq 0.9$, we have $\delta < 10^{-5}$ after at most 8 iterations and $\delta < 10^{-11}$ after at most 9 iterations.)

In general, the point $\hat{x}$ that enters the definition of the proximity measure $\delta$ is an optimal solution to the problem, and hence is not known. Consequently, it is generally not possible to provide a computable test to decide whether the quadratic convergence neighbourhood has been reached. A standard quadratic convergence result simply consists in asserting that "eventually" the iterate will reach a neighbourhood of quadratic convergence. We can call this result an *asymptotic* one. In this chapter we shall prove a result of this type for a particular interior point method named "predictor–corrector". This result states that we may expect a sudden acceleration of the method near the solution, a phenomenon that is often observed in practice.

A distinctive feature of interior point methods is that a stronger quadratic convergence result holds whenever the reference point $\hat{x}$ is an interior feasible solution (and thus not an optimal one). In that case, it turns out that the proximity measure is in effect computable. Moreover a quadratic convergence neighbourhood $\hat{x}$ is often defined by a simple inequality on the proximity measure. One obtains in this way a *local* quadratic convergence result. This result can further be made global if one can provide a bound on the number of iterations required to reach the neighbourhood. This property makes it possible to move in a controlled way the target $\hat{x}$ to approach an optimal point $x^*$. This is essentially the philosophy of the so-called path-following methods. (The other alternative is a potential reduction analysis, an example of which will be given in Section 6 of this chapter).

### 2.3 Notation

The recurring use of some special linear algebra operations has led researchers in the field of interior point methods to introduce shorthand notation. For instance, the vector of all ones, denoted throughout this chapter by the letter $e$, is frequently used. The dimension of the vector $e$ is to be inferred from the context at the mild risk of some ambiguity.

The first common operation in interior point methods is the scaling of a matrix $A$ by a vector $x$: each column of $A$ is to be multiplied by the corresponding component of the vector $x$. To this end we associate the vector $x$ with the diagonal matrix $X = \mathrm{diag}(x)$, whose entries on its diagonal are the components of $x$. The scaling operation amounts to the matrix product $AX$.

The second recurrent linear algebra operation is the component-wise product of two vectors of the same dimension, say $x$ and $s$. The $i$-th component of the product is $(x_i s_i)$. We shall denote it by $xs$. This operation is sometimes known as the Hadamard product. Our notation for the Hadamard product is unambiguous provided the reader keeps in mind the fact that $x$ and $s$ are both vectors. With this notation we can write

$$xs = Xs = Sx,$$

and

$$x^T s = e^T(xs).$$

A third operation of interest consists of taking the component-wise power (positive or negative) of a vector. In the same spirit as in the case of the Hadamard product, we shall denote the vector by $x^p$ whose components are $(x_i)^p$. Note that we could have written $x^p = X^p e$ equivalently.

The last piece of notation is a further extension of the preceding ideas. We shall now define a multidimensional mapping from a one-dimensional function. For instance, let $f : \mathbb{R} \to \mathbb{R}$ and $x \in \mathbb{R}^n$, then $f(x)$ is a shorthand notation for the mapping $g : \mathbb{R}^n \to \mathbb{R}^n$ defined by $(g(x))_i = f(x_i)$. With this in mind, one can write

$$\sum_i \ln(x_i) = e^T \ln(x).$$

More complicated expressions can be built in this way. For example,

$$\|x^{\frac{1}{2}}\|^2 = e^T x,$$

while

$$\frac{x}{e + \sqrt{x}} = (\frac{x_1}{1+\sqrt{x_1}}, \ldots, \frac{x_n}{1+\sqrt{x_n}}).$$

To conclude, let us recall the meaning of the symbols $O$ and $\Omega$ that are useful in the study of convergence. Assume $t$ is a parameter and let $x$ be a variable of interest with a functional dependence with respect to $t$. We say that $x$ is $O(t)$ (of order $t$) if there exists a positive constant $C_1$ such that $|x| \leq C_1|t|$. We say that $x$ is $\Omega(t)$ if there exist two positive constants $C_2$ and $C_3$, such that $C_2|t| \leq |x| \leq C_3|t|$. The constants $C_i$ need not be specified. The notation is particularly useful when $|t|$ tends to 0 or $\infty$; it provides information on the behaviour of the variable $x$ without requiring a full description of the functional dependence. The argument in $O$ (or $\Omega$) can be any function of $t$. The most common one is an exponent, $O(t^p)$. The extreme case with $p = 0$ is often denoted $O(1) = O(t^0)$ (resp. $\Omega(1) = \Omega(t^0)$). The statement $x = O(1)$ simply means that $|x|$ is bounded while $x = \Omega(1)$ implies that $|x|$ is bounded away from zero.

## 3 The theory of linear programming revisited

### 3.1 Introduction

In mathematical programming, algorithms and the study of optimality conditions are strongly interrelated. In linear programming, for instance, the optimality conditions put emphasis on extreme points. The solution method of choice, up until the recent past, was the simplex algorithm, a method that follows a pathway to a solution through extreme points. Reciprocally, the simplex algorithm yields, as a by-product, the basic duality results of linear programming.

Interior point methods have shifted the interest towards interior points, more precisely, towards continuous interior trajectories. The most distinguished trajectory is the so-called *central path*. This is in a way an ideal trajectory that most interior point methods seek to approximate. True enough, the convergence analysis of interior point methods does not formally require a study of the properties of the central path (or other related interior continuous trajectories such as weighted central paths). However, the strong results achieved by these methods suggest that the above trajectories are worth a study of their own. This is the topic of this section.

It appears that the effort is well worthwhile. Not only shall we retrieve, in a rather simple way, the full duality theory for linear programming, but we shall also pay attention to the concept of optimal partition. The latter has the great advantage of being unique and of characterizing the optimal solution set. This is in sharp contrast with the optimal vertices and/or bases which may be exponentially many. One may draw several important consequences. Let us point out two of them. Firstly, one can interpret most interior point algorithms as iterative schemes that strive to identify the optimal partition. Since this partition is unique, whatever the amount of degeneracy, one may expect that interior point methods are not adversely affected by degeneracy. This is an experimental fact, a well established one nowadays. The other consequence is in the realm of sensitivity analysis: if the problem is parameter dependent, we define critical parameter values as values at which the optimal partition changes. Between the critical values, the optimal value varies linearly. This topic is the subject of active research [18].

## 3.2 Problem definition and assumptions

We shall deal with the following asymmetrical pair of dual linear programming (LP) problems

$$\begin{aligned} &(P) \qquad \min\{c^T x \,:\, Ax = b, x \geq 0\}, \\ &(D) \qquad \max\{b^T y \,:\, A^T y + s = c, s \geq 0\}, \end{aligned}$$

where $A$ is an $m \times n$ matrix. $c$, $x$, $s \in \mathbb{R}^n$ and $b$, $y \in \mathbb{R}^m$. The feasible regions of $(P)$ and $(D)$ are denoted by $\mathcal{P}$ and $\mathcal{D}$ respectively. It is convenient to posit:

**Assumption 1** *The matrix $A$ has full row rank $m$.*

This assumption enforces a one-to-one correspondence between $y$ and $s$ in $\mathcal{D}$. Simple algebra yields

$$y = (AA^T)^{-1}A(c - s). \tag{3.1}$$

We shall sometimes use this closed form expression for $y$, but most of the time we shall implicitly refer to it by using $y \in \mathcal{D}$ or $s \in \mathcal{D}$ instead of the correct statement $(y, s) \in \mathcal{D}$. Assumption 1 is not restrictive: if it is not fulfilled, we can simply remove constraints from the primal formulation until the constraint matrix has full rank.

If $x \in \mathcal{P}$ and $x > 0$, we shall say that $x$ is a *positive vector* in $\mathcal{P}$, and if $(y, s) \in \mathcal{D}$ with $s > 0$, then we shall say that $s$ is a *positive vector* in $\mathcal{D}$. The main assumption in this section is that such positive vectors exist in both $\mathcal{P}$ and $\mathcal{D}$.

**Assumption 2** *Both $\mathcal{P}$ and $\mathcal{D}$ contain a positive vector.*

We shall refer to this assumption as the *interior point assumption*. At first sight the interior point assumption looks quite restrictive. This is not so. In a later section, we shall reformulate any general problem as an equivalent problem for which the interior point assumption holds naturally. We shall then extend all the results of the present section to the general case.

If $x \in \mathcal{P}$ and $(y, s) \in \mathcal{D}$, then we call $(x, s)$ a *primal-dual feasible pair.* In that case one has

$$c^T x - b^T y = x^T s \geq 0, \tag{3.2}$$

which implies that all primal objective values $c^T x$ are larger than or equal to all dual objective values $b^T y$. This is known as the *weak duality result.* A consequence is that if the so-called *duality gap* $x^T s$ vanishes, then $x$ is an optimal solution for $(P)$ and $(y, s)$ is an optimal solution for $(D)$.

We now derive a first consequence of the interior point assumption. In doing so, we shall use a simple but useful relationship that we state for further use as a lemma.

**Lemma 3.1** *Assume $\bar{x} \in \mathcal{P}$ and $\bar{s} \in \mathcal{D}$. Then for all primal-dual feasible pairs $(x, s)$*

$$x^T s = \bar{s}^T x + \bar{x}^T s - \bar{x}^T \bar{s}.$$

**Proof** From the feasibility assumption, $x - \bar{x}$ and $s - \bar{s}$ are orthogonal since the first one belongs to the null space of $A$ while the second is in the row space of $A$. Expanding the scalar product $(x - \bar{x})^T (s - \bar{s})$ and equating it to zero yields the result. □

**Corollary 3.2** *For all positive $K$, the set of all primal-dual feasible pairs $(x, s)$ such that $x^T s \leq K$ is bounded.*

**Proof** By Assumption 2, there exists a positive primal-dual feasible pair $(\bar{x}, \bar{s})$. Substituting $K$ for $x^T s$ in Lemma 3.1, we get

$$\bar{s}^T x + \bar{x}^T s \leq K + \bar{x}^T \bar{s}.$$

This implies that both $\bar{s}^T x$ and $\bar{x}^T s$ are bounded. Since $\bar{x} > 0$ and $\bar{s} > 0$, we conclude that all components of $x$ and $s$ must also be bounded. □

A consequence of this result is that the optimal sets of $(P)$ and $(D)$ are bounded. Surprisingly enough, the converse is also true (see e.g. [35]). The interior point assumption is thus equivalent to having bounded optimal solution sets.

### 3.3 The target map and its inverse

In this section we deal with a result whose importance remained unnoticed for a long time, though it is particularly relevant to interior point methods. However, its scope is wider and we consider it to be a fundamental theorem on linear programming.

We can paraphrase this result by saying that there exists a one-to-one mapping between the positive primal-dual feasible pairs and the positive vectors in $\mathbb{R}^n$. This property was probably first proved by Kojima *et al.* [28]. We shall give a simple alternative proof for it. To this end, we introduce the mapping $\psi : \mathbb{R}^n_+ \times \mathbb{R}^n_+ \mapsto \mathbb{R}^n_+$, defined by $\psi(x, s) := xs$.

**Theorem 3.3** *For each $w \in \mathbb{R}^n_+$ there exists a unique positive primal-dual feasible pair $(x, s)$ such that $\psi(x, s) = w$.*

**Proof** The proof uses the function $g : \mathbb{R}_+ \to \mathbb{R}$ defined by

$$g(t) := t - \ln t - 1, \quad t > 0. \tag{3.3}$$

This function is strictly convex and tends to $+\infty$ as $t$ tends to 0 or $+\infty$. Thus, it achieves a unique minimum at $t = 1$. Since $g(1) = 0$, we also have that $g(t) \geq 0$. Now let $w$ be any positive vector in $\mathbb{R}^n$ and consider

$$h(x, s) := \sum_{j=1}^{n} w_j\, g(t_j) = \sum_{j=1}^{n} x_j s_j - \sum_{j=1}^{n} w_j \ln \frac{x_j s_j}{w_j} - \sum_{j=1}^{n} w_j, \tag{3.4}$$

where $t_j = (x_j s_j)/w_j$. Note that by definition $h$ is non-negative.

From Lemma 3.1 the quadratic term $x^T s$ in $h$ is linear, hence convex for all feasible primal-dual pairs. Obviously the second term in $h$ is strictly convex since its Hessian has a positive diagonal and zero elements elsewhere. Hence $h$ is strictly convex on its restriction to the primal-dual feasible set.

We proceed by showing that $h$ has a minimizer. It suffices to prove that the level sets of $h$ are compact. Let us assume that $(x,\, s)$ belongs to some level set $\mathcal{L}$ of $h$. So, $\mathcal{L}$ consists of all positive primal-dual pairs $(x,\, s)$ for which $h(x,\, s) \leq \gamma$ for some non-negative $\gamma$. Since $h(x, s)$ is the sum of the $n$ non-negative terms $w_j g(t_j)$, $1 \leq j \leq n$, and each $w_j$ is positive, it follows that each $g(t_j)$ is bounded from above. In view of the definition of $g$ there exists a positive number $M > 0$ such that

$$\frac{1}{M} \leq x_j s_j \leq M, \quad j = 1, \ldots n. \tag{3.5}$$

From the right-hand side inequality in (3.5) we get $x^T s \leq nM$. Hence, by Corollary 3.2, we obtain that the level set is bounded. From the left-hand side of (3.5), we also obtain that the level set is uniformly bounded away from the boundary of the positive orthant. Since $h$ is continuous, we obtain that the level set is compact. Consequently, $h$ achieves a minimum value at some positive primal-dual pair $(x,\, s)$. Note that the strict convexity of the restriction of $h$ implies that this pair $(x,\, s)$ is unique.

Let us now consider the first-order optimality conditions for a minimizer of $h$. These conditions require that the gradient of $h(x, s)$ be orthogonal to the constraint set. Taking the derivatives with respect to $x$ and $s$, we obtain the conditions

$$s - wx^{-1} \in \text{row space of } A, \tag{3.6}$$

$$x - ws^{-1} \in \text{null space of } A, \tag{3.7}$$

which show that the vectors $s - wx^{-1}$ and $x - ws^{-1}$ are orthogonal.

Let $d$ and $v$ be positive vectors defined by $d = \sqrt{x/s}$ and $v = \sqrt{xs}$. Multiplying the vectors in (3.6) and (3.7) by $d$ and $d^{-1}$ respectively, we get the same vector $v - w/v$. Moreover, this vector must simultaneously belong to the row space and to the null space of $AD$ where $D = \text{diag}(d)$. Thus $v - w/v$ is the zero vector. Equivalently, $xs = v^2 = w$. We have just proved that for all positive

$w$, there exists a unique positive primal-dual pair $(x,\, s)$ such that $xs = w$. This completes the proof □

From the proof of Theorem 3.3 we have that the minimizer of $h$ for a given $w$ is uniquely defined by the equations

$$\begin{aligned} Ax &= b, \quad x \geq 0, \\ A^T y + s &= c, \quad s \geq 0, \\ xs &= w. \end{aligned} \tag{3.8}$$

We shall later make extensive use of this formulation.

The result of Theorem 3.3 is so important for our further analysis that we shall define the *target map* to be the restriction of $\psi$ to the positive primal-dual pairs, and define the *inverse target map* as its inverse mapping. For obvious reasons, we shall often name the set of positive vectors in $\mathbb{R}^n$ the *target space*, or the *w-space*. Though we did not make it explicit, the target map and its inverse clearly depend on the LP problem under consideration. Actually, this dependence is two-way since it has been recently proved in [3] that the problem data, represented by the triple $(A,\, b,\, c)$, can be reconstructed from the inverse target map. Hence the target map and its inverse contain all the information relative to the underlying LP problem.

Let us mention two additional nice properties of the target map (and its inverse). Firstly, one can easily check that the inverse function theorem holds at every definition point, so that the target map and its inverse are infinitely many times differentiable (see, e.g. [35]). Secondly, we note that if $w = \psi(x, s)$, then $x^T s = e^T w$. So in the $w$-space the positive primal-dual pairs with a fixed duality gap are represented by a hyperplane orthogonal to the all-one vector $e$.

### 3.4 The central path

The inverse target map assigns to every positive vector $w$ the positive primal-dual pair $\psi^{-1}(w)$. We will write

$$\psi^{-1}(w) = (x(w),\, s(w))\,.$$

Since the inverse target map has nice continuity and differentiability properties, we can associate a trajectory in the $w$-space with a trajectory in the $(x,\, s)$-space. We can further note that any trajectory in the $w$-space that converges to the origin has a correspondent in the $(x,\, s)$-space for which the associated duality gap converges to zero. This will be a convenient way to design paths to the optimal set.

Among all possible trajectories in the $w$-space, the set of points along the main diagonal play a special role. These points have the form $\mu e$, with $\mu > 0$. The inverse image of the main diagonal constitutes the so-called *central path* of the problem pair consisting of $(P)$ and $(D)$. The name central stems from the fact that the main diagonal is the central line of the positive orthant. It will be convenient to denote the pair $(x(\mu e),\, s(\mu e))$ by $(x(\mu),\, s(\mu))$. The curve in the primal space $\mathcal{P}$, obtained by letting $x(\mu)$ run through all of its values $(\mu > 0)$, is called the central path of $(P)$. Similarly, the curve in the dual space $\mathcal{D}$, obtained by letting $s(\mu)$ run through all of its values, is called the central path of $(D)$. As pointed out earlier, the full rank assumption on $A$ implies that there is a unique $y$ associated with $s(\mu)$. We shall denote it by $y(\mu)$. Recall that $x(\mu)$ and $s(\mu)$ are uniquely determined by the system

$$\begin{aligned} Ax &= b, \quad x \geq 0, \\ A^T y + s &= c, \quad s \geq 0, \\ xs &= \mu e. \end{aligned}$$

The central paths of $(P)$ and $(D)$ are analytic curves.

Since the limit of $x(\mu)^T s(\mu)$ is zero when $\mu$ tends to zero, we see that the limiting points $x(0)$ and $s(0)$, if they exist, must be optimal solutions of $(P)$ and $(D)$ respectively. Actually, those limits do exist and are unique. Let us immediately prove the first part of this statement. We shall postpone the uniqueness part and the characterization of the limiting point to a later section.

**Theorem 3.4** *The pair $(x(\mu), s(\mu))$ has at least one accumulation point when $\mu$ tends to zero.*

**Proof** Consider the set of all centered primal-dual pairs $(x(\mu), s(\mu))$ with $0 < \mu \leq \bar{\mu}$ for some positive $\bar{\mu}$. Since the duality gap for any such pair is $n\mu$ ($n\mu \leq n\bar{\mu}$), this set is bounded by Corollary 3.2. So when $\mu$ approaches zero, this set has at least one accumulation point, say $(x^*, s^*)$. Consequently, there exists a sequence $(x(\mu_k), s(\mu_k))$, $k \in \mathbb{N}$, which converges to $(x^*, s^*)$ when $\mu_k$ goes to zero. Since $x(\mu_k)^T s(\mu_k) = n\mu_k$, letting $k$ go to infinity we obtain that $(x^*)^T s^* = 0$. Since one has

$$c^T x^* - b^T y^* = (x^*)^T s^* = 0,$$

the pair $(x^*, s^*)$ is optimal. □

The central path is a candidate of choice in the design of an algorithm. We can even say that almost all interior point methods use the central path as a guideline to the optimal sets of $(P)$ and $(D)$. At this stage we already know enough about the central path to describe and analyse some interesting interior point algorithms. However, we feel it is worth spending some additional time to elicit some further theoretical issues. Firstly, we will introduce the extremely important notion of the optimal partition of an LP problem. Secondly, we will characterize the limiting points. Finally, we will show how all the results in this chapter are also meaningful for problem pairs which do not satisfy the interior point assumption.

## 3.5 The optimal partition

The set of optimal solutions of $(P)$ and $(D)$ will be denoted by $\mathcal{P}^*$ and $\mathcal{D}^*$ respectively. We introduce two subsets, $B$ and $N$, of the index set $\{1, 2, \cdots, n\}$:

$$B := \{i \, : \, x_i > 0 \text{ for some } x \in \mathcal{P}^*\},$$

$$N := \{i \, : \, s_i > 0 \text{ for some } s \in \mathcal{D}^*\}.$$

Since $x^T s = 0$, whenever $x \in \mathcal{P}^*$ and $s \in \mathcal{D}^*$, the subsets $B$ and $N$ are disjoint. The question arises whether those subsets form a partition. This issue has been resolved by Goldman and Tucker in an early contribution to the theory of linear programming [9]. It is possible to give an alternative proof of this result, based on the interior point methodology, that is both simple and illuminating.

**Theorem 3.5. (Goldman and Tucker)** *There exists an optimal primal-dual pair* $(x^*, s^*)$ *such that* $x^* + s^* > 0$.

**Proof** According to Theorem 3.4 the central path has at least one limiting pair $(x^*,\, s^*)$ as $\mu$ goes to zero and this pair is optimal. We shall further prove that it is strictly complementary.

The optimality of the pair implies that the support $\sigma(x^*)$ of $x^*$ is a subset of $B$ and the support $\sigma(s^*)$ of $s^*$ is a subset of $N$. Let us apply Lemma 3.1 to the pairs $(x^*, s^*)$ and $(x(\mu_k), s(\mu_k))$. Since $x(\mu_k)^T s(\mu_k) = n\mu_k$ and $(x^*)^T s^* = 0$, we arrive at

$$\sum_{j\in\sigma(x^*)} x_j^* s_j(\mu_k) + \sum_{j\in\sigma(s^*)} x_j(\mu_k) s_j^* = n\mu_k.$$

Dividing both sides by $\mu_k$ and recalling that $x_j(\mu_k)s_j(\mu_k) = \mu_k$, we obtain

$$\sum_{j\in\sigma(x^*)} \frac{x_j^*}{x_j(\mu_k)} + \sum_{j\in\sigma(s^*)} \frac{s_j^*}{s_j(\mu_k)} = n.$$

Now let $k$ go to infinity. Then the first sum becomes equal to the number of non-zero coordinates in $x^*$. Similarly, the second sum becomes equal to the number of non-zero coordinates in $s^*$. Hence we find that the cardinalities of the supports of $x^*$ and $s^*$ sum up to $n$. This implies $\sigma(x^*) = B$ and $\sigma(s^*) = N$ (i.e. the pair $(x^*, s^*)$ is strictly complementary). □

An immediate corollary of the above theorem is that $B$ and $N$ form a partition of the index set. This partition, which is uniquely determined by the problem pair $(P)$ and $(D)$, is called the *optimal partition* of the problem pair.

The optimal partition is an important and relevant concept in linear programming because it characterizes the primal and dual optimal sets. Since the central path ends at the strictly complementary pair (we shall show soon that the limit point is unique), we may anticipate that interior point methods will naturally identify the optimal partition and thus the optimal sets.

To prove that the optimal partition $(B,\, N)$ characterizes the optimal sets, we introduce a piece of notation. We shall denote the restriction of a vector $x \in \mathbb{R}^n$ to a subset $I$ of the index set by $x_I$. The optimal sets $\mathcal{P}^*$ and $\mathcal{D}^*$ can alternatively be defined by

$$\mathcal{P}^* = \{x \;:\; x \in \mathcal{P},\, x_N = 0\},$$

and

$$\mathcal{D}^* = \{s \;:\; s \in \mathcal{D},\, s_B = 0\}.$$

Clearly by Theorem 3.5 those sets are non-empty. Moreover, they are orthogonal, implying that their elements are optimal to $(P)$ and $(D)$ respectively. Since every optimal primal solution must be orthogonal to any dual optimal solution, we further obtain that $\mathcal{P}^*$ is the primal optimal set. The same property holds for $\mathcal{D}^*$.

It is worth mentioning that the optimal partition $(B, N)$ can be any partition of the full index set. In particular, it may well happen that either the set $B$ or the set $N$ is empty. In these extreme cases, one has $\mathcal{D}^* = \mathcal{D}$ or $\mathcal{P}^* = \mathcal{P}$ respectively (i.e. all dual feasible points are optimal if $B$ is empty, and all primal feasible points are optimal if $N$ is empty).

It is not surprising that the optimal partition of a problem plays a key role in parametric or sensitivity (postoptimal) analysis. Since these topics go beyond the scope of this chapter, we refer the reader to [18, 21] for further details.

### 3.6 Limiting behaviour of the central path

Having settled the issue of the existence of a limiting point, we shall now address the question of uniqueness. We shall show that the central path converges to a very special optimal primal-dual pair.

If we consider the optimal set $\mathcal{P}^*$ of $(P)$, then we know that it consists of all primal feasible points $x$ for which $x_i = 0,\ i \in N$. From Theorem 3.5 we know that there is an optimal solution $x$ where $x_B > 0$. Therefore, among the optimal solutions we can look for one that maximizes the function

$$\prod_{i \in B} x_i.$$

Since this function is concave and the optimal set is bounded, it will have a unique maximizer. This particular optimal solution will be called the *central solution* of $(P)$.

Similarly, the optimal set $\mathcal{D}^*$ of $(D)$ consists of all dual feasible points $(y,\ s)$ for which the $i$-th slack value $s_i$ vanishes whenever $i \in B$, and we know that there exists an optimal solution such that $s_N > 0$. By maximizing the function

$$\prod_{i \in N} s_i$$

and using the fact that this function is concave and the optimal set is bounded, we again find that there exists a unique maximizer. This particular optimal solution of $(D)$ will be called the *central solution* of $(D)$.

We shall now prove that the limit point of the primal (resp., dual) central path is the central solution of $(P)$ (resp., $(D)$). The idea of this proof is due to McLinden [33, 34].

**Theorem 3.6** *The primal central path converges to the central solution of* $(P)$, *and the dual central path converges to the central solution of* $(D)$.

**Proof** Let $(x^*, s^*)$ be any accumulation point of the central path when $\mu$ approaches zero. Recall from Theorem 3.4 that there is at least one such point, and by Theorem 3.5, this point is strictly complementary. So the support of $x^*$ is $B$ and the support of $s^*$ is $N$. Letting $(x^0, s^0)$ be any optimal primal-dual pair, the support of $x^0$ and $s^0$ will be contained in $B$ and $N$ respectively. Now let $\mu > 0$; we apply Lemma 3.1 to the pairs $(x^0, s^0)$ and $(x(\mu_k), s(\mu_k))$. Since $x(\mu)^T s(\mu) = n\mu$ and $(x^0)^T s^0 = 0$, we get

$$x(\mu)^T s^0 + s(\mu)^T x^0 = n\mu.$$

Dividing by $\mu = x_i(\mu)s_i(\mu)$, $i = 1 \ldots n$, we obtain

$$\sum_{i=1}^{n}\left(\frac{x_i^0}{x_i(\mu)} + \frac{s_i^0}{s_i(\mu)}\right) = n.$$

Since $x_i^0 = 0, \forall i \in N$ and $s_i^0 = 0$, $\forall i \in B$, it follows that

$$\sum_{i\in B} \frac{x_i^0}{x_i(\mu)} + \sum_{i\in N} \frac{s_i^0}{s_i(\mu)} = n.$$

Letting $\mu$ tend to zero, along with an appropriate subsequence if necessary, so that $(x(\mu), (x(\mu))$ converges to $(x^*, s^*)$, we get

$$\sum_{i\in B} \frac{x_i^0}{x_i^*} + \sum_{i\in N} \frac{s_i^0}{s_i^*} = n.$$

Using the arithmetic-geometric mean inequality, we get

$$\left(\prod_{i\in B} \frac{x_i^0}{x_i^*} \prod_{i\in N} \frac{s_i^0}{s_i^*}\right)^{\frac{1}{n}} \leq \frac{1}{n}\left(\sum_{i\in B} \frac{x_i^0}{x_i^*} + \sum_{i\in N} \frac{s_i^0}{s_i^*}\right) = 1.$$

This inequality holds for arbitrary $x^0 \in \mathcal{P}^*$ and $s^0 \in \mathcal{D}^*$. Taking $s^0 = s^*$, we get

$$\prod_{i\in B} x_i^0 \leq \prod_{i\in B} x_i^*, \ \forall x^0 \in \mathcal{P}^*,$$

which proves that $x^*$ is the central solution of $(P)$. In the same way, by taking $x^0 = x^*$ we obtain

$$\prod_{i\in N} s_i^0 \leq \prod_{i\in N} s_i^*$$

for all $s^0$ in $\mathcal{D}^*$, thus proving that $s^*$ is the central solution of $(D)$. Hence the theorem follows. □

## 3.7 Relaxing the interior point assumption

The duality theory for linear programming we have constructed so far is rich but incomplete. The results hold only under the rather restrictive interior point assumption. There are a number of cases where this assumption is not met. For instance, the primal and/or the dual feasible set may be empty. Even worse, those sets may be non-empty but the primal and/or the dual optimal sets may be unbounded, a case that is ruled out by the interior point assumption.

There is a rather standard strategy for handling the general case: it consists of embedding the original problem into a larger one for which the interior point assumption holds. The relationship between the embedding and the original problem must be such that the solutions of the embedding, that are known to exist, can be interpreted in terms of the original problem. Let us point out that the issue we shall deal with is not specific to the non-algorithmic focus of this section. Most interior point algorithms rely heavily on the existence of interior solutions. It is therefore not surprising that most embedding strategies proposed so far were devised for algorithmic purposes. We shall present here an ingenious one that was published quite recently. The most remarkable fact about it is not that the embedding includes both the primal and the dual variables but that the embedding is self-dual.

Let us make two preliminary remarks. First the issue raised by the interior point assumption is not really typical of the interior point approach. The simplex algorithm also requires an initial basic feasible solution and it is well known that various embeddings (like the big "M" and 2-phase method) have been proposed to solve this problem. The second remark is that the embedding which we shall propose below was meant for and can be used for initiating interior point methods. For this reason we shall dispense with the treatment of feasibility issues in the sections dealing with algorithms.

The embedding we shall now discuss is based on a technique from [9]. The given problem pair, say $(P)$ and its dual $(D)$, are embedded into a large problem which satisfies the interior point assumption, and which is homogeneous and self-dual. In this chapter we will outline this approach without going into all the details. We follow the approach of [54, 56].

For any $x^0, s^0, y^0, \kappa^0, \tau^0$ where $x^0, s^0, \kappa^0$ and $\tau^0 > 0$, we consider the following linear programming problem:

$$
\begin{array}{lrrrrl}
\min & & & & \beta\theta, & \\
\text{s.t.} & & Ax & -b\tau & +\bar{b}\theta & = 0, \\
& -A^T y & & +c\tau & -\bar{c}\theta & \geq 0, \\
& b^T y & -c^T x & & +\bar{z}\theta & \geq 0, \\
& -\bar{b}^T y & +\bar{c}^T x & -\bar{z}\tau & & = -\beta,
\end{array}
$$

$$
x, \tau \geq 0, \quad y, \theta \text{ free.}
$$

In that problem, we have

$$
\begin{aligned}
\bar{b} &= b\tau^0 - Ax^0, \\
\bar{c} &= c\tau^0 - A^T y^0 - s^0, \\
\bar{z} &= c^T x^0 - b^T y^0 + \kappa^0, \\
\beta &= (x^0)^T s^0 + \kappa^0 \tau^0,
\end{aligned}
$$

where $s$ and $\kappa$ are the slack vector and the slack scalar respectively that are associated with the second and third constraints.

Let us immediately point out that the problem is not of the standard form because it includes equality and inequality constraints as well as non-negative and free variables. Then, although we may show that the interior point assumption holds in the present case, we cannot directly apply the results of the previous sections. However, there is no difficulty in extending the previous analysis to have it apply to the currently more general format. This work has been done in [54, 56], in particular, the notions of "central path", "optimal partition" and "strictly complementary solution" are extended naturally. We shall not reproduce the proofs here.

The remarkable feature of this problem is that its dual has exactly the same feasible region. Moreover, the dual objective is to maximize $-\beta\theta$, which is the same as the primal objective by switching the sign. The two problems are identical, and we say that the problem is *self-dual.* One can easily check that

$$y = y^0,\ x = x^0,\ \tau = \tau^0,\ \theta = 1,\ s = s^0,\ \kappa = \kappa^0$$

satisfies all the constraints, with equality for the inequality constraints. Since this primal, and thus dual, solution is interior (the non-negative variables take on positive values), the interior point assumption is satisfied. We can apply the results of the previous sections.

Since the problem has at least a feasible solution, and since this solution is also feasible for the dual, the primal and dual objective take on opposite values $\beta\theta$ and $-\beta\theta$. The duality gap is $2\beta\theta \geq 0$. Since $\beta > 0$, we have $\theta \geq 0$. If we multiply the four restrictions in the problem by $y^T$, $x^T$, $\tau$ and $\theta$ respectively, and then take the sum, we find another formula for the duality gap, namely,

$$x^T s + \tau\kappa = \beta\theta.$$

We know from the previous section that the problem has an optimal solution. In that case the associated duality gap $2\beta\theta$ is zero, hence $\theta = 0$. More interestingly, we can also assert that the problem has at least one strictly complementary solution. Let $(y^*, x^*, \tau^*, \theta^*, s^*, \kappa^*)$ be such a solution. We have

$$(x^*)^T s^* + \tau^* \kappa^* = 0$$

and

$$x^* + s^* > 0, \quad \tau^* + \kappa^* > 0.$$

Let $\tilde{B}$ denote the set of $i$ for which $x_i^* > 0$ and $\tilde{N}$ the set of $i$ for which $s_i^* > 0$. We consider two cases, namely $\tau^* > 0$ and $\tau^* = 0$. In the first case, strict complementarity implies that $\kappa^* = 0$. Using this, one can easily verify that $x = x^*/\tau^*$, $y = y^*/\tau^*$ and $s = s^*/\tau^*$ is a strictly complementary solution of the original problem pair, and that $(\tilde{B}, \tilde{N})$ is the optimal partition of the original problem pair.

Let us now consider the second case: $\tau^* = 0$. Since we also have $\theta^* = 0$, it follows that $x^*$, $s^*$, $\kappa^*$ satisfy

$$\begin{array}{rrrl} & Ax^* & & = 0, \\ A^T y^* & & +s^* & = 0, \\ b^T y^* & -c^T x^* & & = \kappa^*, \\ -\bar{b}^T y^* & +\bar{c}^T x^* & & = -\beta, \\ & x^* \geq 0, \; s^* \geq 0, \; \kappa^* \geq 0, & & \end{array}$$

with

$$x^* + s^* > 0, \;\; \kappa^* > 0.$$

Now suppose that $(P)$ is feasible. Then there exists an $\bar{x}, \bar{x} \geq 0$, such that $A\bar{x} = b$. Then we have

$$b^T y^* = (A\bar{x})^T y^* = (\bar{x})^T A^T y^* = -(\bar{x})^T s^* \leq 0.$$

Using $b^T y^* - c^T x^* = \kappa^*$, we find that $c^T x^* \leq -\kappa^*$. Note that $\kappa^* > 0$ implies $x^* \neq 0$. Thus we obtain a ray $\bar{x} + \alpha x^*, \alpha \geq 0$, in the feasible region of $(P)$ along which the objective value goes to $-\infty$. Therefore, the primal problem $(P)$ is unbounded and hence the dual problem $(D)$ is infeasible.

The case where $(D)$ is feasible follows in the same way. We omit the proof here. We conclude that if $\tau^* = 0$, then $(P)$ is unbounded and $(D)$ is infeasible or vice versa, or both problems are infeasible. The embedding technique has thus made it possible to find out whether the original problem and its dual are both feasible, or at least one of them is infeasible. In the former case, the duality theory developed earlier applies to the embedding problem and, as a by-product, to the original problem.

# 4 Target-following methods

The aim of a solution method for linear programming is to find a solution to the optimality condition

$$\begin{array}{rcl} Ax & = & b, \;\; x \geq 0, \\ A^T y + s & = & c, \;\; s \geq 0, \\ xs & = & 0. \end{array}$$

An acceptable alternative, as we discussed in the introductory section, would be to find a solution of the system

$$\begin{array}{rcl} Ax & = & b, \;\; x \geq 0, \\ A^T y + s & = & c, \;\; s \geq 0, \\ xs & = & \bar{w}, \end{array} \tag{4.1}$$

for some small $\bar{w}$. Then the duality gap $x^T s = e^T \bar{w}$ would be small and the solution of (4.1) is an approximate optimal solution. The question is: does it help in any way to have to solve (4.1) instead of the optimality conditions? As we put the question in those general terms, the answer is obviously no. However, we do not need to tackle this difficult problem at once, instead, we may consider a succession of easier problems that gradually lead to the global solution.

Let us paraphrase this iterative process. Firstly, we may assume that some primal-dual feasible pair is at hand. (We know that this assumption is not really restrictive, in particular if we resort to a self-symmetric primal-dual reformulation.) Next, if this initial pair, say $(x^0, s^0)$, were by chance close to $\bar{w}$ (in the sense that the complementary product $x^0 s^0$ would be close to $\bar{w}$), it is likely that an accurate solution may be computed at a very low computational cost. Unfortunately there is no reason to have $(x^0, s^0)$ close to a very small target (or the problem would be solved just as it is posed). However we are free to set a first target $\bar{w}^0$ close to $(x^0, s^0)$. Then an excellent approximation of the inverse target map $\psi^{-1}(\bar{w})$ is computed and the target is moved slightly in some suitable way. If the resulting target sequence $\bar{w}^k$ converges to zero, the associated sequence of approximate solutions $(x^k, s^k)$ of (4.1) converges to an optimal solution of the linear programming problem.

We have just paraphrased the target-following scheme, a concept that was informally introduced by Mizuno [36, 37] and systematized in [25]. This concept captures the basic ideas that underlie some of the most prominent interior point methods that appeared in the literature. As we shall see, it can be simply specialized to each of them, thereby providing easy and short convergence analyses.

We chose to restrict our discussion to primal-dual methods. As the previous section shows, the primal-dual framework is the natural one. However, there exist many pure primal (or pure dual) interior point algorithms. It can be shown that the target-following scheme can be applied to this more restricted framework. To avoid excessive length we shall omit those developments here. We shall further restrict our discussion to feasible methods, i.e. methods for which an initial feasible starting point is at hand. The issue of an infeasible start will be considered in the last section of this chapter.

The presentation of target-following methods is organized as follows. In Section 4.1, we shall be concerned with efficient computation of the image of the inverse target map when the target is close to a given primal-dual feasible pair. In Section 4.2, we shall formalize the target-following scheme and apply it to a variety of interior point methods. Although we will omit some technical details in the analysis, we shall compute bounds on the number of iterations in each case. In most instances, those bounds are polynomial. In Section 4.3, we shall discuss a modification of the basic scheme to allow longer steps between successive targets.

### 4.1 Approximating target points

Before dealing with algorithms, we must address the question of how to solve the system (4.1) for a target close to a given initial primal-dual feasible pair. We will be concerned here with means of drastically improving the initial approximation of (4.1). The main ingredient is Newton's method. Newton's method is well known to be quadratically convergent in a neighbourhood of the solution. Once this neighbourhood is attained, a nearly exact solution may be obtained in a matter of a few Newton steps. The strong result of interior point methods is that it is possible to describe this neighbourhood analytically and to perform an easily computable test to check if the iterate belongs to this neighbourhood.

### 4.1.1 *A primal-dual Newton process*

We now assume that we are given some primal-dual pair $(x, s)$, and $w$ is the corresponding point in the $w$-space. So $w = xs$. Moreover, let $\bar{w}$ be the current target point in the $w$-space. Our aim is to find an approximate solution of the system (4.1). Ideally we would like to find displacements $\Delta x, \Delta y, \Delta s$ and $\Delta w$ in the $x$, $y$, $s$ and $w$-space respectively, such that

$$\begin{aligned} \Delta x &= x(\bar{w}) - x, \\ \Delta y &= y(\bar{w}) - y, \\ \Delta s &= s(\bar{w}) - s, \\ \Delta w &= \bar{w} - w. \end{aligned}$$

So by substitution, we want these displacements to satisfy the following system:

$$\begin{aligned} A &= 0, \\ A^T\Delta y + \Delta s &= 0, \\ s\Delta x + x\Delta s &= \bar{w} - w - \Delta x\Delta s. \end{aligned} \tag{4.2}$$

Due to the non-linear term $\Delta x\Delta s$ in (4.2), it is not possible to give a closed form solution to the above system. To overcome this difficulty, one considers the linear term and neglects all terms of higher order. This results in the linear system

$$\begin{aligned} A\Delta x &= 0, \\ A^T\Delta y + \Delta s &= 0, \\ s\Delta x + x\Delta s &= \Delta w, \end{aligned} \tag{4.3}$$

for which one can easily compute a solution. This solution is itself an approximation of the solution of the initial system. This approximation procedure is nothing other than Newton's method.

Since $A$ has full row rank, one can easily show that the linear system of equations to be solved using Newton's method is non-singular. The triple of displacements $\Delta x, \Delta y$ and $\Delta s$ obtained in this way is the search direction we will use. We call it the primal-dual Newton direction. The result of a step along this direction is denoted by $x^*, y^*$ and $s^*$ and we let $w^* = x^*s^*$. Then we have

$$w^* = (x + \Delta x)(s + \Delta s) = xs + x\Delta s + s\Delta x + \Delta x\Delta s = w + \Delta w + \Delta x\Delta s.$$

So we obtain the useful relation

$$w^* = \bar{w} + \Delta x\Delta s \tag{4.4}$$

which makes it clear that the error made by neglecting the second-order term is given by $\Delta x \Delta s$. This product plays a significant role in the analysis of all primal-dual interior point methods. It represents the second-order effect in the Newton step. Clearly we want this product to be small in order to make Newton's method efficient. Below we will present some estimates for the Euclidean norm and the infinity norm of $\Delta x \Delta s$. Similar results were obtained by Mizuno *et al.* [40] in a different way for the case where $\bar{w}$ is on the central path.

One can easily obtain explicit expressions for $\Delta x, \Delta y$ and $\Delta s$, but we will not need them in our analysis. Instead it will be convenient to use a different formulation of the system. To this end we introduce the vectors

$$d := \sqrt{xs^{-1}}, \qquad v := \sqrt{w} = \sqrt{xs}.$$

Using $d$ we can rescale both $x$ and $s$ to the same vector, namely $v$,

$$d^{-1}x = ds = v.$$

We also use $d$ to rescale $\Delta x$ and $\Delta s$,

$$p_x := d^{-1}\Delta x, \qquad p_s := d\Delta s.$$

Note that

$$\Delta x \Delta s = p_x p_s$$

and the orthogonality of $\Delta x$ and $\Delta s$ implies that $p_x$ and $p_s$ are also orthogonal. Now we may write

$$x\Delta s + s\Delta x = xd^{-1}d\Delta s + sdd^{-1}\Delta x = v(p_x + p_s).$$

Hence, Newton's direction is determined by the following linear system:

$$\begin{aligned} ADp_x &= 0, \\ DA^T p_y + p_s &= 0, \\ p_x + p_s &= v^{-1}\Delta w. \end{aligned}$$

Letting

$$p_v := v^{-1}\Delta w,$$

we have

$$p_x + p_s = p_v$$

and $p_x$ and $p_s$ are simply the orthogonal components of the decomposition of $p_v$ in the null space of $AD$ and the row space of $AD$ respectively.

We also use the vector $q_v$, defined by

$$q_v := p_x - p_s.$$

Note that the orthogonality of $p_x$ and $p_s$ implies that

$$\|q_v\| = \|p_v\|.$$

We also have

$$\begin{aligned} p_x &= \tfrac{1}{2}(p_v + q_v), \\ p_s &= \tfrac{1}{2}(p_v - q_v), \end{aligned}$$

and therefore

$$p_x p_s = \tfrac{1}{4}(p_v^2 - q_v^2). \tag{4.5}$$

**Lemma 4.1** *One has* $\|p_x p_s\|_\infty \le \frac{1}{4}\|p_v\|^2$ *and* $\|p_x p_s\| \le \frac{1}{2\sqrt{2}}\|p_v\|^2$.

**Proof** Using (4.5) we may write

$$\|p_x p_s\|_\infty \le \tfrac{1}{4}\max\left(\|p_v\|_\infty^2, \|q_v\|_\infty^2\right) \le \tfrac{1}{4}\max\left(\|p_v\|^2, \|q_v\|^2\right) = \tfrac{1}{4}\|p_v\|^2.$$

Using (4.5) again we obtain

$$\begin{aligned} \|p_x p_s\|^2 &= e^T(p_x p_s)^2 = \tfrac{1}{16}e^T\left(p_v^2 - q_v^2\right)^2 = \tfrac{1}{16}\left\|p_v^2 - q_v^2\right\|^2 \\ &\le \tfrac{1}{16}\left(\left\|p_v^2\right\|^2 + \left\|q_v^2\right\|^2\right) \le \tfrac{1}{16}\left(\|p_v\|^4 + \|q_v\|^4\right) = \tfrac{1}{8}\|p_v\|^4. \end{aligned}$$

This implies Lemma 4.1. □

### 4.1.2 *Proximity and feasibility*

In the analysis of Newton's method we need a measure for the proximity of the current iterate $w = xs$ to the target point $\bar{w}$. For this purpose we introduce the following proximity measure:

$$\delta(w, \bar{w}) := \frac{1}{2\sqrt{\min(\bar{w})}}\|p_v\|.$$

Since $p_v = v^{-1}\Delta w = v^{-1}(\bar{w} - w)$ and $v = \sqrt{w}$, this can be rewritten as

$$\delta(w, \bar{w}) := \frac{1}{2\sqrt{\min(\bar{w})}}\left\|\frac{\bar{w} - w}{\sqrt{w}}\right\|.$$

We first investigate when the (full) Newton step to the target point $\bar{w}$ can be taken without becoming infeasible. The next lemma gives a simple condition on $\delta(w, \bar{w})$ that guarantees that the new iterates $x^* := x + \Delta x$ and $s^* := s + \Delta s$ are positive.

**Lemma 4.2** *The Newton step is feasible if* $\left\|\frac{p_x p_s}{\bar{w}}\right\|_\infty < 1$. *This condition is satisfied if*

$$\delta(w, \bar{w}) < 1.$$

**Proof** Let $0 \le \alpha \le 1$ be a steplength along the Newton direction. We define $x(\alpha) = x + \alpha\Delta x$ and $s(\alpha) = s + \alpha\Delta s$. The new iterate, after a full Newton step, is given by $x^* := x(1)$ and $s^* := s(1)$. We have

$$\begin{aligned} x(\alpha)s(\alpha) &= (v + \alpha p_x)(v + \alpha p_s) \\ &= v^2 + \alpha v(p_x + p_s) + \alpha^2 p_x p_s \end{aligned}$$

$$\begin{aligned} &= w + \alpha(\bar{w} - w) + \alpha^2 p_x p_s \\ &= w(1-\alpha) + \alpha\bar{w}\left(e + \alpha\frac{p_x p_s}{\bar{w}}\right). \end{aligned} \tag{4.6}$$

As $\left\|\frac{p_x p_s}{\bar{w}}\right\|_\infty < 1$ holds by assumption, we have, for all $\alpha, 0 \leq \alpha \leq 1$, that $(e + \alpha\frac{p_x p_s}{\bar{w}}) > 0$ and hence $x(\alpha)s(\alpha) > 0$. This implies that $x(\alpha) \neq 0$ and $s(\alpha) \neq 0$ for all $\alpha$, $0 \leq \alpha \leq 1$. Consequently, the linear functions of $\alpha$, $x(\alpha)$ and $s(\alpha)$ do not change sign on the interval $[0, 1]$. Since $x(0)$ and $s(0)$ are positive, $x(1)$ and $s(1)$ are positive as well. The first statement in the lemma is proved. The condition on $\delta$ follows from the inequalities

$$\left\|\frac{p_x p_s}{\bar{w}}\right\|_\infty \leq \frac{\|p_x p_s\|_\infty}{\min(\bar{w})} \leq \frac{\|p_v\|^2}{4\min(\bar{w})} = \delta^2$$

where the last inequality follows from Lemma 4.1. □

The Newton step has another important property that it sets the duality gap at the same value as the gap in the target point $\bar{w}$. This is the content of the following lemma.

**Lemma 4.3** *Let the primal-dual feasible pair $(x^*, s^*)$ be obtained after a full Newton step with respect to $\bar{w}$. Then the duality gap of this pair achieves the target value, i.e. $(x^*)^T s^* = e^T\bar{w}$.*

**Proof** Recall from (4.4) that $w^* = \bar{w} + \Delta x\Delta s$. Using the fact that $\Delta x$ and $\Delta s$ are orthogonal, we may write

$$(x^*)^T s^* = e^T w^* = e^T\bar{w} + \Delta x^T\Delta s = e^T\bar{w}.$$

□

This lemma has two important implications. The first is that if subsequent Newton steps are taken with $\bar{w}$ being fixed, then the duality gap remains constant! The other implication concerns the case where one takes a full Newton step at each iteration but also adjusts the target between each step in a suitable way. Then the convergence of the method can be evaluated simply by considering the targets (and their associated duality gaps) without having to bother about the iterates themselves.

### 4.1.3 *Local quadratic convergence*

We are interested in how close a Newton step brings us to the target point $\bar{w}$. In this respect the following lemma is of much interest. It shows that if the current iterate $w$ is close enough to the target point $\bar{w}$, then the Newton step ensures quadratic convergence of the proximity measure.

**Lemma 4.4** *Assume that $\delta := \delta(w, \bar{w}) < 1$ and let $w^*$ result from a Newton step at $w$ with respect to $\bar{w}$. Then one has*

$$\delta(w^*, \bar{w})^2 \leq \frac{\delta^4}{2(1-\delta^2)}.$$

**Proof** From Lemma 4.2 we know that $x^*$ and $s^*$ are feasible. By definition,

$$\|p_v\| = 2\,\delta\sqrt{\min(\bar{w})}.$$

Using (4.4) and Lemma 4.1, we get

$$\min(w^*) \geq \min(\bar{w}) - \|p_x p_s\|_\infty \geq \min(\bar{w}) - \frac{1}{4}\|p_v\|^2 = \min(\bar{w})(1-\delta^2). \quad (4.7)$$

Again using (4.4) we may write

$$\begin{aligned} \delta(w^*,\, \bar{w})^2 &= \frac{1}{4\min(\bar{w})}\left\|(w^*)^{-\frac{1}{2}}(\bar{w}-w^*)\right\|^2 \\ &= \frac{1}{4\min(\bar{w})}\left\|(w^*)^{-\frac{1}{2}}(p_x p_s)\right\|^2 \\ &\leq \frac{1}{4\min(\bar{w})}\frac{\|p_x p_s\|^2}{\min(w^*)}. \end{aligned}$$

Substituting the bounds derived in Lemma 4.1 and (4.7) yields

$$\delta(w^*,\, \bar{w})^2 \leq \frac{1}{32\min(\bar{w})}\frac{\|p_v\|^4}{\min(\bar{w})(1-\delta^2)}.$$

Using $\|p_v\| = 2\,\delta\sqrt{\min(\bar{w})}$ again, we eliminate $\|p_v\|^4$ and get

$$\delta(w^*,\, \bar{w})^2 \leq \frac{\delta^4}{2\,(1-\delta^2)}$$

which completes the proof of the lemma. □

One can easily check that if $\delta \leq 1/\sqrt{2}$ holds, then $\delta(w^*, \bar{w}) \leq \delta^2$, and thus guarantee the quadratic convergence of the sequence of Newton steps. So the primal-dual Newton method provides a very efficient tool for getting accurate approximations of the primal-dual pair corresponding to some target point in the $w$-space.

## 4.2 Short-step methods

Now we describe the main idea of the target-following approach. Assume we are given some positive primal-dual feasible pair $(x^0,\, s^0)$. Because of the results in Section 3.7, we can make this assumption without loss of generality. Now put $w^0 := x^0 s^0$ and assume that we have a sequence

$$w_0,\, w_1,\, \cdots,\, w_k,\, \cdots,\, w_K \quad (4.8)$$

of points $w_k$ in the $w$-space with the following property:

> Given the primal-dual pair for $w_k$, it is "easy" to construct the primal-dual pair for $w_{k+1}$, $0 \leq k < K$.

We call such a sequence a *traceable target sequence* of length $K$, its elements *target points* and the corresponding primal-dual pairs *target pairs.*

If a traceable sequence is available, then we can solve the given problem pair $(P)$ and $(D)$ up to the precision level $e^T w_K$ in $K$ iterations. The $k$-th iteration in this method would consist of finding the primal-dual target-pair corresponding to the target point $w_k$. We know from Section 4.1 that Newton's method can be highly efficient for that purpose (i.e. local quadratic convergence), provided that the proximity of the primal-dual pair $(x_{k-1}, s_{k-1})$ to $w_k$ satisfies

$$\delta(x_{k-1}s_{k-1}, w_k) \leq \frac{1}{\sqrt{2}}, \quad k = 1, 2, \ldots, K. \tag{4.9}$$

Then in a matter of one or a few Newton iterations, we would get a new primal-dual feasible pair $(x_k, s_k)$ which approximates the target $w_k$ very accurately.

Assume that (4.9) holds for some $k$, i.e. $(x_{k-1}, s_{k-1})$ is close to $w_k$. Since $(x_k, s_k)$ nearly achieves the target $w_k$, the pair $(x_k, s_k)$ will be close to $w_{k+1}$ if $w_k$ is itself close to $w_{k+1}$. So to build a traceable target sequence, we only have to make sure that the successive targets are close enough to one another; hence the qualifier *short-step* for these methods. The target-following approach is a little more general than the standard path-following schemes that appear in the literature. The vast majority of the latter correspond to target sequences on the central path. Examples of short-step methods along the central path can be found in [11, 25, 30, 42, 43, 45, 47]. See also [14].

To keep our presentation simple, we shall assume that possibly more than one Newton step is taken for a given target, so the primal-dual feasible pair eventually achieves a nearly perfect fit with the target. In that way we only have to concentrate on the proximity between successive targets. This choice is not really restrictive. One must realize that quadratic convergence is very fast: if (4.9) holds, in no more than 6 Newton iterations we are sure to have a primal-dual feasible pair $(x, s)$ such that $\delta(xs, w) < 10^{-10}$ which is a near-perfect match. So we can safely state that after $O(K)$ Newton steps, we will have found the primal-dual pair corresponding to the duality gap $e^T w_K$.

If we insisted that exactly one Newton step be taken between two successive targets, the corresponding primal-dual feasible pairs would never exactly match their respective targets and the closeness of the targets would not be enough to prove that the sequence is traceable. However it is possible to obtain the same convergence result at the cost of a little more involved analysis. For a detailed analysis of this case, the reader is referred to [25].

In the next subsections we present some example of traceable target sequences. The examples are chosen to cover the most prominent primal-dual methods and results in the literature.

### 4.2.1 *Short-step path-following methods*

Under this heading the literature generally refers to short-step methods for which the target sequence lies on the central path. The best-known iteration bound for interior point methods for linear programming is of the form

$$O\left(\sqrt{n}\ln\frac{(x^0)^T s^0}{\epsilon}\right). \tag{4.10}$$

Here $(x^0)^T s^0$ denotes the duality gap for the initial primal-dual feasible pair, and $\epsilon$ is the duality gap we want to achieve. This iteration bound has been obtained by several authors. Without trying to be complete we mention the papers [11, 42, 45, 47]. In these methods the target points have the form $w = \mu e$, $\mu > 0$. Let the next target point be

$$\bar{w} = (1-\theta)w,\ 0 < \theta < 1.$$

Then some straightforward calculations yield $\delta(w, \bar{w})$ where

$$\begin{aligned}\delta(w, \bar{w}) &= \frac{1}{2\sqrt{\min(\bar{w})}}\left\|\frac{\bar{w}-w}{\sqrt{w}}\right\| \\ &= \frac{\theta}{2\sqrt{\min(\bar{w})}}\left\|\sqrt{w}\right\| \\ &= \frac{\theta\sqrt{n}}{2\sqrt{1-\theta}}.\end{aligned}$$

Assuming that $n \geq 4$ we find that

$$\delta(w, \bar{w}) \leq \frac{1}{\sqrt{2}}, \quad \text{if} \quad \theta = \frac{1}{\sqrt{n}}.$$

Obviously, each iteration reduces the duality gap by the factor $1-\theta$. So after $k$ iterations the duality gap is smaller than $\epsilon$, provided that

$$(1-\theta)^k (x^0)^T s^0 \leq \epsilon.$$

Taking logarithms we obtain

$$-k\ln(1-\theta) \geq \ln(x^0)^T s^0 - \ln\epsilon.$$

Using the elementary inequality $-\ln(1-\theta) \geq \theta$, $\theta > 0$, this will certainly hold if

$$k \geq \frac{1}{\theta}\ln\frac{(x^0)^T s^0}{\epsilon}.$$

Replacing $\theta$ by its value, we obtain the iteration bound (4.10).

### 4.2.2 *Weighted path-following methods*

With a little extra effort, we can also analyse the case where the target sequence lies on the line segment $w = \mu w^0$, $\mu > 0$, for some fixed positive vector $w^0$. When $w^0 = \left\|w^0\right\| e$, the sequence is on the central path. For an arbitrary $w^0$, the sequence is on the so-called *weighted central path.* This sequence allows the procedure to start away from the central path. Let the next target point be given by

$$\bar{w} = (1-\theta)w, \; 0 < \theta < 1.$$

Straightforward calculations yield $\delta(w, \bar{w})$ where

$$\begin{aligned}
\delta(w, \bar{w}) &= \frac{1}{2\sqrt{\min(\bar{w})}} \left\| \frac{\bar{w} - w}{\sqrt{w}} \right\| \\
&= \frac{1}{2\sqrt{\min(\bar{w})}} \left\| \theta \sqrt{w} \right\| \\
&= \frac{\theta}{2\sqrt{1-\theta}} \left\| \sqrt{\frac{w^0}{\min(w^0)}} \right\| \\
&\leq \frac{\theta\sqrt{n}}{2\sqrt{1-\theta}} \sqrt{\frac{\max(w^0)}{\min(w^0)}}.
\end{aligned}$$

Defining

$$\rho(w) := \sqrt{\frac{\min(w)}{\max(w)}}, \; w > 0,$$

we find

$$\delta(w, \bar{w}) \leq \frac{\theta\sqrt{n}}{2\rho(w^0)\sqrt{1-\theta}}.$$

Assuming again that $n \geq 4$ we find that

$$\delta(w, \bar{w}) \leq \frac{1}{\sqrt{2}} \quad \text{if} \quad \theta = \frac{\rho(w^0)}{\sqrt{n}}.$$

Each iteration reduces the duality gap by the factor $1 - \theta$. So using the same arguments as in the previous section, we are sure that the duality gap is smaller than $\epsilon$ after at most

$$\frac{\sqrt{n}}{\rho(w^0)} \ln \frac{(x^0)^T s^0}{\epsilon}$$

iterations. Comparing this result with the iteration bound of the previous section, we observe that one introduces a factor of $1/\rho(w^0) > 1$ by not using the central path.

The above result indicates that in some sense the central path is the best path to follow to the optimal set. There is a price to pay for not starting on the central path, and this price is high since we lose polynomiality (recall that $\rho(w^0)$ is a quantity that depends on the problem data and as such it should appear under a logarithm). This result gives us evidence of the very special status of the central path among all possible weighted central paths.

### 4.2.3 *Centering method*

In the next example we address a different problem. Given an initial primal-dual feasible pair corresponding to $w^0$, $\rho(w^0) < 1$, we want to reach a primal-dual pair on the central path as fast as possible. We shall propose a target sequence that converges in

$$O\left(\sqrt{n}\ln\frac{1}{\rho(w^0)}\right) \tag{4.11}$$

iterations. If we have this target sequence followed by a sequence along the central path, as in our first example, we find the following iteration bound:

$$\sqrt{n}\left(\ln\frac{1}{\rho(w^0)} + \ln\frac{(x^0)^T s^0}{\epsilon}\right).$$

Note that this bound, for a strategy that first centralizes and then optimizes, is better than the one we obtained for the more direct strategy of following a sequence along the weighted path. In fact, this bound is currently the best one known when the starting point lies away from the central path.

To prove (4.11), let $\bar{w}$ be obtained from some point $w$ outside the central path by replacing each $w_i$ such that

$$w_i < (1+\theta)\min(w)$$

by $(1+\theta)\min(w)$, where $\theta$ is some positive constant satisfying $(1+\theta)\rho(w) \leq 1$. It then follows that

$$\rho(\bar{w}) = (1+\theta)\rho(w)$$

and

$$\begin{aligned}
\delta(w, \bar{w}) &= \frac{1}{2\sqrt{\min(\bar{w})}}\left\|\frac{\bar{w}-w}{\sqrt{w}}\right\| \\
&\leq \frac{1}{2\sqrt{(1+\theta)\min(w)}}\left\|\frac{\theta\min(w)\,e}{\sqrt{w}}\right\| \\
&\leq \frac{\theta\sqrt{n}}{2\sqrt{1+\theta}} \\
&\leq \frac{\theta\sqrt{n}}{2}.
\end{aligned}$$

Thus we have

$$\delta(v, \bar{v}) \leq \frac{1}{\sqrt{2}} \quad \text{if} \quad \theta = \frac{\sqrt{2}}{\sqrt{n}}.$$

At each iteration, $\rho(w)$ increases by the factor $1+\theta$. So we will certainly have reached the central path if the iteration number $k$ satisfies

$$(1+\theta)^k \rho(w^0) \geq 1.$$

Substituting the value of $\theta$ and then taking logarithms, we get

$$k \ln\left(1+\frac{\sqrt{2}}{\sqrt{n}}\right) \geq \ln\frac{1}{\rho(w^0)}.$$

One may easily verify that for $n \geq 3$ this inequality is satisfied if

$$\frac{k}{\sqrt{n}} \geq \ln\frac{1}{\rho(w^0)}.$$

Thus we find that no more than

$$\sqrt{n}\ln\frac{1}{\rho(w^0)}$$

iterations will bring the iterate on the central path. Note that this result is in accordance with (4.11).

Let us mention here the converse of the centering problem: it consists of finding a primal-dual pair with prescribed weights when the initial point lies on the central path. This problem is known as the *weighted centering problem.* It has been considered by Atkinson [2] and also by den Hertog [19]. Both authors obtain an iteration bound similar to (4.11).

#### 4.2.4 *Centering and optimizing together*

In the previous subsection we discussed a two-phase strategy for the case where the initial primal-dual feasible pair $(x^0, s^0)$ is not on the central path: Phase I is devoted to centralizing and Phase II is devoted to optimizing. Although this strategy achieves the best iteration bound, it is worth considering an alternative strategy that would combine the two phases into one. Let $w^0 := x^0 s^0$ and consider the function $f : \mathbb{R}_+ \longrightarrow \mathbb{R}^n$ defined by

$$f(\theta) = \frac{w^0}{e + \theta w^0}, \ \theta \geq 0.$$

The image of $f$ defines a path in the $w$-space starting at $f(0) = w^0$ and converging to the origin when $\theta$ goes to infinity. Asymptotically, the path becomes tangent to the central path because for very large values of $\theta$, one has

$$f(\theta) \approx \frac{e}{\theta}.$$

The direction of the tangent is obtained by differentiating with respect to $\theta$. This yields

$$\frac{df(\theta)}{d\theta} = \frac{-(w^0)^2}{(e+\theta w^0)^2} = -f(\theta)^2.$$

This shows that the path is a trajectory of the vector field $-w^2$ in the $w$-space. Without going further into it we refer to [22] where this field was used to obtain the primal-dual analogue of the so-called primal affine scaling direction of Dikin [4]. Because of this relation we will refer to the path as the *Dikin path* in the $w$-space starting at $w^0$. Note that if $w^1$ lies on it, then the Dikin path starting at $w^1$ is just the continuation of the path starting at $w^0$.

One can easily establish that along the path the proximity to the central path is improving. This goes as follows. Let $w := f(\theta)$. Then using the fact that $f$ preserves the ordering of the coordinates, we may write

$$\rho(w)^2 = \frac{\frac{\min(w^0)}{1+\theta\min(w^0)}}{\frac{\max(w^0)}{1+\theta\max(w^0)}} = \rho(w^0)^2\,\frac{1+\theta\max(w^0)}{1+\theta\min(w^0)} \geq \rho(w^0)^2.$$

This ratio goes to 1 as $\theta$ goes to $\infty$. Consequently, the Dikin path achieves both goals that were assigned to it: centralize and optimize.

Let us now try to devise a traceable target sequence along the Dikin path. Suppose that $w$ is a point on this path. Without loss of generality, we may assume that $w = f(0) = w^0$. Let $\bar{w} := f(\theta)$ for some positive $\theta$. Then we have

$$\begin{aligned}
\delta(w,\bar{w}) &= \frac{1}{2\sqrt{\min(\bar{w})}}\left\|\frac{\bar{w}-w}{\sqrt{w}}\right\| \\
&= \frac{1}{2\sqrt{\min(\bar{w})}}\left\|\frac{\frac{w}{e+\theta w}-w}{\sqrt{w}}\right\| \\
&= \frac{1}{2\sqrt{\min(\bar{w})}}\left\|\frac{\theta(w)^{\frac{3}{2}}}{e+\theta w}\right\| \\
&= \frac{\sqrt{1+\theta\min(w)}}{2\sqrt{\min(w)}}\left\|\frac{\theta(w)^{\frac{3}{2}}}{e+\theta w}\right\| \\
&\leq \frac{\sqrt{1+\theta\min(w)}}{2\rho(w)}\left\|\frac{\theta w}{e+\theta w}\right\| \\
&\leq \frac{\theta\,\|w\|}{2\rho(w)}.
\end{aligned}$$

The last inequality follows from $e + \theta w > e$. So we have

$$\delta(w,\bar{w}) \leq \frac{1}{\sqrt{2}} \quad \text{if} \quad \theta = \frac{\rho(w)\sqrt{2}}{\|w\|}.$$

The duality gap is reduced at least by the factor $1 + \theta\min(w)$, because

$$e^T\bar{w} := e^T\frac{w}{e+\theta w} \leq \frac{e^T w}{1+\theta\min(w)}.$$

Replacing $\theta$ by its value defined above, one has

$$1+\theta\min(w) = 1 + \frac{\rho(w)\sqrt{2}}{\|w\|}\min(w) \geq 1 + \frac{\rho(w)^3\sqrt{2}}{\sqrt{n}}.$$

Using $\rho(w) > \rho(w^0)$, we deduce in the usual way that after

$$\frac{\sqrt{n}}{\rho(w^0)^3\sqrt{2}} \ln \frac{(x^0)^T s^0}{\epsilon} \tag{4.12}$$

iterations the duality gap is smaller than $\epsilon$.

For small values of $\rho(w^0)$ this bound is significantly worse than the bounds obtained in the previous section when starting off the central path. This weak result follows since we only use the fact that $\rho$ is increasing, but not that it tends to 1 (i.e. we did not take account of the centralizing effect of the Dikin path). To improve the bound, we shall look at the process in a different way. Instead of directly estimating the number of target moves until a suitable duality gap is achieved, we shall concentrate on the number of steps that are required to achieve near centrality, a state that can be measured by, for example, $\rho > \frac{1}{2}$.

In view of the above definition of $w$ and $\bar{w}$, we have, by simple calculation,

$$\rho(\bar{w})^2 = \rho(w)^2 \left(1 + \frac{\theta(1-\rho(w)^2)\max(w)}{1+\theta\min(w)}\right).$$

Hence, as long as $\rho(w)^2 \leq \frac{1}{2}$, one has

$$\rho(\bar{w})^2 \geq \rho(w)^2 \left(1 + \frac{\theta\max(w)}{2(1+\theta\min(w))}\right).$$

Taking

$$\theta = \frac{\rho(w)^3}{\min(w)\sqrt{n}},$$

one has $\delta(w, \bar{w}) \leq \frac{1}{2}$, and moreover,

$$\theta\min(w) = \frac{\rho(w)^3}{\sqrt{n}} \leq 1, \qquad \theta\max(w) = \frac{\rho(w)}{\sqrt{n}} \geq \frac{\rho(w^0)}{\sqrt{n}}.$$

Hence

$$\rho(\bar{w})^2 \geq \rho(w)^2 \left(1 + \frac{\theta\max(w)}{4}\right) \geq \rho(w)^2 \left(1 + \frac{\rho(w^0)}{4\sqrt{n}}\right).$$

So after $k$ iterations, we have

$$\rho(\bar{w})^2 \geq \left(1 + \frac{\rho(w^0)}{4\sqrt{n}}\right)^k \rho(w^0)^2.$$

Now using

$$\ln\left(1 + \frac{\rho(w^0)}{4\sqrt{n}}\right) \geq \frac{\rho(w^0)}{5\sqrt{n}},$$

one can easily deduce that after

$$\frac{5\sqrt{n}}{\rho(w^0)} \ln \frac{1}{2(\rho(w^0))^2}$$

iterations one will have $\rho(\bar{w})^2 \geq 1/2$. If we consider the Dikin path initiating at that point, by replacing $\rho(w^0)$ by $1/\sqrt{2}$, we obtain from (4.12) the bound of

$$2\sqrt{n}\ln\frac{(x^0)^T s^0}{\epsilon}$$

additional iterations to achieve convergence. By adding the two iteration numbers, we obtain the iteration bound

$$\sqrt{n}\left(2\ln\frac{(x^0)^T s^0}{\epsilon} + \frac{5}{\rho(w^0)}\ln\frac{1}{2(\rho(w^0))^2}\right).$$

Note that this bound, as in the case of the weighted central path, is worse than the two-phase strategy studied earlier.

## 4.3 Long-step target-following methods

The complexity bounds derived in the previous sections are based on a worst case analysis: each target step is chosen to be short enough so that, in any possible instance, proximity will remain under control. Moreover, the target step is not related to the particular primal-dual feasible pair. As a consequence, when using an implementation of a short-step target-following method, the required running time may give rise to some disappointment. It then becomes tempting to take longer target steps. An obvious improvement would be to relate the target move to the primal-dual feasible pair and to make the move as large as possible while maintaining sufficient proximity to the primal-dual feasible pair. In that case a Newton step still yields a new primal-dual feasible pair closer to the target, and the process may be repeated. This enhancement of the short-step strategy into the so-called maximal step strategy does not improve the overall theoretical complexity bound, but it has a dramatic effect on the asymptotic convergence rate. In a recent paper [15], it is shown that the maximal step method is asymptotically quadratically convergent, i.e. in the final iterations the duality gap approaches zero quadratically fast.

Despite this nice asymptotic result, in general, the maximal step method may take too short steps to be really efficient. In practical applications it is wise to achieve longer steps. As already pointed out, long-target-steps almost surely imply that proximity between the iterate and the target becomes larger than 1. One obvious shortcoming of a long-target-step is that the full Newton step may cause infeasibility. To overcome his difficulty one must use a damped Newton step. However, it is not at all clear how a damped step will contribute to an improvement of the proximity measure in any controllable manner. Fortunately, the lack of control on the proximity measure is compensated by a judicious use of the barrier function (3.4). Remember that the inverse target was defined as being the unique minimizer of the weighted barrier function corresponding to the target. The idea is then to make sure that a damped Newton step will produce a significant decrease of this function. Since the barrier function is bounded below by its minimum, the cumulative decreases force convergence. The strong result here is that as long as the proximity measure is larger than some fixed but arbitrary value, the damped Newton step induces a decrease of the barrier function by at least some fixed amount that is independent of the current iterate and the problem data. From that property and the bound on the cumulative decreases, one gets a bound on the number of iterations required to enter the vicinity of the target. It goes beyond the scope of this chapter to go into a detailed report of the results obtained in this direction. We conclude this section by mentioning a few papers which the reader may consult for further information.

The first long-step results were those of Gonzaga [12, 13] and Roos and Vial [46]. We also mention den Hertog, Roos and Vial [20] and Mizuno *et al.* [39]. The results in these papers were obtained in a pure primal format. Other papers deal with the primal-dual framework, like Kojima, Mizuno and Yoshise [30] who use a so-called potential function and Jansen, Roos, Terlaky and Vial [25] who present an analysis based on the primal-dual logarithmic barrier function. Finally, [24] extends the target-following scheme described in this chapter to the long-step case.

## 5 Asymptotic behaviour of the predictor–corrector method

It is an experimental observation that for almost all interior point methods for linear programming one experiences a dramatic acceleration of convergence shortly before termination. Order of magnitudes are gained at each iteration and in a matter of two or three additional iterations the process ends with a solution. In 1993, Mizuno, Todd and Ye [40] published their predictor–corrector method. It soon appeared that one could prove asymptotic quadratic convergence for this method (see, e.g. [16, 17, 55]). Since then, this type of analysis has been applied with success to other algorithms (see e.g. [15]). In this section we would like to give an example of this type of analysis, so that the reader may gain a better understanding of what happens in the limit. We choose to apply the analysis to the predictor–corrector method because it is simpler than for other methods.

Our first goal is to recast the primal-dual predictor–corrector method into a target-following format and also to give a bound on the iteration number till convergence. Then we shall provide a simple geometric proof of the asymptotic convergence rate. Let us mention an additional reason for devoting a full section to this particular method. In its original formulation the method is not extremely efficient for a simple reason that we shall mention in due course in this section. However, a clever variant of the method significantly cuts down the computational effort to achieve the greatest practical efficiency among all interior point methods. So the method has become very popular and is worth the detour.

### 5.1 The predictor–corrector method

Let us revisit the standard target-following method along the central path. Recall that the target sequence $w_1, w_2, \ldots, w_K$ is of the form $w_k = \mu_k e$ and $\mu_{k+1} = (1-\theta)\mu_k$ for some fixed $0 < \theta < 1$. The Newton step is the solution of a system of linear equations. In that system, the first two sets of constraints are homogeneous: they aim at preserving feasibility. The last equation in the system has the form

$$s\Delta x + x\Delta s = \bar{w} - xs, \tag{5.1}$$

and thus expresses the aim of reaching the target $\bar{w} = (1-\theta)\mu e$ from the present iterate $(x, s)$. We will let $w = xs$. Since the Newton step produces a point with the same duality gap as the target, we may assume (if the process can be repeated) that the current point $(x, s)$ achieves the same duality gap as $\mu e$. In other words, $x^T s = e^T w = n\mu$. Let us rewrite the right-hand side of equation (5.1) as

$$\bar{w} - w = (1-\theta)\frac{e^T w}{n}e - w = (1-\theta)(\frac{e^T w}{n}e - w) + \theta(0 - w).$$

Using the linearity of the defining system, we can interpret the Newton direction associated with the target $\bar{w} = (1-\theta)\mu e$ as the convex combination of two Newton directions corresponding to the targets $\mu e = e^T we/n$ and 0 respectively.

The first direction has the interesting property that it leaves the duality gap unchanged. Its effect is pure centering. For this reason it is often called a *centering step*. The displacements in the $x$- and $s$-space follow from

$$s\Delta x + x\Delta s = \mu e - w, \quad \mu = \frac{e^T w}{n}.$$

The second direction, on the contrary, aims at closing the duality gap. Its target is the zero vector and the displacements in $x$-space and $s$-space follow from

$$s\Delta x + x\Delta s = -w.$$

For historical reasons the name for this is *affine scaling*. Thus it becomes clear that the standard path-following method arises by judiciously combining the two directions.

In the predictor–corrector method, instead of combining the two directions in a single step, one breaks down this single step into two steps: (1) an affine scaling step and (2) a pure centering step. It is easy to see that the full affine scaling step will produce an infeasible point (otherwise a single step would solve the problem). Moreover, due to the second-order effect, the step is done at the expense of centering. These shortcomings of the full affine scaling step can be circumvented by using a damped affine scaling step. So we change the affine scaling target from 0 to $(1-\theta)w$. The Newton step with respect to this new target is the same as a $\theta$ fraction of the full affine scaling step, as appears when looking at the equation

$$s\Delta x + x\Delta s = (1-\theta)w - w = \theta(0 - w).$$

By taking $\theta$ small enough we can now enforce feasibility of the Newton step, and at the same time get control on the loss of centering. The centering step that ensues will maintain the duality gap at its new value $(1-\theta)e^T w/n$. If we globalize those two steps (i.e. affine scaling and centering) in what we may call a major iteration, we obtain the predictor–corrector method, and it will now be

**Table 1** Targets and iterates in the $w$-space

| stage | target | iterate |
|---|---|---|
| 0 | $\bar{w} = \frac{e^T w}{n} e$ | $w$ |
| 1 | $\bar{w}^+ = (1-\theta)w$ | $w^+$ |
| 2 | $\bar{w}^{++} = (1-\theta)\bar{w}$ | $w^{++}$ |

clear that this is very much like the standard path-following method.

The global convergence analysis of the predictor–corrector method will exhibit a bound on the number of major iterations that is comparable to the one obtained for the standard path-following method. The theory does not hint that the number of major iterations will be in any way smaller than the total number of iterations of another target-following method. Since each major iteration costs two Newton iterations, it is even possible that the predictor–corrector method behaves worse. To make the method computationally efficient, one has to substitute the exact calculation of the centering step by an approximation of it. With this modification, the predictor–corrector method is now considered to be the most efficient interior point computational scheme for large-scale linear programming.

### 5.2 Convergence analysis

To ease the formal analysis of the algorithm, we summarize in Table 1 the components of a major iteration to view the process as a target-following scheme. Each row of the table displays the past target and the current point that approximates this target. The first and last row corresponds to the status at the beginning and at the end of one major iteration. In both cases the target lies on the central path. The middle row corresponds to the affine scaling step with a target lying on the ray joining the origin to $w$.

Our aim will be to maintain the proximity of the iterate to the target in Stage 0 to less than or equal to some fixed constant $\delta$. Since Stage 2 will be Stage 0 of the next major iteration, the critical issue is to make sure that the proximity measure of the last point $w^{++}$ to the target $\bar{w}^{++}$ is at most $\delta$. If this property holds the process can be repeated. The question is: which values should be given to $\delta$ and $\theta$ to achieve this result?

To simplify the presentation, we will use $\delta = \frac{1}{2}$ below, but any other constant in the interval $(0,1)$ could also be used. Since $w^{++}$ is obtained from $w^+$ by a full Newton step at $w^+$ with respect to the target $\bar{w}^{++}$, it follows from the local quadratic convergence result in Lemma 4.4 that we will have $\delta(w^{++}, \bar{w}^{++}) \leq \frac{1}{2}$ if

$$\delta(w^+, \bar{w}^{++}) \leq \frac{1}{\sqrt{2}}. \tag{5.2}$$

Let us give first an expression for $\delta = \delta(w, \bar{w})$. Replacing $\bar{w}$ by its value $\mu e$, with

$\mu = e^T w/n$, one gets

$$\delta = \frac{1}{2}\left\|\sqrt{\frac{w}{\mu}} - \sqrt{\frac{\mu}{w}}\right\|.$$

From this one can derive useful bounds on $w$:

$$\frac{1}{\rho^2} \leq \frac{w}{\mu} \leq \rho^2, \tag{5.3}$$

where

$$\rho = \delta + \sqrt{1+\delta^2} \leq \frac{1}{2}\left(1+\sqrt{5}\right).$$

The (elementary) proof of this result can be found in [25].

Let us also recall from (4.4) that $w^+$ can be written as

$$w^+ = (1-\theta)w + \theta^2 \Delta x \Delta s, \tag{5.4}$$

where $\Delta x$ and $\Delta s$ denote the scaled displacements in the $x$- and the $s$-space respectively for a (full) affine scaling step at $w$. The target after the intermediate step is

$$\bar{w}^{++} = (1-\theta)\mu e.$$

Since the iterate achieves the same duality gap as the target after a Newton step, we can assume

$$e^T w = n\mu.$$

Using this we now derive a bound on the proximity measure of the intermediate step $w^+$ to the final target $\bar{w}^{++}$ as follows.

$$\begin{aligned}
\delta(w^+, \bar{w}^{++}) &= \frac{1}{2\min\sqrt{\bar{w}^{++}}}\left\|\frac{w^+ - \bar{w}^{++}}{\sqrt{w^+}}\right\| \\
&= \frac{1}{2\sqrt{(1-\theta)\mu}}\left\|\frac{(1-\theta)w + \theta^2\Delta x\Delta s - (1-\theta)\mu e}{\sqrt{w^+}}\right\| \\
&\leq \frac{\sqrt{1-\theta}}{2\sqrt{\mu}}\left\|\frac{w-\mu e}{\sqrt{w}}\frac{\sqrt{w}}{\sqrt{w^+}}\right\| + \frac{\theta^2}{2\sqrt{(1-\theta)\mu}}\left\|\frac{\Delta x\Delta s}{\sqrt{w^+}}\right\|.
\end{aligned}$$

Using simple inequalities on the norms, we have

$$\delta(w^+, \bar{w}^{++}) \leq \delta\sqrt{(1-\theta)}\left\|\sqrt{\frac{w}{w^+}}\right\|_\infty + \frac{\theta^2\left\|\Delta x\Delta s\right\|}{2\sqrt{(1-\theta)\mu}}\left\|\frac{1}{\sqrt{w^+}}\right\|_\infty. \tag{5.5}$$

In our search for a bound on the various terms appearing in the right-hand side, we find it convenient to introduce the notation

$$\omega = \frac{\left\|\Delta x\Delta s\right\|\theta^2\rho^2}{(1-\theta)\mu}. \tag{5.6}$$

We have

$$\begin{aligned}
\min\left(\frac{w^+}{w}\right) &= \min\left((1-\theta)e - \frac{\theta^2 \Delta x \Delta s}{w}\right) \\
&\geq (1-\theta) - \theta^2 \left\|\frac{\Delta x \Delta s}{w}\right\|_\infty \\
&\geq (1-\theta) - \frac{\theta^2 \rho^2}{\mu} \left\|\Delta x \Delta s\right\|_\infty \\
&\geq (1-\theta)(1-\omega).
\end{aligned}$$

In a similar way, we obtain

$$\begin{aligned}
\min\left(w^+\right) &\geq \min(w) \min\left(\frac{w^+}{w}\right) \\
&\geq \frac{\mu}{\rho^2}(1-\theta)(1-\omega).
\end{aligned}$$

Introducing these bounds into (5.5) yields

$$\begin{aligned}
\delta(w^+, \bar{w}^{++}) &\leq \frac{\delta}{\sqrt{1-\omega}} + \frac{1}{2\rho\sqrt{1-\omega}} \frac{\|\Delta x \Delta s\| \, \theta^2 \rho^2}{(1-\theta)\mu} \\
&\leq \frac{1}{\sqrt{1-\omega}}\left(\delta + \frac{\omega}{2\rho}\right). \qquad (5.7)
\end{aligned}$$

Recall from (5.2) that we need to have $\delta(w^+, \bar{w}^{++}) \leq 1/\sqrt{2}$. This will certainly hold if

$$\frac{1}{\sqrt{1-\omega}}\left(\delta + \frac{\omega}{2\rho}\right) \leq \frac{1}{\sqrt{2}}.$$

Since $\delta$ and $\rho$ are known, this inequality yields an upper bound on $\omega$, which depends monotonically on $\delta$. One may easily verify that for $\delta = 1/2$, the inequality is satisfied if $\omega \leq 1/4$. Using the definition (5.6) of $\omega$, we obtain the following condition on $\theta$:

$$4\,\|\Delta x \Delta s\|\, \theta^2 \rho^2 \leq (1-\theta)\mu.$$

From (5.3) one easily deduces that $\rho^2 \leq 3$. Hence, it will be sufficient if

$$12\,\|\Delta x \Delta s\|\, \theta^2 \leq (1-\theta)\mu. \qquad (5.8)$$

Recall that $\Delta x$ and $\Delta s$ are the scaled displacements for the full affine scaling step. Using (4.1) we get

$$\|\Delta x \Delta s\| \leq \frac{1}{2\sqrt{2}} \|p_v\|^2 = \frac{1}{2\sqrt{2}} \left\|\frac{w}{\sqrt{w}}\right\|^2 = \frac{1}{2\sqrt{2}} e^T w = \frac{n\mu}{2\sqrt{2}} < \frac{n\mu}{2}.$$

Substituting this bound, we conclude that (5.8) will certainly be satisfied if

$$6n\theta^2 \leq 1-\theta.$$

**Table 2** Asymptotic orders of magnitude of some relevant variables

| | $B$ | $N$ |
|---|---|---|
| $x$ | $\Omega(1)^*$ | $\Omega(\mu)$ |
| $s$ | $\Omega(\mu)$ | $\Omega(1)^*$ |
| $w$ | $\Omega(\mu)$ | $\Omega(\mu)$ |
| $d$ | $\Omega(\frac{1}{\sqrt{\mu}})$ | $\Omega(\sqrt{\mu})$ |
| $\Delta x$ | $O(\mu)^*$ | $O(\mu)$ |
| $\Delta s$ | $O(\mu)$ | $O(\mu)^*$ |

If we take $\theta = 1/\left(3\sqrt{n}\right)$, this inequality holds, and we will have $\delta(w^{++}, \bar{w}^{++}) \leq 1/2$. So this value of $\theta$ guarantees that the process can repeat itself. Each predictor–corrector iteration reduces the duality gap by $(1-\theta)$. This implies convergence in $O(\sqrt{n})$ major iterations. Recall again that each major iteration requires two steps of the same complexity as a standard iteration in any interior point method.

## 5.3 Asymptotic convergence

The basic issue in studying the rate of convergence is to see whether the term $\theta$ can be made closer and closer to 1 in the limit, i.e. when $\mu$ is getting very small. On the other hand the global convergence analysis of the previous section showed that the important point was to make sure that the parameter $\theta$, which acts as the damping factor in the affine scaling step, respects the inequality (5.8). To achieve the goal of quadratic convergence, we must require that in the limit the term $\|\Delta x \Delta s\|$ is of the order $O(\mu^2)$, so that one can have $(1-\theta) = O(\mu)$.

To obtain this result, we will have to concentrate on the magnitude of the components of the iterate $(x,\, s)$ at the beginning of each major iteration as well as those of the (full) affine scaling direction $(\Delta x,\, \Delta s)$. The limiting behaviour of these terms is strongly dependent on whether an individual component belongs to one set of the optimal partition or to the complementary one. Table 2 summarizes the fundamental result concerning the order of the various terms of interest. In that table, $(B,\, N)$ denotes the optimal partition of the problem. From this table we read that, e.g. $x_B = \Omega(1)$ and $\Delta x_N = O(\mu)$. The symbols $\Omega$ and $O$ imply that there exist positive constants $C_1$,$C_2$ and $C_3$ such that $C_1 e \leq x_B \leq C_2 e$ and $\Delta x_N \leq C_3 \mu$. Some of the statements in the table are almost trivial; the more difficult ones are indicated by a star.

Let us temporarily postpone the proof of the statements in Table 2 and show that one easily proves quadratic convergence simply using the order estimates given in the table. We write

$$\|p_x p_s\| = \|\Delta x \Delta s\| = \|\Delta x_B \Delta s_B + \Delta x_N \Delta s_N\| \leq \|\Delta x_B \Delta s_B\| + \|\Delta x_N \Delta s_N\| .$$

From the table we deduce that each of the last two terms is $O(\mu^2)$. Hence we have $\|p_x p_s\| = O(\mu^2)$. Substitution in (5.8) gives the following condition on $\theta$:

$$12\theta^2 \|p_x p_s\| = \theta^2 O(\mu^2) \leq \mu(1-\theta).$$

Clearly this allows us to take values of $\theta$ for which $1 - \theta = O(\mu)$. This justifies our claim, namely that the predictor–corrector method is globally quadratically convergent.

The rest of this section is devoted to proving the estimates in Table 2. Note that the third line, which concerns $w = xs$, is immediate from (5.3). Note also that the fourth line follows directly from the definition of $d$ and the estimates for $x$ and $s$ (first and second lines). So let us formally prove the first two lines here. We start with the estimates for $x_B$ and $s_B$. (The estimates for $x_N$ and $s_N$ are obtained in exactly the same way.) We need the following two positive numbers:

$$\begin{aligned} \rho_p &:= \min_{i \in B} : \max\{x_i \ : \ x \in \mathcal{P}^*\}, \\ \rho_d &:= \min_{i \in N} : \max\{s_i \ : \ (y,\, s) \in \mathcal{D}^*\}. \end{aligned}$$

Note that these numbers only depend on the data of the problem and not on the iterates. These quantities were introduced earlier in [51]. Now let $i \in B$ and let $\bar{x} \in \mathcal{P}^*$ be such that $\bar{x}_i$ is maximal. Then we may write

$$s_i = s_i \bar{x}_i \frac{1}{\bar{x}_i} \leq \frac{s^T \bar{x}}{\rho_p} \leq \frac{n\mu}{\rho_p}.$$

This shows that $s_B = O(\mu)$. Since $x_B s_B = w_B = \Omega(\mu)$, it follows that $x_B$ is bounded away from zero. On the other hand, since the pair $(x,\, s)$ has duality gap $n\mu$ and hence, by Corollary 3.2, belongs to a bounded set, we have $x_B = O(1)$. Thus we may conclude that $x_B = \Omega(1)$ and consequently, $s_B = \Omega(\mu)$. In exactly the same way, one derives that $s_N = \Omega(1)$ and $x_N = \Omega(\mu)$. Hence we have proved the statement in the first four lines of Table 2.

We proceed with the last two lines of Table 2. We obtain an order estimate for $\Delta x_N$ by the following simple argument. We have

$$\left\|d_N^{-1} \Delta x_N\right\| \leq \left\|d^{-1} \Delta x\right\| = \|p_x\| \leq \|p_v\|.$$

Since $p_v = -v$, and $\|v\| = \sqrt{n\mu}$, we obtain $\left\|d_N^{-1} \Delta x_N\right\| = O(\sqrt{\mu})$. Now using $d_N = \Omega(\sqrt{\mu})$, it follows that $\Delta x_N = O(\mu)$. In the same way, we can prove $\Delta s_B = O(\mu)$.

As we indicated earlier, the last two star-marked statements in Table 2, namely the estimates for $\Delta x_B$ and $\Delta s_N$, are more difficult to obtain.[1] We will only deal with the estimate for $\Delta x_B$. The result for $\Delta s_N$ can be obtained in a similar way.

[1]We kindly acknowledge that the basic idea of the analysis below was communicated privately to us by our colleague Gonzaga, and also refer to some papers dealing with the global quadratic convergence of the predictor–corrector method [17, 55].

Using the notation introduced in Section 4.1.1, we may write

$$d^{-1}\Delta x = p_x = P_{AD}(p_v) = -P_{AD}(v) = -P_{AD}(ds).$$

Letting $(\tilde{y}, \tilde{s})$ denote any dual optimal pair, we have

$$s = c - A^T y = A^T \tilde{y} + \tilde{s} - A^T y = \tilde{s} + A^T(\tilde{y} - y).$$

So we have

$$ds = d\tilde{s} + (AD)^T(\tilde{y} - y).$$

This means that $ds - d\tilde{s}$ belongs to the row space of $AD$. The row space being orthogonal to the null space of $AD$, it follows that

$$P_{AD}(ds) = P_{AD}(d\tilde{s}).$$

Hence we obtain

$$d^{-1}\Delta x = -P_{AD}(d\tilde{s}).$$

This means that

$$d^{-1}\Delta x = \text{argmin}_h \left( \|d\tilde{s} - h\|^2 \; : \; ADh = 0 \right),$$

or equivalently,

$$d^{-1}\Delta x = \text{argmin}_h \Big( \|d_B\tilde{s}_B - h_B\|^2 + \|d_N\tilde{s}_N - h_N\|^2 \; : \; A_B D_B h_B + A_N D_N h_N = 0 \Big).$$

Note that the estimate for $h_N = d_N^{-1}\Delta x_N$ is already known. This vector is $O(\sqrt{\mu})$. What is needed now is an estimate for $h_B$. To this end we shall express $h_B$ as an indirect function of $h_N = d_N^{-1}\Delta x_N$. Using $\tilde{s}_B = 0$ and taking $h_N$ as above, we obtain

$$d_B^{-1}\Delta x_B = \text{argmin}_{h_B} \left( \|h_B\|^2 \; : \; A_B D_B h_B = -A_N \Delta x_N \right).$$

Stated otherwise, the vector $d_B^{-1}\Delta x_B$ is the vector of smallest norm in the affine space

$$\mathcal{S} := \left\{ z \in \mathrm{I\!R}^{|B|} \; : \; A_B D_B z = -A_N \Delta x_N \right\}.$$

Since $\Delta x_N = O(\mu)$, we also have $A_N \Delta x_N = O(\mu)$. Thus we can assert that there exists a vector $z \in \mathcal{S}$ such that $D_B z = O(\mu)$ where the constant after the order symbol $O$ only depends on $A_B$ as it should. Using $d_B^{-1} = \Omega(\sqrt{\mu})$, we have $z = O(\mu\sqrt{\mu})$. Recall that $h_B = d_B^{-1}\Delta x_B$ is the vector of smallest norm in $\mathcal{S}$. Hence we have

$$\left\|d_B^{-1}\Delta x_B\right\| \leq \|z\| = O(\mu\sqrt{\mu}).$$

Using $d_B^{-1} = \Omega(\sqrt{\mu})$ again, we prove that $\|\Delta x_B\| = O(\mu)$, and hence $\Delta x_B = O(\mu)$. This completes the proof.

## 6 An infeasible start version of Karmarkar's algorithm

In the previous sections on target-following algorithms and the predictor–corrector method, we alluded to the problem of finding an initial primal-dual interior feasible pair. We invoked the possibility of recasting the problem in a skew symmetric self-dual format, but it is a valid question to know whether or not the feasibility issue can be directly handled within the original problem formulation. For interior point methods, the answer is positive.

The first infeasible start algorithm [7] was given in the framework of the primal projective algorithm of Karmarkar [26]. More recently, several papers dealing with infeasible start in primal-dual methods have appeared, see, e.g. [29, 38, 41, 44, 53, 57]. Those papers provide a theoretical justification of the actual implementation of primal-dual algorithms (see [10]). However, we shall not discuss those extensions. We shall rather concentrate on Karmarkar's projective algorithm. The reasons for our choice are twofold. Firstly, Karmarkar's projective algorithm yields a nice and relatively simple handling of the infeasible start issue. Secondly, Karmarkar's algorithm is the first interior point method for linear programming with a proven polynomial complexity. Discussing the infeasible start issue within the framework of the projective algorithm provides an opportunity to review some of the ideas that appeared in Karmarkar's seminal paper [26].

To conclude this short presentation, let us mention that the infeasible start issue is not specific to interior point methods. The simplex method faces the very same difficulty. The classical answer in that case is to use either a "big M" approach or a two-phase method. There also exist less well-known one-phase methods that do not require a "big M" nor extra variables. In these methods, called "criss-cross" methods, infeasible basic solutions are allowed, and both primal and dual pivoting operations are performed. For an exposition of the method, we refer to [58]. Finite convergence is proved in [48].

### 6.1 Karmarkar's canonical problem

Karmarkar [26] used the following canonical formulation

$$\begin{array}{lrcl} \min & c^T x, & & \\ \text{s.t.} & Ax & = & 0, \\ & e^T x & = & n, \\ & x & \geq & 0. \end{array} \tag{6.1}$$

Note that the constraints are homogeneous, except the last one which can be viewed as a normalizing constraint. This formulation implicitly assumes boundedness of the feasible set. Under this boundedness assumption, it is possible to recast the standard constraints $Ax = b$, $x \geq 0$ into Karmarkar's framework. The boundedness assumption states that the system $\{x \geq 0 \,:\, Ax = 0\}$ has no solution other than $x = 0$. It is easy to show, e.g. using duality theory for linear programming, that there must exist a vector $\bar{y}$ such that $A^T\bar{y} > 0$. For any $x$ such that $Ax = b$, one has $(A^T\bar{y})^T x = b^T\bar{y}$. This redundant constraint can be added to the definition of the feasible set $Ax = b$, $x \geq 0$. By subtracting suitable multiples of this constraint to $Ax = b$, and properly rescaling the variables, one easily obtains an equivalent problem in Karmarkar's format. Unfortunately, it is not always easy to elicit $\bar{y}$. An alternative is to introduce the brute force constraint $e^T x + x_{n+1} = M$, where $M$ is some large number and $x_{n+1} \geq 0$ is a slack variable. Proceeding as before, we can recast the problem in Karmarkar's format.

Karmarkar makes two further explicit assumptions:

**Assumption 3** *The problem has an interior feasible solution.*

**Assumption 4** *The optimal value is zero.*

From Section 3 we know that Assumption 3, together with the boundedness assumption implied by Karmarkar's formulation, is nothing other than the interior point assumption of the first section of this chapter. As to Assumption 4, it states, in a disguised way, that the problem at hand is not really an optimization problem. The real issue is to find a feasible solution such that $c^T x = 0$.

Karmarkar used geometric arguments to design the search direction and then resorted to a potential reduction analysis to prove polynomial convergence. He relied on the concept of projective transformation, hence the name *projective algorithm*. This elegant but indirect approach is not mandatory. One can also interpret the search direction as the projected Newton method with respect to the potential. The algorithm is thus a standard damped Newton method. We shall use this approach and proceed in three steps. In the first step, we shall deal with the pure feasibility problem as in [8]. In doing so we shall use Karmarkar's canonical format which differs from the one used up to now in this chapter. In the second step, we shall generalize the method to handle an infeasible initial point and an objective with an unknown optimal value. In the last step, we shall show how the method can be naturally applied to problems in the standard form.

## 6.2 Projective feasibility algorithm

Let us begin with Karmarkar's original problem: that is, an interior feasible solution is known and the optimal value is 0. As in [8], let us not distinguish anymore between the objective and the constraints. The problem of interest is to find a feasible solution to

$$\begin{aligned} \tilde{A}x &= 0, \\ e^T x &= n, \\ x &\geq 0, \end{aligned} \tag{6.2}$$

where $\tilde{A} = \begin{pmatrix} c^T \\ A \end{pmatrix}$. The projective algorithm, as in all interior point methods, does not compute the exact solutions. Instead, it produces points such that

$$\|\tilde{A}x\|_\infty \leq \varepsilon.$$

It can be proved that obtaining an $\varepsilon$-solution, with $\varepsilon$ small enough, is the same as solving the problem exactly [8]. To measure the progress towards feasibility, we shall use a generalization of Karmarkar's potential, namely

$$\varphi(x) = n \ln \|\tilde{A}x\|_\infty - e^T \ln x.$$

It is important to note that the potential $\varphi(x)$ is positively homogeneous of degree zero. Also observe that if $x > 0$ is feasible, then $\|\tilde{A}x\|_\infty = c^T x$. Furthermore, if $x$ is not optimal, then $c^T x > 0$ and the potential becomes

$$\varphi(x) = n \ln c^T x - e^T \ln x,$$

as defined by Karmarkar in [26].

**Lemma 6.1** *If $x > 0$ satisfies $\varphi(x) \leq n \ln \varepsilon$, then the point $(n/e^T x)x$ is an $\varepsilon$-solution.*

**Proof** Since $\varphi$ is positively homogeneous of degree zero, we have $\varphi((n/e^T x)x) = \varphi(x) \leq n \ln \varepsilon$. So we can assume that $x$ is normalized, i.e. $e^T x = n$. Using the standard inequality between the geometric and the arithmetic means

$$e^T \ln x \leq n \ln \frac{e^T x}{n} = 0,$$

we obtain for normalized points $(e^T x = n)$,

$$n \ln \|\tilde{A}x\|_\infty \leq \varphi(x) \leq n \ln \varepsilon.$$

Thus $\|\tilde{A}x\|_\infty \leq \varepsilon$. □

By the above lemma, the original feasibility problem can be replaced by the unconstrained non-linear programming problem of minimizing $\varphi$, or more precisely, of finding a small enough value for the potential. Another practical consequence is that we may disregard the normalization constraint $e^T x = n$ in all intermediate steps.

The point now is to devise an iterative procedure that decreases the potential. For strongly convex unconstrained problems, the method of choice is Newton's method: the iterative step is the one that minimizes the local second-order approximation of the potential. Unfortunately, we face two problems here. The first one is simply that the term $n \ln \|\tilde{A}x\|_\infty$ in the potential $\varphi$ is not convex (not even differentiable!). To get around this difficulty, we consider the restriction of the potential to the affine subspace $x + \Delta x$, with $\tilde{A}\Delta x = 0$. In that case, the first term, $n \ln \|\tilde{A}x\|_\infty$, is constant on that subspace and the potential reduces to $-e^T \ln x$ (a strongly convex function). If it turns out that the restriction imposed by the constraint of staying on the given subspace does not prevent significant decreases of the potential, we will have a viable alternative for the minimization of $\varphi$.

Let us postpone the proof of this last statement, but now formally define the Newton step with respect to the restriction of the potential to the affine subspace. Taking into account the fact that the first term of $\varphi$ is constant, we get for the derivatives $\nabla\varphi = -x^{-1}$ and $\nabla^2\varphi = X^{-2}$. The Newton step $\Delta x$ is thus the solution of

$$\min\{\frac{1}{2}\Delta x^T X^{-2}\Delta x - \Delta x^T x^{-1} : \tilde{A}\Delta x = 0\}.$$

It follows that the Newton direction is the unique solution of the linear system in $\Delta x$ and an auxiliary vector $u$:

$$\begin{aligned} X^{-2}\Delta x \quad -x^{-1} + \tilde{A}^T u &= 0, \\ \tilde{A}\Delta x \qquad\qquad\qquad &= 0. \end{aligned}$$

Using the standard primal scaling $\Delta x = xp_x$ we obtain the equivalent system

$$\begin{aligned} p_x \quad +X\tilde{A}^T u &= e, \\ \tilde{A}Xp_x \qquad\quad &= 0, \end{aligned}$$

whereby it appears that $p_x$ is the projection of the vector $e$ on the null space of the matrix $\tilde{A}X$. We shall denote it by $P_{\tilde{A}X}e$. For this reason the Newton step with respect to the restriction of $\varphi$ is called the *projected* Newton step.

The second difficulty is that $x + \Delta x$ may not be positive and the potential at that point would not be defined. So we may have to take a damped projected Newton step, i.e. an iterative step of the form $x + \alpha\Delta x$, where $\tilde{A}\Delta x = 0$ and $0 < \alpha \leq 1$ is a damping factor.

Prior to stating the main result concerning the projective algorithm, let us mention a simple lemma on logarithms:

**Lemma 6.2** *Let $h$ be a vector such that $||h|| < 1$. Then*

$$e^T \ln(e + h) \geq e^T h + ||h|| + \ln(1 - ||h||).$$

For a proof, we refer to [23]. The above lemma will be used in the proof of the next fundamental lemma.

**Lemma 6.3** *Assume $p_x = P_{\tilde{A}X}e$. Then*

(1) *if $p_x \neq 0$ and $p_x \geq 0$, then the vector $n(xp_x)/(x^T p_x)$ solves the problem given by (6.1);*

(2) *if $p_x < e$, then the problem given by (6.1) is infeasible;*

(3) *if $p_x \not< e$, then the point $x^+ = x + (xp_x)/(1 + ||p_x||)$ is interior feasible and $\varphi(x^+) - \varphi(x) \leq -0.31$.*

**Proof** Statement 1 is immediate. Assume case 2 holds: $p_x < e$. Then $\tilde{A}^T u = x^{-1}(e - p_x) > 0$ and for all non-trivial $\hat{x} \geq 0$, we have $\hat{x}^T\tilde{A}^T u > 0$, ruling out the existence of a feasible solution with $\tilde{A}\hat{x} = 0$. If case 3 holds, we have

$$x^+ = x(e + \alpha p_x), \quad \alpha = \frac{1}{1 + ||p_x||}.$$

Clearly $x^+ > 0$. Using the definition of the potential and Lemma 6.2 with $h = \alpha p_x$, we get

$$\begin{aligned}\varphi(x^+) &= n\ln\|\tilde{A}x^+\|_\infty - e^T\ln x^+\\ &= n\ln\|\tilde{A}x\|_\infty - e^T\ln x - e^T\ln(e+\alpha p_x)\\ &\leq \varphi(x) - \alpha e^T p_x - \alpha\|p_x\| - \ln(1-\alpha\|p_x\|).\end{aligned}$$

One has $e^T p_x = \|p_x\|^2$, as can be easily seen. Hence we obtain

$$\varphi(x^+) \leq \varphi(x) - \alpha\|p_x\|(1+\|p_x\|) - \ln(1-\alpha\|p_x\|).$$

Substituting the value of $\alpha$, we get

$$\varphi(x^+) \leq \varphi(x) - \|p_x\| + \ln(1+\|p_x\|).$$

Since the right hand side is decreasing with $\|p_x\|$, we can replace $\|p_x\|$ by a lower bound. By assumption $p_x \not< e$, we have $\|p_x\| \geq 1$. We obtain $\varphi(x^+) - \varphi(x) \leq -1 + \ln 2 \leq -0.31$. □

In view of the above lemma, we can describe one step of the projective algorithm. Let $x^0 = e$ be the initial point (any other point with positive coordinates would do) and let $x > 0$ be the current iterate. Compute $p_x$. If case 1 or case 2 occurs, the algorithm stops with a relevant conclusion. If case 3 holds, we define the damped Newton step by

$$\hat{x} = x + \alpha x p_x,$$

with

$$\alpha = \frac{1}{1+\|p_x\|}.$$

Then, the iterate is scaled by

$$x^+ := \frac{n}{e^T\hat{x}}\hat{x} = \frac{1}{1+\alpha\frac{x^T p_x}{n}}\hat{x}.$$

This step is feasible and it decreases the potential by at least $\sigma = 1 - \ln 2 \approx 0.31$.

We can use case 3 of the above lemma to bound the number of iterations. Indeed, after $k$ iterations of the projective algorithm, we have

$$\varphi(x^k) \leq \varphi(x^0) - k\sigma.$$

Recall that

$$\varphi(x^k) \geq n\ln\|\tilde{A}x^k\|_\infty \quad (\text{since } -e^T\ln x^k \geq -n\ln\tfrac{e^T x^k}{n} = 0)$$

and

$$\varphi(x^0) = n\ln\|\tilde{A}e\|_\infty \quad (\text{since } e^T\ln x^0 = e^T\ln e = 0).$$

It follows that

$$n\ln\|\tilde{A}x^k\|_\infty \leq \varphi(x^k) \leq n\ln\|\tilde{A}e\|_\infty - k\sigma.$$

If $k$ is large enough to have

$$n \ln \|\tilde{A}e\|_\infty - k\sigma \leq n \ln \varepsilon, \tag{6.3}$$

we certainly have $\ln \|\tilde{A}x^k\|_\infty \leq n \ln \varepsilon$, and the current iterate is $\varepsilon$-feasible. So the algorithm performs no more than

$$\frac{n}{\sigma} \ln \frac{\|\tilde{A}e\|_\infty}{\varepsilon}$$

iterations.

We would like to stress a distinctive feature of the projective algorithm. We have

$$\tilde{A}x^+ = \frac{1}{1 + \alpha\frac{x^T p_x}{n}} \tilde{A}(x + \alpha x p_x) = \frac{1}{1 + \alpha\frac{x^T p_x}{n}} \tilde{A}x.$$

The new constraint violation $\tilde{A}x^+$ is a multiple $\gamma = (1 + \alpha x^T p_x / n)^{-1}$ of the former violation. This multiplicative factor is positive (but not necessarily less than one). Consequently, the signs of the constraint violations do not change. Note also that the same multiplicative factor applies to all constraints. Finally, if some constraint is satisfied at the beginning, it remains satisfied till the end. Let $\gamma_i$, $i = 1, \ldots, k$, be the successive factors in $k$ iterations. Then we have

$$\tilde{A}x^k = \Big(\prod_{i=1}^{k} \gamma_i\Big) \tilde{A}e.$$

In view of (6.3), we can reformulate the convergence result as: the cumulative effect of the multiplicative factors is tending exponentially fast to zero:

$$\prod_{i=1}^{k} \gamma_i \leq \exp\left(\frac{-k\sigma}{n}\right). \tag{6.4}$$

To conclude this first step of our analysis, let us look at Karmarkar's canonical problem, i.e. an optimization problem with a known interior feasible solution $x^0 > 0$ and a known optimal value equal to zero. Since the projective algorithm preserves feasibility and $Ax^0 = 0$, we shall have $Ax = 0$, i.e. the constraints in the original optimization problem are always satisfied throughout the iterations. Consequently, $\|\tilde{A}x\|_\infty = c^T x > 0$ throughout the iterations. The bound on the number of iterations is thus

$$\frac{n}{\sigma} \ln \frac{c^T x^0}{\varepsilon}.$$

## 6.3 The infeasible start projective algorithm

Let us now consider the case where the optimal value is unknown and no interior feasible solution is given a priori. The traditional approach in linear programming is to solve the problem in two phases: in phase I the algorithm constructs a feasible solution; in phase II the algorithm seeks optimality. The projective algorithm allows a joint treatment of feasibility and optimality. It is sometimes named a combined phase I – phase II algorithm. Our presentation follows [7]. The same parametrization of the objective is used by Todd and Burrell [50] under Assumption 3. We refer to [1, 6, 49] for alternative infeasible start approaches in a primal framework.

From now on, we shall only assume

**Assumption 5** *There is a known lower bound $z$ to the optimal value of the problem.*

With the lower bound $z$ we associate the feasibility problem

$$\begin{aligned} c^T x - z &= 0, \\ Ax &= 0, \\ e^T x &= n, \\ x &\geq 0. \end{aligned}$$

Obviously, if $z$ is a strict lower bound, the feasibility problem has no solution. But we know that the projective algorithm will discover this. The crux is that when the blocking case $p_x < e$ occurs, there is a way to improve the bound until $p_x \not< e$. So the algorithm will go on with a guaranteed decrease of the potential at each step. Let us see how it works.

First, using the normalizing constraint, we transform the first feasibility constraint into a homogeneous one:

$$c^T x - z = (c - \frac{z}{n} e)^T x.$$

We let

$$c_z = c - \frac{z}{n} e \qquad \text{and} \qquad \tilde{A} = \begin{pmatrix} c_z^T \\ A \end{pmatrix}.$$

To measure progress towards a solution, we shall use the same potential as before. This potential depends on $z$ through the first row of $\tilde{A}$, so we shall denote it as

$$\varphi(x; z) = n \ln \|\tilde{A}x\|_\infty - e^T \ln x.$$

In our analysis, we shall also use the potential that is associated with the simple feasibility problem $\{x \geq 0 \,:\, Ax = 0,\, e^T x = n\}$. We shall denote it

$$\psi(x) = n \ln \|Ax\|_\infty - e^T \ln x.$$

In the algorithm, we shall use the same search direction $p_x = P_{AX} e$ as before. To express its dependence with respect to $z$, we introduce some notation. Let $u_e$, $u_c$ and $u_x$ be such that

$$\begin{aligned} P_{AX} e &= e - XA^T u_e, \\ P_{AX} Xc &= Xc - XA^T u_c, \\ P_{AX} Xe &= Xe - XA^T u_x. \end{aligned}$$

We can now give the formula for $p_x$.

**Lemma 6.4** *The projected Newton direction* $p_x = P_{\tilde{A}X}e$ *at* $x$ *is given by the formula:*

(1) *if* $P_{AX}Xc_z = 0$,

$$p_x = P_{AX}e; \tag{6.5}$$

(2) *otherwise*

$$p_x = P_{AX}e - \rho(z)P_{AX}Xc_z, \tag{6.6}$$

*where*

$$\rho(z) = \frac{e^T P_{AX}Xc_z}{\|P_{AX}Xc_z\|^2}.$$

**Proof** If $P_{AX}Xc_z = 0$, then $Xc_z$ is in the row space of $XA^T$. Therefore the null space of $\tilde{A}X$ is equal to the null space of $AX$. Hence

$$p_x = P_{\tilde{A}X}e = P_{AX}e.$$

This proves the first part of the lemma. To prove the second part, we note first that $\rho(z)$ is well defined. Then we use a uniqueness property of the projection: the projection of $e$ on the null space of $\tilde{A}X$ is the only vector in the null space of $\tilde{A}X$ such that $e - p_x$ belongs to the row space of that matrix. Let us check whether these properties hold when $p_x$ is given by (6.6). That $p_x$ belongs to the null space of $AX$ follows directly from the formula. We also have

$$c_z^T Xp_x = c_z^T XP_{AX}e - \rho(z)c_z^T XP_{AX}Xc_z.$$

Since $c_z^T XP_{AX}e = e^T P_{AX}Xc_z$ and $c_z^T XP_{AX}Xc_z = \|P_{AX}Xc_z\|^2$, we have

$$c_z^T Xp_x(z) = 0,$$

which proves that $p_x$ is indeed in the null space of $\tilde{A}X$. On the other hand, we have

$$e - p_x(z) = X\left(\rho(z)(c - \frac{z}{n}e) - A^T\left(\rho(z)(u_c - \frac{z}{n}u_x) - u_e\right)\right). \tag{6.7}$$

The right-hand side is clearly in the column space of $(XA^T, Xc_z)$, i.e. the row space of $\tilde{A}X$. Hence, (6.6) rightly defines the projection. The proof is complete. □

**Remark** The first case in the above lemma has the following interpretation. Let $\bar{x}$ be any feasible point. Since

$$\begin{aligned}(\bar{x}x^{-1})^T P_{AX}Xc_z &= (\bar{x}x^{-1})^T(X(c - \frac{z}{n}e) - XA^T(u_c - \frac{z}{n}u_x)) \\ &= \bar{x}^T(c - \frac{z}{n}e) - \bar{x}^T A^T(u_c - \frac{z}{n}u_x) = c^T\bar{x} - z,\end{aligned}$$

we see that $P_{AX}Xc_z = 0$ implies that $c^T\bar{x} - z = 0$. The objective function of the linear programming problem takes the constant value $z$ on the feasible set. If this extremely degenerate case occurs, then the problem reduces to finding a point in the feasible set, a case that has been dealt with in the preceding section.

**Remark** The second case in the lemma shows that $p_x$ is a function of $z$. To emphasize this relationship we shall use, when necessary, the notation $p_x(z)$. We also note that the function $p_x(z)$ is continuous for all $z$ such that $P_{AX}Xc_z \neq e$.

As long as $p_x(z) \not< e$ we know from the analysis of the previous section that a regular step of the projective algorithm decreases the potential by at least $\sigma = -0.31$. However, $z$ being a lower bound, the system $\tilde{A}x = 0$, $x \geq 0$ has no non-zero solution and we must expect that $p_x(z) < e$ at some iteration. The key result, which we shall now prove, is that $p_x(z) < e$ guarantees that $z$ can be raised to yield a better lower bound. This increase of $z$ is pursued until the condition $p_x(z) \not< e$ is met again. The limit point $z^+$ is defined by

$$z^+ = \inf\{\tilde{z} \geq z : e - p_x(\tilde{z}) \not> 0\}. \tag{6.8}$$

Using the continuity of $p_x$, we can assert that $z^+ > z$. We may have $z^+ = \infty$.

The key point is to show that $z^+$ is indeed a lower bound. It is important to note that in this analysis the point $x$ is fixed; only $z$ varies. The vectors $u_e$, $u_c$ and $u_e$ do not change in this process.

**Lemma 6.5** *Let $x > 0$ be given and assume that there exists no $\tilde{z}$ such that $P_{AX}Xc_{\tilde{z}} = 0$. Assume $e - p_x(z) > 0$ and let $z^+$ be as defined in (6.8).*

(1) *If $\rho(\tilde{z}) \leq 0$ for some $z \leq \tilde{z} < z^+$, then the problem given by (6.1) is infeasible.*

(2) *If $\rho(\tilde{z}) > 0$ for all $z \leq \tilde{z} < z^+$ and $z^+ = \infty$, then the problem given by (6.1) is infeasible.*

(3) *If $\rho(\tilde{z}) > 0$ for all $z \leq \tilde{z} < z^+$ and $z^+ < \infty$, then $z^+$ is a lower bound with $p_x(z^+) \not< e$.*

**Proof** Assume case 1 and let $\bar{z}$ be the smallest value larger than $z$ satisfying $\rho(\bar{z}) \leq 0$. Assume first $\rho(\bar{z}) = 0$. For any non-trivial $\bar{x} \geq 0$, one has from (6.7)

$$0 < \bar{x}^T X^{-1}(e - p_x(\bar{z})) = \bar{x}^T A^T u_e.$$

Thus $A\bar{x} \neq 0$ and the problem has no feasible solution. Assume now $\rho(\bar{z}) < 0$. Since $\rho$ is a continuous function, it must be that $\bar{z} = z$ and thus $\rho(z) < 0$. Assume that $\bar{x} \geq 0$ is a feasible solution: $A\bar{x} = 0$, $e^T\bar{x} = n$. Then we get

$$0 < \bar{x}^T X^{-1}(e - p_x(z)) = \rho(z)(c^T\bar{x} - z).$$

Since $\rho(z) < 0$, the inequality implies $c^T\bar{x} < z$, contradicting the assumption that $z$ is a lower bound. Thus we must conclude that the problem is infeasible.

Assume now case 2. For any $z \leq \tilde{z} < z^+$ we simultaneously have $\rho(\tilde{z}) > 0$ and $e - p_x(\tilde{z}) > 0$. Let $s(\tilde{z})$ be defined by

$$s(\tilde{z}) := c - \frac{\tilde{z}}{n}e - A^T\left(u_c - \frac{\tilde{z}}{n}u_x - \rho(\tilde{z})^{-1}u_e\right). \tag{6.9}$$

Multiplying definition (6.9) by $\rho(\tilde{z})x > 0$, we obtain by (6.7)

$$\rho(\tilde{z})xs(\tilde{z}) = (e - p_x(\tilde{z})) > 0. \tag{6.10}$$

It follows that $s(\tilde{z}) > 0$. Pick any feasible (hence possibly optimal) solution $\hat{x}$; we have by (6.9) and (6.10) that

$$c^T\hat{x} - \tilde{z} = \hat{x}^T s(\tilde{z}) > 0. \tag{6.11}$$

Consequently, $\tilde{z}$ is a lower bound. The value $z^+$ is the supremum of a family of lower bounds: it is thus a lower bound itself. Since $z^+ = \infty$, we immediately conclude that the problem is infeasible.

Finally, assume case 3. Since the conditions are the same as in case 2, except $z^+ < \infty$, we have that $z^+ > z$ is a lower bound. By construction we also have $p_x(z^+) \not< e$ (though $p_x(z^+) \leq e$). This concludes the proof of Lemma 6.5. □

Lemma 6.5 does not say how to compute the solution $z^+$ of (6.8). This can be done by a coordinate-wise analysis of the behaviour of $e - p_x(z)$ as $z$ increases. Recall that in the lemma we excluded the degenerate case $P_{AX}Xc_z = 0$ and proved $\rho(\tilde{z}) > 0$ for all $z \leq \tilde{z} < z^+$. Hence $e^T P_{AX}Xc_{\tilde{z}} = \rho(\tilde{z})\|P_{AX}Xc_{\tilde{z}}\|^2 > 0$. Multiplying (6.7) by $\rho(\tilde{z})^{-1}e^T P_{AX}Xc_{\tilde{z}} > 0$, we see that

$$\begin{aligned}\rho(\tilde{z})^{-1}(e^T P_{AX}Xc_{\tilde{z}})(e - p_x(\tilde{z})) = \\ (e^T P_{AX}Xc_{\tilde{z}})P_{AX}Xc_{\tilde{z}} + \|P_{AX}Xc_{\tilde{z}}\|^2 A^T u_e < 0.\end{aligned}$$

Since $P_{AX}Xc_{\tilde{z}}$ is a linear function of $\tilde{z}$, each coordinate in the right-hand side is a quadratic function of $\tilde{z}$. By solving at most $n$ quadratic equations, one can find the smallest value for which all coordinates are non-positive, and at least one is zero. This is the approach followed in [7].

A more powerful update and also an easier one to compute was introduced in [5]. Keep in mind that we are interested in finding an improved bound. To this end consider the linear programming problem

$$\min\{(c - \frac{z}{n}e)^T x \,:\, Ax = 0,\, e^T x = n,\, x \geq 0\}$$

and its dual

$$\max\{\xi \,:\, A^T u + \xi e \leq c - \frac{z}{n}e\}.$$

For all primal feasible $x$ and dual feasible $\xi$, one has $(c - (z/n)e)^T x \geq \xi$. Therefore, if for some $z$ we find a feasible $\xi > 0$, i.e. a vector

$$s = (c - \frac{z}{n}e) - A^T u$$

such that $s > 0$, we have that $z$ is a lower bound. The same argument was implicitly used in the proof of the lemma to show that $\tilde{z}$ was a lower bound. We did it by introducing the vector $s(\tilde{z})$ defined by (6.9)

$$s(\tilde{z}) := c - \frac{\tilde{z}}{n}e - A^T\left(u_c - \frac{\tilde{z}}{n}u_x - \rho(\tilde{z})^{-1}u_e\right).$$

Let us replace $\rho(\tilde{z})^{-1}$ in this expression by a free variable $\tau$ and solve the two-dimensional linear programming problem (in the variables $\tilde{z}$ and $\tau$)

$$\max\{\tilde{z} \; : \; s = c - \frac{\tilde{z}}{n}e - A^T\left(u_c - \frac{\tilde{z}}{n}u_x - \tau u_e\right) \geq 0\}.$$

Its solution $\hat{z}$ is such that $s(\hat{z}) \geq 0$, implying that $\hat{z}$ is a lower bound; this bound also satisfies $\hat{z} \geq z^+$. If $p_x(\hat{z}) < e$, we can apply Lemma 6.5 and either conclude that the problem is infeasible or exhibit an improved bound $\hat{z}^+$ and a corresponding dual variable $\hat{s}(z^+) \geq 0$ of the form given by (6.9). This contradicts the optimality of $\hat{z}$. Hence we have $p_x(\hat{z}) \not< e$.

In view of the above results, let us paraphrase an iteration of the infeasible start projective algorithm. Let $x > 0$ be the current iterate and let $z$ be a lower bound.

- If $P_{AX}Xc_{\tilde{z}} = (P_{AX}Xc) - (\tilde{z}/n)(P_{AX}Xe) = 0$ for some $\tilde{z} \geq z$, then use the projective feasibility algorithm to solve $\{x \geq 0 \; : \; Ax = 0, \; e^Tx = n\}$. If the latter problem has a solution, then $\tilde{z}$ is the optimal value.
- Compute a lower bound $z^+ \geq z$ such that $e - p_x(z^+) \not> 0$. (If $e - p_x(z) \not> 0$, then take $z^+ = z$.)
- Find $\alpha > 0$ such that

  $$\varphi(x(e + \alpha p_x); z^+) - \varphi(x; z^+) < -\sigma = -0.31,$$

  where a default value for $\alpha$ is $1/(1 + \|p_x\|)$.
- Normalize $x$ by $x := (n/e^Tx)x$.

We shall now define an $\varepsilon$-solution for the infeasible start problem as a normalized point ($x > 0$ and $e^Tx = n$) such that

$$\|Ax\|_\infty \leq \varepsilon \qquad \text{and} \qquad c^Tx - z \leq \varepsilon.$$

Any point that satisfies this criterion is $\varepsilon$-feasible and its objective value satisfies $c^Tx \leq z^* + \varepsilon$. This does not exclude ($\varepsilon$-infeasible) points such that $c^Tx < z^*$.

Pick, as in the feasibility problem, the initial point $x^0 = e$ and assume $z^0$ is a known lower bound. Without loss of generality, we shall assume that $c^Tx^0 > z^0$. To prove convergence we must analyse separately the effects of the two basic steps, the projective step and the update of $z$. For the projective step, the condition $p_x \not< e$ always holds. As a result the potential $\varphi(x; z)$ decreases by at least $\sigma = 0.31$. Note that in the meantime the potential $\psi(x)$, associated with the pure feasibility problem, also decreases by the same amount. This is due to the fact that $p_x$ is in the null space of $\tilde{A}X$ and also of $AX$: the contribution to the decrease of the potential comes from the term $e^T \ln x$ that is common to the two potentials. The projective step has the additional effect that each row of $\tilde{A}X$ is multiplied by a positive factor. In particular, the term $c^Tx - z$ does not change sign in this process.

The effect of an update of $z$ is more ambiguous. The term $|c^Tx - z|$ may either increase or decrease, depending on the sign of $c^Tx - z$ at the beginning and at the end of the updating process. As a consequence, the potential may either increase or decrease.

To cope with this difficulty, we break down the analysis into two parts. Assume first that the updating of $z$ never generates a negative term $c^T x - z$. Consequently, the update of $z$ always decreases $\varphi(x; z)$. We can apply the analysis of the first section to show that in at most

$$\frac{n}{\sigma} \ln \left( \frac{\max\{\|Ae\|_\infty, |c^T e - z^0|\}}{\varepsilon} \right) \tag{6.12}$$

iterations, we reach a situation where $\tilde{\varphi}(x; z) \leq n \ln \varepsilon$ and the convergence criterion is met.

Assume now that $c^T x - z$ becomes non-positive after some update. As pointed out earlier, the sign of $c^T x - z$ will not change in subsequent iterations. However, recall that the potential $\psi(x)$ associated with the simple feasibility problem keeps decreasing at each projective iteration and remains unchanged during an update of $z$. Consequently, in at most

$$\frac{n}{\sigma} \ln \left( \frac{\|Ae\|_\infty}{\varepsilon} \right) \leq \frac{n}{\sigma} \ln \left( \frac{\max\{\|Ae\|_\infty, |c^T e - z^0|\}}{\varepsilon} \right)$$

iterations, we reach a situation where $\psi(x) \leq n \ln \varepsilon$. Since $c^T x - z \leq 0$ the stopping criterion is met.

**Remark** The convergence analysis is considerably simpler if we assume that the iterate $x$ is feasible. This implies that $e$ is in the null space of $AX$ and thus $u_e = 0$ and $P_{AX} e = e$. The reader can check that $\rho(z)$ reduces to $(c^T x - z)/\|P_{AX} X c_z\|^2 > 0$. This automatically rules out case 1 of Lemma 6.5 and also makes easy the proof of case 2. The update of $z$ is given by a one-dimensional linear program, i.e. by a simple ratio test. This is the approach of [50].

**Remark** The whole analysis of this section can be adapted to fit directly the standard formulation, i.e. constraints of the form $Ax = b$, $x \geq 0$. This has been done in [7] under the assumption of a bounded feasible set. This boundedness condition is the same as the one we assumed to perform the transformation from the standard form to Karmarkar's formulation. The direct approach, which keeps closer to the original data, is preferable: we chose however to carry out the analysis in Karmarkar's framework for the sake of an easier presentation. The boundedness of the feasible set assumption has been relaxed in [52]. In that paper the author uses the condition that the zero vector is the only solution of the system $\{x \geq 0 : Ax = 0, c^T x = 0\}$. This is nearly equivalent to saying that the optimal set is bounded.

## Bibliography


1. K.M. Anstreicher. A combined phase I – phase II scaled potential algorithm for linear programming. *Mathematical Programming*, 52:429–439, 1991.
2. D.S. Atkinson and P.M. Vaidya. A scaling technique for finding the weighted analytic center of a polytope. *Mathematical Programming*, 57:163–192, 1992.

3. J.P. Crouzeix and C. Roos. On the inverse target map of a linear programming problem. Working Paper, University of Clermont, France, 1994.
4. I.I. Dikin. Iterative solution of problems of linear and quadratic programming. *Doklady Akademii Nauk SSSR*, 174:747–748, 1967. Translated into English in *Soviet Mathematics Doklady*, 8:674–675, 1967.
5. C. Fraley. Linear updates for a single – phase projective method. *Operations Research Letters*, 9:169–174, 1990.
6. R.M. Freund. A potential–function reduction algorithm for solving a linear program directly from an infeasible "warm start". *Mathematical Programming*, 52:441–466, 1991.
7. G. de Ghellinck and J.P. Vial. A polynomial Newton method for linear programming. *Algorithmica*, 1:425–453, 1986.
8. G. de Ghellinck and J.P. Vial. An extension of Karmarkar's algorithm for solving a system of linear homogeneous equations on the simplex. *Mathematical Programming*, 39:79–92, 1987.
9. A.J. Goldman and A.W. Tucker. Theory of linear programming. In H.W. Kuhn and A.W. Tucker, (eds.), *Linear Inequalities and Related Systems*, 53–97. Princeton University Press, Princeton, New Jersey, 1956.
10. J. Gondzio and T. Terlaky. A computational view of interior point methods. In J.E. Beasley, (ed.), *Advances in Linear and Integer Programming.* Oxford University Press, Oxford, Great Britain, 1996.
11. C.C. Gonzaga. An algorithm for solving linear programming problems in $O(n^3L)$ operations. In N. Megiddo, (ed.), *Progress in Mathematical Programming: Interior Point and Related Methods*, 1–28. Springer Verlag, New York, 1989.
12. C.C. Gonzaga. Large steps path-following methods for linear programming, Part I: Barrier function method. *SIAM Journal on Optimization*, 1:268–279, 1991.
13. C.C. Gonzaga. Large steps path-following methods for linear programming, Part II: Potential reduction method. *SIAM Journal on Optimization*, 1:280–292, 1991.
14. C.C. Gonzaga. Path following methods for linear programming. *SIAM Review*, 34:167–227, 1992.
15. C.C. Gonzaga. The largest step path following algorithm for monotone linear complementarity problems. Technical Report 94–07, Faculty of Technical Mathematics and Informatics, TU Delft, NL–2628 CD Delft, The Netherlands, 1994.
16. C.C. Gonzaga and R.A. Tapia. On the convergence of the Mizuno–Todd–Ye algorithm to the analytic center of the solution set. Technical Report 92–36, Dept. of Mathematical Sciences, Rice University, Houston, TX 77251, USA, 1992.
17. C.C. Gonzaga and R.A. Tapia. On the quadratic convergence of the simpli-

fied Mizuno–Todd–Ye algorithm for linear programming. Technical Report 91–41, Dept. of Mathematical Sciences, Rice University, Houston, TX 77251, USA, June 1992. Revised November 1992.

18. H.J. Greenberg. The use of the optimal partition in a linear programming solution for postoptimal analysis. *Operations Research Letters*, 15:179–184, 1994.
19. D. den Hertog. *Interior Point Approach to Linear, Quadratic and Convex Programming*, volume 277 of *Mathematics and its Applications*. Kluwer Academic Publishers, Dordrecht, The Netherlands, 1994.
20. D. den Hertog, C. Roos and J.P. Vial. A complexity reduction for the long-step path-following algorithm for linear programming. *SIAM Journal on Optimization*, 2:71–87, 1992.
21. B. Jansen, C. Roos and T. Terlaky. An interior point approach to postoptimal and parametric analysis in linear programming. Technical Report 92–21, Faculty of Technical Mathematics and Informatics, TU Delft, NL–2628 CD Delft, The Netherlands, April 1992.
22. B. Jansen, C. Roos and T. Terlaky. A polynomial primal-dual Dikin-type algorithm for linear programming. Technical Report 93–36, Faculty of Technical Mathematics and Informatics, TU Delft, NL–2600 GA Delft, The Netherlands, May 1993.
23. B. Jansen, C. Roos, T. Terlaky and J.P. Vial. Primal-dual algorithms for linear programming based on the logarithmic barrier method. Technical Report 92–104, Faculty of Technical Mathematics and Informatics, TU Delft, NL–2600 GA Delft, The Netherlands, 1992.
24. B. Jansen, C. Roos, T. Terlaky and J.P. Vial. Long-step primal-dual target-following algorithms for linear programming. Working Paper, Faculty of Technical Mathematics and Informatics, TU Delft, NL–2600 GA Delft, The Netherlands, 1993.
25. B. Jansen, C. Roos, T. Terlaky and J.P. Vial. Primal-dual target-following algorithms for linear programming. Technical Report 93–107, Faculty of Technical Mathematics and Informatics, TU Delft, NL–2600 GA Delft, The Netherlands, 1993.
26. N.K. Karmarkar. A new polynomial-time algorithm for linear programming. *Combinatorica*, 4:373–395, 1984.
27. L.G. Khachiyan. A polynomial algorithm in linear programming. *Doklady Akademii Nauk SSSR*, 244:1093–1096, 1979. Translated into English in *Soviet Mathematics Doklady*, 20:191–194, 1979.
28. M. Kojima, N. Megiddo, T. Noma and A. Yoshise. *A Unified Approach to Interior Point Algorithms for Linear Complementarity Problems*, volume 538 of *Lecture Notes in Computer Science*. Springer Verlag, Berlin, Germany, 1991.
29. M. Kojima, S. Mizuno and M.J. Todd. Infeasible-interior-point primal-dual potential-reduction algorithms for linear programming. *SIAM Journal on*

*Optimization*, 5:52–67, 1995.

30. M. Kojima, S. Mizuno and A. Yoshise. A primal-dual interior point algorithm for linear programming. In N. Megiddo, (ed.), *Progress in Mathematical Programming: Interior Point and Related Methods*, 29–47. Springer Verlag, New York, 1989.
31. E. Kranich. Interior point methods for mathematical programming: A bibliography. Discussion Paper 171, Institute of Economy and Operations Research, FernUniversität Hagen, P.O. Box 940, D–5800 Hagen 1, West–Germany, May 1991. Available through *netlib*, see Kranich [32].
32. E. Kranich. Interior-point methods bibliography. *SIAG/OPT Views-and-News, A Forum for the SIAM Activity Group on Optimization*, 1:11, 1992.
33. L. McLinden. An analogue of Moreau's approximation theorem with application to the non-linear complementarity problem. *Pacific Journal of Mathematics*, 88:101–161, 1980.
34. L. McLinden. The complementarity problem for maximal monotone multifunctions. In R.W. Cottle, F. Giannessi and J.L. Lions, (eds.), *Variational Inequalities and Complementarity Problems*, 251–270. John Wiley and Sons, New York, 1980.
35. N. Megiddo. Pathways to the optimal set in linear programming. In N. Megiddo, (ed.), *Progress in Mathematical Programming: Interior Point and Related Methods*, 131–158. Springer Verlag, New York, 1989. Identical version in: *Proceedings of the 6th Mathematical Programming Symposium of Japan, Nagoya, Japan*, 1–35, 1986.
36. S. Mizuno. An $O(n^3L)$ algorithm using a sequence for linear complementarity problems. *Journal of the Operations Research Society of Japan*, 33:66–75, 1990.
37. S. Mizuno. A new polynomial time method for a linear complementarity problem. *Mathematical Programming*, 56:31–43, 1992.
38. S. Mizuno. Polynomiality of infeasible-interior-point algorithms for linear programming. *Mathematical Programming*, 67:109–119, 1994.
39. S. Mizuno, M.J. Todd and Y. Ye. Anticipated behaviour of long-step algorithms for linear programming. Technical Report 882, School of Operations Research and Industrial Engineering, Cornell University, Ithaca, NY 14853–3801, USA, 1990.
40. S. Mizuno, M.J. Todd and Y. Ye. On adaptive step primal-dual interior-point algorithms for linear programming. *Mathematics of Operations Research*, 18:945–963, 1993.
41. S. Mizuno, M.J. Todd and Y. Ye. A surface of analytic centers and infeasible-interior-point algorithms for linear programming. Technical Report, School of Operations Research and Industrial Engineering, Cornell University, Ithaca, NY 14853–3801, USA, January 1993. To appear in *Mathematics of Operations Research*.

42. R.D.C. Monteiro and I. Adler. An $O(n^3L)$ primal-dual interior point algorithm for linear programming. Technical Report ORC 87–4, Dept. of Industrial Engineering and Operations Research, University of California, Berkeley, CA 94720, USA, 1987.

43. R.D.C. Monteiro and I. Adler. Interior path following primal-dual algorithms: Part I: Linear programming. *Mathematical Programming*, 44:27–41, 1989.

44. F.A. Potra. An infeasible interior-point predictor–corrector algorithm for linear programming. Reports on Computational Mathematics 26, Dept. of Mathematics, The University of Iowa, Iowa City, IA 52242, USA, June 1992.

45. J. Renegar. A polynomial-time algorithm, based on Newton's method, for linear programming. *Mathematical Programming*, 40:59–93, 1988.

46. C. Roos and J.P. Vial. Long steps with the logarithmic penalty barrier function in linear programming. In J. Gabszevwicz, J.F. Richard and L. Wolsey, (eds.), *Economic Decision-Making: Games, Economics and Optimization, dedicated to J. H. Dreze*, 433–441. Elsevier Science Publisher B.V., Amsterdam, The Netherlands, 1989.

47. C. Roos and J.P. Vial. A polynomial method of approximate centers for linear programming. *Mathematical Programming*, 54:295–305, 1992.

48. T. Terlaky. A convergent criss-cross method. *Mathemathische Operationsforschung und Statistics ser. Optimization*, 16:683–690, 1985.

49. M.J. Todd. Combining phase I and phase II in a potential reduction algorithm for linear programming. *Mathematical Programming*, 59:133–278, 1993.

50. M.J. Todd and B.P. Burrell. An extension of Karmarkar's algorithm for linear programming using dual variables. *Algorithmica*, 1:409–424, 1986.

51. S. Vavasis and Y. Ye. Condition numbers for polyhedra with real number data. Technical Report TR 93–1398, Department of Computer Science, Cornell University, Ithaca, New York 14853–7501, USA, 1993.

52. J.P. Vial. A unified approach to projective algorithms for linear programming. In S. Dolecki, (ed.), *Optimization: Proceedings of the 5th French-German Conference in Castel-Novel, Varetz, France, October 1988, Lecture Notes in Mathematics*, 1405:191–220. Springer Verlag, Berlin, West-Germany, 1989.

53. S.J. Wright. A path-following infeasible-interior-point algorithm for linear complementarity problems. *Optimization Methods and Software*, 2:79–106, 1993.

54. F. Wu, S. Wu and Y. Ye. On quadratic convergence of the $O(\sqrt{n}L)$-iteration homogeneous and self-dual linear programming algorithm. Technical Report, Dept. of Management Science, University of Iowa, Iowa City, IA 52242, USA, December 1992.

55. Y.Ye, O. Güler, R.A. Tapia and Y. Zhang. A quadratically convergent

$O(\sqrt{n}L)$ iteration algorithm for linear programming. *Mathematical Programming*, 59:151–162, 1993.

56. Y. Ye, M.J. Todd and S. Mizuno. An $O(\sqrt{n}L)$-iteration homogeneous and self-dual linear programming algorithm. *Mathematics of Operations Research*, 19:53–67, 1994.
57. Y. Zhang and D. Zhang. Superlinear convergence of infeasible interior-point methods for linear programming. Research Report 92–15, Dept. of Mathematics and Statistics, University of Maryland Baltimore County, Baltimore, MD 21228, USA, 1992. To appear in *Mathematical Programming.*
58. S. Zionts. The criss-cross method for solving linear programming problems. *Management Science*, 15:426–445, 1969.

# 3

# A computational view of interior point methods

**Jacek Gondzio**

*Logilab, HEC Geneva,*
*Section of Management Studies, University of Geneva,*
*102 Bd Carl Vogt, CH-1211 Geneva 4, Switzerland*

**Tamás Terlaky**

*Faculty of Technical Mathematics and Informatics,*
*Delft University of Technology,*
*P.O. Box 5031, 2600 GA Delft, The Netherlands*

## 1 Overview

Many issues that are crucial for an efficient implementation of an interior point algorithm are addressed in this chapter. To start with, a prototype primal-dual algorithm is presented. Next, many tricks that make it efficient in practice are discussed in detail. Those include: preprocessing techniques, initialization approaches, methods for computing search directions (and the underlying linear algebra techniques), centering strategies and methods of stepsize selection.

Several reasons for the manifestation of numerical difficulties, for example the primal degeneracy of optimal solutions or the lack of feasible solutions, are explained in a comprehensive way.

A motivation for obtaining an optimal basis is given and a practicable algorithm to perform this task is presented. Advantages of different methods to perform postoptimal analysis (applicable to interior point optimal solutions) are discussed.

Important questions that still remain open in the implementation of interior point methods are also addressed, e.g. performing correct postoptimal analysis, detecting infeasibility or resolving difficulties arising in the presence of unbounded optimal faces. The challenging practical problem of warm start is recalled and two potentially attractive approaches to it are suggested.

To facilitate understanding of different implementation strategies, some illustrative numerical results on a subset of problems from the NETLIB collection are presented.

---

The research of Jacek Gondzio has been partially supported by the Committee for Scientific Research of Poland, grant No PB 8 S505 015 05 and by the Fonds National de la Recherche Scientifique Suisse, grant No 12-34002.92.

The research of Tamás Terlaky has been partially supported by OTKA No 2116.

## 2 Introduction

Karmarkar's publication in 1984 of a new polynomial time algorithm for linear programming (LP) [40] drew enormous attention from the mathematical programming community and has led to great activity, resulting in a flood of papers (see, e.g. [45]).

The idea of crossing the interior of the feasible region in search of an optimum of the linear program was present at least since the 1960s. These methods were for example: an affine-scaling method of Dikin [18] and a logarithmic barrier method SUMT of Fiacco and McCormick [22]. For at least two reasons, however, these methods could not at the time be shown to be competitive to simplex. First, due to the storage limitations, the size of the problems solved in the late 1960s never exceeded several hundred rows and columns and for such sizes the simplex method is practically unbeatable. Secondly, there were no sparse symmetric solvers available at that time (they appeared at the beginning of the 1970s) so the orthogonal projections must have killed the efficiency of interior point methods (IPMs). IPMs need significantly more memory than the simplex method which was an unacceptable requirement at that time.

Clearly, the situation was quite different in 1984, which encouraged Karmarkar to claim an excellent efficiency for his new approach. In fact, these claims still had to wait several years to be confirmed by computational results [1, 2, 59].

Soon after Karmarkar's publication, Gill, Murray, Saunders, Tomlin and Wright [27] built the bridge between this new interior point method and the logarithmic barrier approach. Barrier methods were developed for the primal and for the dual LP formulation (see, e.g. the surveys [32, 65, 67]). Early implementations that were based on pure primal or dual methods gave competitive results with simplex implementations. Nowadays all the state-of-the-art IPM implementations are those of primal-dual methods, hence in this chapter we concentrate only on primal-dual methods.

First Megiddo [52] proposed applying a logarithmic barrier method to the primal and the dual problems at the same time. Independently, Kojima, Mizuno and Yoshise [44] developed the theoretical background of this method and gave complexity results. Its early implementations [15, 51] showed much promise and encouraged further research. For extensions that represent current state-of-the-art primal-dual implementations, see Lustig, Marsten and Shanno [49, 47, 48] and Mehrotra [54, 55].

A primal-dual algorithm is a feasible IPM if all the iterates are primal and dual feasible. If the iterates are positive but infeasible then the primal-dual algorithm is called an *infeasible IPM*. This algorithm attains feasibility at the same time as optimality is reached. It had been successfully implemented that way [47] and had shown very good practical convergence long before a theoretical justification for such behaviour was found by Kojima, Megiddo and Mizuno [43]. The method has proven polynomial complexity: $O(n^2L)$ in [79] and $O(nL)$ in [57, 62].

Although the complexity of the infeasible primal-dual algorithm is worse than the best-known complexity $O(\sqrt{n}L)$ of most feasible IPMs (see, e.g. the surveys [32, 65, 67]), it is now widely accepted that primal-dual infeasible IPMs are more efficient in implementations. Since infeasible IPMs are the methods of choice to date for "state of the art" implementations, throughout the whole chapter we mean a *primal-dual infeasible IPM* when we speak about a primal-dual algorithm. To facilitate this, in Section 3 we shall introduce a prototype primal-dual infeasible IPM algorithm.

A common feature of almost all IPMs is that they can be interpreted in terms of following the path of centers [68] that leads to the optimal solution (see, e.g. [32, 65]). With some abuse of mathematics, a basic iteration of a path-following algorithm consists of moving from one point in a neighbourhood of the central path to another one called the *target* that preserves the property of lying in a neighbourhood of the central path and reduces the distance to optimality measured with some estimation of the duality gap. Such a movement can in principle involve more than one step towards the target. Depending on how significant is the update of the target (and consequently, whether just one or more Newton steps are needed to reach the vicinity of the new target) one distinguishes between short-step and long-step methods. Due to the considerable cost of every Newton step, usually (at least in implementations) one Newton step is allowed before a new target is defined.

Every Newton step requires computing at least one orthogonal projection onto the null space of a scaled linear operator $AD$, where $A$ is the LP constraint matrix and $D$ is a positive diagonal scaling matrix that changes in subsequent iterations. Primal, dual and primal-dual variants of IPMs differ on the way matrix $D$ is defined, but the effort to compute Karmarkar's projection is always the same. Every orthogonal projection involves inversion of the matrix $AD^2A^T$ – the most time consuming linear algebra operation which takes about 60–90% of the computation time of a single interior point iteration. Unless the linear program is specially structured and this structure can be exploited to determine an easily invertible preconditioner for an iterative method (such as e.g. a conjugate gradient algorithm, implemented successfully for network problems [61, 64]), direct methods [19] that compute a sparse symmetric factorization (Cholesky decomposition of the positive definite system $AD^2A^T$ or Bunch–Parlett [7, 13] decomposition of the indefinite augmented system $\begin{bmatrix} D^{-2} & A^T \\ A & 0 \end{bmatrix}$) are the methods of choice. Computing projections onto affine spaces seems crucial for the efficiency of any interior point algorithm. We shall thus discuss it in detail in Section 4 which also addresses other issues of implementation of the IPM such as the role of presolve analysis, the choice of the starting point, the choice of the stepsizes in the primal and in the dual spaces, the role of centering, higher-order methods, the termination conditions and, finally, the comparison of theoretical and practical complexity.

In Section 5 we shall add some remarks relating to manifestations of degeneracy and ill-conditioning in the computation of projections.

For about forty years, the simplex method (starting from its discovery in 1947 [16] until Karmarkar's breakthrough [40]) was the only effective algorithm for linear programs. Hence, operations research practitioners got used to seeing linear programming from the simplex perspective. This, in particular, applies to the use of the postoptimality analysis available from the optimal basis solution. In fact, such a postoptimality analysis is almost always incomplete (frequently incorrect), see e.g. [17, 33, 37]. Nevertheless, there exist many applications in which an optimal basis is necessary, e.g. reoptimization in integer programming. In such a case a need arises for identifying an optimal basis from the interior point optimal solution. Fortunately, this can be done in strongly polynomial time [53]. We shall address the problem of optimal basis identification in Section 6.

Section 7 will be devoted to some crucial questions that still remain open. Sensitivity analysis based on interior point optimal solution is generally more expensive but produces correct information. We discuss how to handle problems with unbounded level sets, how to detect infeasibility and how to implement efficient warm start in interior point algorithms.

Most relevant issues of interior point method implementations will be illustrated by solving a subset of the NETLIB LP problem test collection using version 2.0 of the HOPDM (Higher Order Primal Dual Method) code [4, 31]. All of our computations are made on a SUN SPARC-10 workstation.

Later on in the chapter we will frequently speak about stability, robustness and efficiency of different methods. On stability, the usual numerical stability is meant. Talking about robustness, one thinks about whether the algorithm gives reliable answers on a wide range (optimally all) of problem instances. Finally, efficiency relates to the speed of the algorithm, and the speed of the implementation.

## 3 A prototype primal-dual algorithm

Let us consider a primal linear programming problem:

$$\begin{array}{ll} \text{minimize} & c^T x, \\ \text{subject to} & Ax = b, \\ & x + s = u, \\ & x, s \geq 0, \end{array} \tag{3.1}$$

where $c, x, s, u \in \mathbb{R}^n, b \in \mathbb{R}^m, A \in \mathbb{R}^{m*n}$; and its dual:

$$\begin{array}{ll} \text{maximize} & b^T y - u^T t, \\ \text{subject to} & A^T y - t + z = c, \\ & z, t \geq 0, \end{array} \tag{3.2}$$

where $y \in \mathbb{R}^m$ and $z, t \in \mathbb{R}^n$. Here we assume that rank $A = m$ (in Section 4.1.3 we will see that this assumption is not restrictive).

To derive the primal-dual algorithm, let us replace non-negativity of constraints in the primal formulation with the logarithmic barrier penalty terms in the objective function, which gives the following logarithmic barrier function:

$$L(x, s, \mu) = c^T x - \mu \sum_{j=1}^{n} \ln x_j - \mu \sum_{j=1}^{n} \ln s_j . \tag{3.3}$$

The first-order optimality conditions for (3.1) with the objective replaced by (3.3) are

$$\begin{aligned} Ax &= b, \\ x + s &= u, \\ A^T y + \mu X^{-1} e - t &= c, \\ \mu S^{-1} e - t &= 0, \\ x &> 0, \\ s &> 0. \end{aligned} \tag{3.4}$$

Substituting $z = \mu X^{-1} e$, the first-order optimality conditions (3.4) give

$$\begin{aligned} Ax &= b, \\ x + s &= u, \\ A^T y + z - t &= c, \\ XZe &= \mu e, \\ STe &= \mu e, \\ x &> 0, \\ s &> 0, \end{aligned} \tag{3.5}$$

where $X, S, Z$ and $T$ are diagonal matrices with the elements $x_j, s_j, z_j$ and $t_j$ respectively, $e$ is the $n$-vector of all ones and $\mu$ is a barrier parameter.

The set of solutions of (3.5) $(x(\mu), s(\mu))$ and $(y(\mu), z(\mu), t(\mu))$ defines the central path of the primal and dual problem, respectively. Having any primal and dual feasible solutions $(x, s)$ and $(y, z, t)$ the quality of centrality is measured by

$$\delta^2(x, s, y, z, t, \mu) = \sum_{j=1}^{n} \left( \sqrt{\frac{x_j z_j}{\mu}} - \sqrt{\frac{\mu}{x_j z_j}} \right)^2 + \sum_{j=1}^{n} \left( \sqrt{\frac{s_j t_j}{\mu}} - \sqrt{\frac{\mu}{s_j t_j}} \right)^2 . \tag{3.6}$$

Note that $(x, s)$ and $(y, z, t)$ are on the *central path* if and only if $\delta(x, s, y, z, t, \mu) = 0$. The smaller $\delta$ is, the better the points are centered.

Observe that the first three of the above equations are linear and force primal and dual feasibility of the solution. The last two equations are non-linear and become the complementarity conditions for $\mu = 0$, which together with the feasibility constraints provides optimality of the solutions. If some solution with a certain $\mu$ is available, then for the complementarity gap one has

$$x^T z + s^T t = 2\mu e^T e = 2n\mu.$$

In any IPM, this quantity measures the error in complementarity. Observe that in feasible IPMs the complementarity gap reduces to the usual duality gap. Clearly it vanishes at an optimal solution.

A single iteration of the basic primal-dual algorithm makes one step of Newton's method applied to the first-order optimality conditions (3.5) with a given $\mu$ and then $\mu$ is updated (usually decreased). The algorithm terminates when infeasibility and the complementarity gap is reduced below a predetermined tolerance.

Having an $x, s, z, t \in \mathbb{R}^n_+, y \in \mathbb{R}^m$, Newton's direction is obtained by solving the following system of linear equations

$$\begin{bmatrix} A & 0 & 0 & 0 & 0 \\ I & 0 & I & 0 & 0 \\ 0 & A^T & 0 & I & -I \\ Z & 0 & 0 & X & 0 \\ 0 & 0 & T & 0 & S \end{bmatrix} \begin{bmatrix} \Delta x \\ \Delta y \\ \Delta s \\ \Delta z \\ \Delta t \end{bmatrix} = \begin{bmatrix} \xi_b \\ \xi_u \\ \xi_c \\ \mu e - XZe \\ \mu e - STe \end{bmatrix}, \tag{3.7}$$

where

$$\begin{aligned} \xi_b &= b - Ax, \\ \xi_u &= u - x - s, \\ \text{and} \qquad \xi_c &= c - A^T y - z + t, \end{aligned}$$

denote the violations of the primal and the dual constraints respectively. Primal-dual infeasible IPMs do not require the feasibility of the solutions ($\xi_b, \xi_u$ and $\xi_c$ might be non-zero) during the optimization process. Feasibility is attained during the process as optimality is reached. It is easy to verify that if a step of length one is made in the Newton direction (3.7), then feasibility is reached immediately. This is seldom the case as a smaller stepsize is usually chosen (a damped Newton iteration is taken) to preserve non-negativity of $x, s, z$ and $t$. If this is the case and a stepsize $\alpha < 1$ is applied, then infeasibilities $\xi_b, \xi_u$ and $\xi_c$ are reduced by a factor of $(1 - \alpha)$.

Let us now look closer at the system of linear equations (3.7). After elimination

$$\begin{aligned} \Delta z &= X^{-1}(\mu e - XZe - Z\Delta x), \\ \Delta s &= \xi_u - \Delta x, \\ \Delta t &= S^{-1}(\mu e - STe - T\Delta s) = S^{-1}(\mu e - STe - T\xi_u + T\Delta x), \end{aligned} \tag{3.8}$$

it reduces to

$$\begin{bmatrix} -D^{-2} & A^T \\ A & 0 \end{bmatrix} \begin{bmatrix} \Delta x \\ \Delta y \end{bmatrix} = \begin{bmatrix} r \\ h \end{bmatrix}, \tag{3.9}$$

where

$$
\begin{aligned}
D^2 &= (X^{-1}Z + S^{-1}T)^{-1}, \\
r &= \xi_c - X^{-1}(\mu e - XZe) + S^{-1}(\mu e - STe) - S^{-1}T\xi_u, \\
h &= \xi_b.
\end{aligned} \tag{3.10}
$$

The matrix in the *augmented system* (3.9) is sparse, symmetric but indefinite. This system of linear equations can be solved directly by the Bunch–Parlett factorization [7, 13], or after multiplying the first equation by $AD^2$ and substituting from the second equation, the system can be reduced to the sparse, symmetric and positive definite *normal equations system*

$$(AD^2A^T)\Delta y = AD^2r + h. \tag{3.11}$$

Advantages of both approaches will be discussed in the next section. Here, we only conclude that once $\Delta y$ is found from (3.11), it uniquely determines

$$\Delta x = D^2A^T\Delta y - D^2r,$$

and $\Delta s, \Delta z$ and $\Delta t$ by (3.8).

Computing $(\Delta x, \Delta y)$ from (3.9) or $\Delta y$ from (3.11) is usually (when a direct approach [19] is applied) divided into two phases: *factorization* of the matrix to some easily invertible form and the following *solve* that exploits this factorization. Usually, the second step is at least an order of magnitude cheaper than the first step. This observation led [41, 54] to the introduction of higher-order terms when computing direction. Computing different corrector terms resolves into multiple solution of the same linear system for several right-hand sides (it thus reuses the same factorization and is relatively inexpensive). The most successful technique that incorporates higher-order information into the primal-dual algorithm comes from Mehrotra [54, 55]. We shall address this technique in more detail in the next section.

For the sake of a brief presentation of the method, we assume that some direction $(\Delta x, \Delta y, \Delta s, \Delta z, \Delta t)$ is computed, the maximum stepsizes in it for primal, $\alpha_P$, and dual, $\alpha_D$, spaces that maintain non-negativity of variables are found and, after being slightly reduced by a factor $\alpha_0$, a new iterate is computed

$$
\begin{aligned}
x^{k+1} &= x^k + \alpha_0\alpha_P\Delta x, \\
s^{k+1} &= s^k + \alpha_0\alpha_P\Delta s, \\
y^{k+1} &= y^k + \alpha_0\alpha_D\Delta y, \\
z^{k+1} &= z^k + \alpha_0\alpha_D\Delta z, \\
t^{k+1} &= t^k + \alpha_0\alpha_D\Delta t.
\end{aligned} \tag{3.12}
$$

After making the step, the barrier parameter $\mu$ is reduced and the process is repeated. Formally, the algorithm can be summarized in Fig. 1.

---

**Input**

$(x^0, s^0) > 0$ and $y^0$, $(z^0, t^0) > 0$ the initial pair of primal and dual solutions respectively;

**Parameters**

$\varepsilon$ is the accuracy parameter;

$\alpha_0$ is the stepsize;

**begin**

$x := x^{(0)}$; $s := s^{(0)}$; $y := y^{(0)}$; $z := z^{(0)}$; $t := t^{(0)}$;

**while** stopping criteria is not satisfied **do**

calculate the search directions by (3.8), (3.9) and (3.10);

calculate the new iterates by (3.12);

**end**

**end.**

---

FIG. 1. Prototype algorithm

## 4 Some tricks that make it work

In this section we shall concentrate on several issues that seem to be crucial for an efficient implementation of interior point algorithms.

### 4.1 Presolve

Linear programs that are solved nowadays often have very large sizes and are usually generated automatically by some modelling support tools. They are thus often formulated in a way that is not necessarily the most suitable for a direct application of an LP solver. Hence it is advantageous to analyse and, if possible, simplify their formulation before passing them to a solver. The important role of presolving in linear programming was recognized [12] a long time ago. It is worth mentioning that presolve is strongly recommended for any solver independently of the method used.

The importance of presolve is even more pronounced when an interior point based solver is applied [1, 31] due to the more involved linear algebra operations and the need for full rank of the matrix $A$. Preprocessing aims at three main goals, namely problem size reduction, problem density reduction and ensuring full rank of the constraint matrix $A$.

#### 4.1.1 *Problem size reduction*

A general-purpose presolve of [31] repeats the logical analysis of the LP problem formulation until no further reduction is obtained (every single reduction creates the possibility of further model simplifications). The following simple operations are applied.

(1) Empty rows and columns are removed.

(2) Singleton inequality constraints are replaced by bounds on the variables (the variable with an entry in a singleton equality row is fixed and removed).

(3) Lower and upper limits for every constraint $i$ are determined by

$$\underline{b_i} = \sum_{\{j:a_{ij}<0\}} a_{ij} u_j \quad \text{and} \quad \overline{b_i} = \sum_{\{j:a_{ij}>0\}} a_{ij} u_j, \tag{4.1}$$

which clearly satisfy

$$\underline{b_i} \le \sum_j a_{ij} x_j \le \overline{b_i}. \tag{4.2}$$

Observe that due to the non-negativity of $x$, the limits $\underline{b_i}$ and $\overline{b_i}$ are non-positive and non-negative respectively. If the inequalities (4.2) are at least as tight as the original (inequality type) LP constraint, then the constraint $i$ is *redundant*. If one of the limits contradicts the LP constraint, then the problem is infeasible. Finally, in some special cases (e.g. "less than or equal to" row with $\underline{b_i} = b_i$, "greater than or equal to" row with $\overline{b_i} = b_i$, or equality type row for which $b_i$ equals one of the limits $\underline{b_i}$ or $\overline{b_i}$), the LP constraint becomes a *forcing* one. This means that the only way to satisfy the constraint is to fix all variables that appear in it to their appropriate bounds.

(4) Constraint limits (4.1) are used to generate implied variable bounds. (Note that LP variables were transformed to the standard form $0 \le x \le u$, before.) This technique makes use of the original form of an LP constraint (i.e. its form before a slack variable has been added to it to transform it to the "standard" equality row of (3.1)). Assume, for example, that a non-redundant "less than or equal to" (LE) type constraint is given, i.e.

$$\underline{b_i} < \sum_j a_{ij} x_j \le b_i.$$

Then

$$\begin{array}{rcccl} \underline{b_i} + a_{ik} x_k & \le & \sum_j a_{ij} x_j & \le & b_i, \quad \forall k,\ a_{ik} > 0, \\ \underline{b_i} + a_{ik}(x_k - u_k) & \le & \sum_j a_{ij} x_j & \le & b_i, \quad \forall k,\ a_{ik} < 0, \end{array}$$

and new implied bounds are given for all variables involved by row $i$

$$\begin{array}{lclcll} x_k & \le & u_k' & = & (b_i - \underline{b_i})/a_{ik}, & \forall k,\ a_{ik} > 0, \\ x_k & \ge & l_k' & = & u_k + (b_i - \underline{b_i})/a_{ik}, & \forall k,\ a_{ik} < 0. \end{array}$$

If these bounds are tighter than the original ones, then the variable bounds are improved. Note that a technique similar to that described above is particularly useful when it imposes finite bounds on free variables. Free variables do not, in such a case, have to be split and represented as the difference of two non-negative variables.

(5) For every singleton column, a row with an entry in it is used to generate implied bounds on a variable referring to it. If these bounds are at least as tight as the original ones, then the variable becomes an implied free. Consequently, both the row (implied free constraint) and the singleton free column are eliminated.

(6) Non-negative unbounded variables ($0 \le x \le +\infty$) referring to singleton columns are used to generate bounds on dual variables $y$. Namely, if the variable $j$ refers to a singleton column with an entry $a_{ij}$ and $u_j = +\infty$ (i.e. $t_j = 0$), then the dual constraint (3.2) becomes an inequality

$$a_{ij} y_i \le c_j.$$

This inequality can be solved and, depending on the sign of $a_{ij}$, produces a lower or upper bound on $y_i$.

(7) Once all dual variables have explicit (possibly infinite) bounds

$$p_i \le y_i \le q_i, \qquad i = 1, 2, \cdots, m, \tag{4.3}$$

lower and upper limits for every dual constraint $j$ are generated

$$\begin{aligned} \underline{c_j} &= \textstyle\sum_{\{i: a_{ij}<0\}} a_{ij} q_i + \sum_{\{i: a_{ij}>0\}} a_{ij} p_i, \\ \text{and} \quad \overline{c_j} &= \textstyle\sum_{\{i: a_{ij}<0\}} a_{ij} p_i + \sum_{\{i: a_{ij}>0\}} a_{ij} q_i. \end{aligned} \tag{4.4}$$

These limits are compared with the cost coefficient $c_j$ and applied to identify variables for which the sign of the reduced cost $z_j - t_j$ can be restricted (*dominated variables*). The dual constraint $j$ becomes redundant if the above holds, so the dominated variable is fixed at its appropriate bound and eliminated from the problem. If the reduced cost $z_j - t_j$ has a weak sign restriction (it is non-negative or non-positive), then the variable is a *weakly dominated* one. Surprisingly, if this is the case, then (under some additional conditions [31]) the variable can also be eliminated.

(8) Dual constraint limits (4.4) are used to generate implied bounds on dual variables. A technique similar to that of point (4) is applied. Implied bounds tighter than the original ones replace those of (4.3).

All techniques mentioned above are worth applying before an LP solver is invoked. They usually reduce the problem size considerably. Sometimes they identify primal or dual infeasibility or unboundedness.

#### 4.1.2 *Improving the sparsity of the problem*

The way in which the LP constraint matrix is involved in interior point iterations justifies further presolve effort that aims at decreasing the cost of calculating the solution of the equations (3.7) and improving the accuracy of solution. Both (whichever form of equation is used to compute the search directions (3.9) or (3.11)) are strongly influenced by the sparsity structure of $A$. The latter depends very much on the conditioning of matrix $A$ and requires at least that it has full row rank.

Sparsity of $A$ can usually be improved. In general, one can look for a non-singular matrix $M \in \mathbb{R}^{m*m}$ such that the matrix $MA$ is the sparsest possible. Primal feasibility constraints can in such a case be replaced by an equivalent formulation

$$MAx = Mb, \tag{4.5}$$

which is much more suitable for direct application of an interior point solver. Exact solution of this *sparsity problem* [14] is an NP-complete problem but efficient heuristics [1, 14, 31] usually produce satisfactory non-zero reductions in $A$. The algorithm of [31], for example, looks for a row of $A$ that has a sparsity pattern being the subset of the sparsity pattern of other rows and uses it to pivot out non-zero elements from other rows.

Even a very sparse $A$ can sometimes produce relatively dense factors in (3.9) or (3.11). Additionally, the latter fills dramatically if dense columns are present in $A$. If the number of dense columns is not excessive, then the technique of splitting them into shorter pieces [30, 72] might be a remedy. Note that the augmented system approach (3.9) suffers less from the presence of dense columns (see Section 4.3).

#### 4.1.3 *Full rank of matrix A*

Theoretically, detecting rank deficiency of $A$ is not a problem. One can continue Gaussian elimination on $A$ until a zero submatrix is obtained. To make this process reliable, complete pivoting [29] should be used, but it is prohibitively expensive due to the destruction of the sparsity structure. A practical generalized Markowitz pivoting [63] can be used to eliminate most linear dependencies with a reasonable computational cost. However, not all codes offer such an option. On the other hand, a lot of linear dependencies can be identified with a search for duplicate rows and with heuristics to make $A$ sparser.

Let us observe that, due to primal degeneracy which is practically always present, the normal equations matrix (3.11) becomes rank deficient when the optimum is approached even in the case when the original $A$ has full row rank (for more details see Section 5.1). All codes are thus equipped with safeguards against such problems and usually recover from small rank deficiency of $A$.

#### 4.1.4 *Some numerical experiments*

Tables 1 and 2 illustrate the advantages of presolve effort. They report problem sizes, normal equations statistics (the number of off-diagonal non-zeros in the adjacency structure $AA^T$, the number of off-diagonal non-zeros in Cholesky factor $L$ and the number of flops required to compute $L$), and the iterations and CPU time to solve some NETLIB problems to 8-digit accuracy in two cases: without presolve analysis and after it respectively. Additionally, Table 2 reports the CPU time for presolve analysis.

These and all the following computational results reported in this chapter have been obtained with the HOPDM code [4, 31]. That code is written in FORTRAN. It has been compiled with the F77 compiler with default optimization option ($-O$) and has been run on a (one processor) SUN SPARC-10 workstation.

**Table 1** Original problem sizes and solution statistics

| | Orig. prob. size | | | Normal equations | | | Solution | |
|---|---|---|---|---|---|---|---|---|
| Problem | $m$ | $n$ | nonz | nz($AA^T$) | nz($L$) | flops | iters | time |
| 25fv47 | 820 | 1571 | 10400 | 11074 | 33217 | 2.42e6 | 26 | 21.41 |
| 80bau3b | 2235 | 9301 | 20413 | 9972 | 40038 | 2.25e6 | 48 | 78.23 |
| bnl2 | 2280 | 3489 | 13999 | 13457 | 84184 | 1.32e7 | 36 | 103.33 |
| cycle | 1886 | 2857 | 20720 | 27714 | 87322 | 8.59e6 | 27 | 66.88 |
| d2q06c | 2171 | 5167 | 32417 | 26991 | 163407 | 3.46e7 | 31 | 237.77 |
| ganges | 1309 | 1681 | 6912 | 7656 | 30563 | 1.67e6 | 19 | 15.19 |
| pilot | 1440 | 3449 | 41092 | 59540 | 204512 | 5.25e7 | 44 | 513.15 |
| pilot87 | 2029 | 4663 | 70682 | 115951 | 489267 | 2.33e8 | 44 | 1935.24 |
| pilotnov | 951 | 1968 | 12186 | 10857 | 48992 | 5.18e6 | 19 | 27.81 |
| sctap3 | 1480 | 2480 | 8874 | 7386 | 17973 | 6.61e5 | 11 | 5.47 |
| ship12s | 1042 | 2763 | 8178 | 5345 | 6457 | 8.08e4 | 13 | 3.48 |
| sierra | 1222 | 2016 | 7252 | 4896 | 11635 | 2.73e5 | 18 | 6.53 |
| stocfor2 | 2157 | 2031 | 8343 | 12738 | 26282 | 4.57e5 | 18 | 9.00 |
| woodw | 1098 | 8405 | 37474 | 20421 | 48988 | 3.44e6 | 28 | 44.50 |

**Table 2** Advantages of presolve analysis

| | Size after presolve | | | Normal equations | | | Solution | | Presolve |
|---|---|---|---|---|---|---|---|---|---|
| Problem | $m$ | $n$ | nonz | nz($AA^T$) | nz($L$) | flops | iters | time | time |
| 25fv47 | 769 | 1535 | 9959 | 10456 | 33377 | 2.55e6 | 28 | 24.05 | 1.32 |
| 80bau3b | 1965 | 8736 | 19048 | 9176 | 37320 | 2.13e6 | 43 | 65.55 | 2.99 |
| bnl2 | 1848 | 3007 | 12458 | 12242 | 79538 | 1.22e7 | 30 | 78.49 | 2.44 |
| cycle | 1400 | 2403 | 14111 | 18376 | 53182 | 3.54e6 | 41 | 57.81 | 2.57 |
| d2q06c | 2012 | 4964 | 30263 | 24860 | 139274 | 2.54e7 | 41 | 254.37 | 5.06 |
| ganges | 840 | 1172 | 5487 | 6627 | 11765 | 2.43e5 | 20 | 6.90 | 1.17 |
| pilot | 1350 | 3329 | 40506 | 58265 | 199640 | 5.18e7 | 44 | 501.63 | 7.98 |
| pilot87 | 1968 | 4595 | 70361 | 115290 | 469878 | 2.18e8 | 43 | 1696.19 | 9.54 |
| pilotnov | 830 | 1861 | 11466 | 9175 | 38896 | 3.64e6 | 19 | 23.89 | 2.63 |
| sctap3 | 1346 | 2356 | 8229 | 6694 | 14767 | 3.89e5 | 13 | 6.54 | 1.03 |
| ship12s | 340 | 1919 | 4273 | 2139 | 2191 | 1.71e4 | 14 | 2.96 | 0.49 |
| sierra | 1129 | 2008 | 6956 | 4418 | 10147 | 2.07e5 | 18 | 7.05 | 1.27 |
| stocfor2 | 1968 | 1854 | 7064 | 9806 | 21180 | 3.30e5 | 19 | 8.70 | 0.71 |
| woodw | 703 | 5347 | 19727 | 12611 | 31237 | 1.90e6 | 33 | 38.28 | 3.02 |

The analysis of results presented in Tables 1 and 2 shows that presolve analysis is a valuable technique that often leads to considerable savings in solution time. These savings, however, are very much problem dependent (they are significant for GANGES but only marginal for PILOT).

### 4.2 Starting point

The choice of a good starting point for an interior point algorithm is still not solved with full satisfaction. Surprisingly, points that are relatively close to the optimal solution (but are not well centered) often lead to bad performance and/or numerical difficulties.

IPMs (also infeasible IPMs) are quite sensitive to the choice of an initial point. A fortunate guess of $(x^0, s^0, y^0, z^0, t^0)$ (possible for some well-understood linear programs) can reduce the computational effort considerably. On the other hand, a bad choice may be disastrous for the efficiency as many iterations will have to be done before the iterates get reasonably centered and the algorithm allows large steps.

The starting points in most implementations of primal-dual infeasible IPMs [4, 47, 55] are some variation of the approximate solution of the following auxiliary QP problem:

$$\begin{array}{ll} \text{minimize} & c^T x + \frac{\varrho}{2}(x^T x + s^T s), \\ \text{subject to} & Ax = b, \\ & x + s = u, \end{array} \tag{4.6}$$

where $\varrho$ is a predetermined weight parameter. A solution of (4.6) can be given by an explicit formula and can be computed at a cost comparable to a single interior point iteration. It is supposed to minimize the norm of the primal solution $(x, s)$ and promotes points that are better in the sense of the LP objective. As the solution of (4.6) may have negative components in $x$ and $s$, those negative components are pushed towards positive values sufficiently bounded away from zero (all elements smaller than $\delta$ are replaced by $\delta$, say, $\delta = 1$). Independently, an initial dual solution $(y, z, t)$ is chosen similarly to satisfy $y = 0$ and the dual constraint (3.2). Again, all elements of $z$ and $t$ smaller than $\delta$ are replaced by $\delta$.

It is worth noting that the code of [76] shows consistently good efficiency for a simple starting point $(x^0, s^0, y^0, z^0, t^0) = (e, e, 0, e, e)$, a phenomenon not known for a primal-dual method (for more details see Sections 7.2 and 7.3).

### 4.3 The linear algebra

Every iteration of an interior point method for linear programming requires computing at least one Newton direction for the first-order optimality conditions. This, in turn, is equivalent to computing the projection of some vector of $\mathbb{R}^n$ onto the null space of the linear operator $AD$. The diagonal scaling matrix $D$ depends on the variant of the algorithm, but the computational effort remains practically the same for all interior point algorithms. This explains why a comparison of the efficiency of different algorithms is often limited to a comparison of iteration numbers to reach the desired accuracy.

As was shown in Section 3, the large sparse system of Newton's equations reduces to the so-called augmented system (3.9). After scaling the primal component of the search direction $\Delta\tilde{x} = -D^{-1}\Delta x$, and substituting $g = Dr$, this system becomes

$$\begin{bmatrix} I & \tilde{A}^T \\ \tilde{A} & 0 \end{bmatrix} \begin{bmatrix} \Delta\tilde{x} \\ \Delta y \end{bmatrix} = \begin{bmatrix} g \\ -h \end{bmatrix}, \tag{4.7}$$

where $\tilde{A} = AD$. Usually this system is referred to as the *augmented system*. Observe that equations (4.7) define the unique orthogonal decomposition of $g$ into $\Delta\tilde{x} \in Ker(\tilde{A})$ and $\tilde{A}^T \Delta y \in Im(\tilde{A}^T)$ when $h = 0$.

The system (4.7) has much better stability properties [6, 7] than its reduced form obtained after eliminating $\Delta\tilde{x}$

$$(\tilde{A}\tilde{A}^T)\Delta y = \tilde{A}g + h, \tag{4.8}$$

which is called the *normal equation*. Direct solution of (4.7) needs Bunch–Parlett [13] factorization of the large, sparse and symmetric but not positive definite matrix

$$\hat{L}\Lambda\hat{L}^T = \begin{bmatrix} I & \tilde{A}^T \\ \tilde{A} & 0 \end{bmatrix}, \tag{4.9}$$

where $\hat{L}$ is a unit lower triangular matrix and $\Lambda$ is a block diagonal matrix with diagonal blocks of size one or two. As the matrix is indefinite, there is no guarantee that non-zero diagonal elements can be found in all intermediate steps of (symmetric) Gaussian elimination. Cholesky decomposition thus cannot, in general, be found and a possible way to overcome the difficulties is to allow indefinite $2 \times 2$ block pivots in $\Lambda$. The augmented system approach has several advantages:

(1) It has good accuracy properties.

(2) It is easily generalizable to exploit the sparsity of KKT (Karush-Kuhn-Tucker) systems arising in non-separable quadratic programming and linear complementarity problems.

(3) It naturally handles free variables. If $x_j \in (-\infty, +\infty)$, then $D_j^{-2}$ in (3.9) is replaced by zero.

(4) Dense columns of $A$ do not degrade its efficiency and do not lead to significant fill-in.

It has two important disadvantages: it is more complicated to implement and it remains, on average, less efficient (about 40%) than its counterpart which applies Cholesky decomposition to the normal equations matrix (4.8)

$$LL^T = \tilde{A}\tilde{A}^T. \tag{4.10}$$

The reason for the efficiency of the Cholesky decomposition comes from the fact that the matrix $\tilde{A}\tilde{A}^T$ is always positive definite. Therefore, the sparsity preserving sequence of pivots (ordering) in which the symmetric Gaussian elimination is performed can be defined in advance, i.e. before the numerical operations start. Thus the *analyse* phase is completely separated from the *factorize* phase [19]. Even more importantly, the sparsity pattern of $\tilde{A}\tilde{A}^T$ is the same in all interior point iterations so the analyse phase has to be done only once before optimization.

If we choose to solve the augmented system (4.7) by Gaussian elimination with $1 \times 1$ pivots from the upper left diagonal block, we end up with the normal equations. Due to stability and fill-in reasons, other pivot sequences, possibly including $2 \times 2$ pivots, are preferable. These sequences are influenced by the numerical values in (4.7). They cannot be determined in advance on the sole basis of the sparsity structure analysis. To choose stable pivots (to get an improvement over normal equations) one has to inspect actual numerical values at the same time. Hence the analyse and factorize phases cannot be separated: the factorization is more expensive than in the normal equations approach. Additionally, due to changes in the diagonal scaling matrix $D$ in subsequent IPM iterations, the pivot sequence should be redefined. In practice, it suffices to update it once every several iterations (only if numerical stability deteriorates) and as the experience of [24] shows, rarely more than one such update is needed. An advantage of the augmented system approach is that more freedom in the pivot choice opens the possibility of getting sparser factors than in the normal equations approach (e.g. the degrading influence of dense columns can be avoided).

Before we pass to a brief description of the Cholesky decomposition, let us comment again on the accuracy of the two competitive approaches. System (4.7) is definitely more stable than (4.8) [7]. To further improve stability both approaches can be equipped with easy to implement techniques that improve their accuracy, e.g. iterative refinement or the use of a regularizing term added to $\tilde{A}\tilde{A}^T$ (which results in bounding all pivots away from zero). Computational experience proves that the normal equations approach also produces sufficiently accurate directions to reach the desired 8-digit correct solutions for practical linear programs [23, 31, 49].

Although this is the case, several researchers [20, 24, 71, 73] have decided to incorporate the augmented system approach [7] in their IPM implementations. We thus find it interesting to give a bit of insight into the computational effort of the two competitive approaches. Table 3 reports the results of running them for several NETLIB problems. It shows the number of non-zeros in triangular factors and the effort (flops) to compute them. In the case of normal equations, we distinguish the effort to form $AA^T$ and to factorize it, and we report the total effort being the sum of the former two. The results for normal equations come from HOPDM code, while the results for the augmented system approach have been reproduced from Table A.3 of [24]. Note that the problem PILOT87 was not solved in [24].

Undoubtedly, it would be advantageous to be able to pick up the preferable approach after the preliminary analysis of the sparsity structure of the LP constraint matrix $A$. This needs both methods to be incorporated into the implementation, which is not usual yet.

Theoretically the most difficult phase in an efficient implementation of the Cholesky decomposition is the analyse phase, i.e. reordering for sparsity. The goal of it is to find a permutation matrix $P$ such that the Cholesky factor of

**Table 3** Augmented system versus normal equations

| | Augmented system | | Normal equations | | | | |
|---|---|---|---|---|---|---|---|
| Problem | nz(L) | flops | nz($AA^T$) | flops($AA^T$) | nz(L) | flops(L) | tot. flops |
| 25fv47 | 48183 | 3.41e6 | 10456 | 8.56e4 | 33377 | 2.55e6 | 2.64e6 |
| 80bau3b | 60965 | 2.36e6 | 9176 | 5.76e4 | 37320 | 2.13e6 | 2.19e6 |
| bnl2 | 93113 | 1.24e7 | 12242 | 5.87e4 | 79538 | 1.22e7 | 1.23e7 |
| cycle | 65687 | 2.72e6 | 18376 | 1.33e5 | 53182 | 3.54e6 | 3.68e6 |
| d2q06c | 177415 | 2.41e7 | 24860 | 3.77e5 | 139274 | 2.54e7 | 2.58e7 |
| ganges | 22005 | 4.97e5 | 6627 | 4.71e4 | 11765 | 2.43e5 | 2.90e5 |
| pilot | 221610 | 4.20e7 | 58265 | 1.09e6 | 199640 | 5.18e7 | 5.29e7 |
| pilot87 | – | – | 115290 | 2.55e6 | 469878 | 2.18e8 | 2.21e8 |
| pilotnov | 57614 | 5.35e6 | 9175 | 1.31e5 | 38896 | 3.64e6 | 3.77e6 |
| sctap3 | 24405 | 5.69e5 | 6694 | 3.58e4 | 14767 | 3.89e5 | 4.25e5 |
| ship12s | 10656 | 8.44e4 | 2139 | 9.94e3 | 2191 | 1.71e4 | 2.70e4 |
| sierra | 20886 | 3.12e5 | 4418 | 2.52e4 | 10147 | 2.07e5 | 2.32e5 |
| stocfor2 | 28119 | 4.11e5 | 9806 | 3.90e4 | 21180 | 3.30e5 | 3.69e5 |
| woodw | 56641 | 2.30e6 | 12611 | 1.06e5 | 31237 | 1.90e6 | 2.01e6 |

$P\tilde{A}\tilde{A}^T P^T$ is the sparsest possible. Unfortunately, this is an NP-complete problem [77]. In practice, a sub-optimal solution is usually found by applying some reordering heuristics. These heuristics work very well in practice. Within a marginal time a spectacular density reduction of the factors can be reached. The most popular heuristics are the *minimum degree* and the *minimum fill-in* orderings which we shall now briefly describe.

## Minimum degree ordering

Markowitz [50] observed that in the $k$-th step of an (unsymmetrical) sparse Gaussian elimination locally, the best pivot candidate $a_{ij}$ is the one that minimizes

$$f_{ij} = (r_i - 1)(c_j - 1),$$

where $r_i$ and $c_j$ are the numbers of non-zero entries of row $i$ and column $j$ respectively in the $k$-th Schur complement. The value $f_{ij}$ gives the number of flops required by the $k$-th step of Gaussian elimination, and at the same time estimates the potential fill-in caused by this step. The pivot sequence found in that way should thus prevent excessive fill-ins and ensure a low cost for the factorization. Tinney and Walker [70] applied this strategy to symmetric matrices. When symmetric positive definite systems are considered, pivot selection is restricted to diagonal elements, hence the Markowitz merit function simplifies to

$$f_j = (c_j - 1)^2, \tag{4.11}$$

and leads to a simple rule that the best candidate for the pivot column is the one with the minimum number of non-zero entries. Interpreting this process in terms of the elimination graph [26], one sees that this is equivalent to the choice of the node that has the minimum degree, which gave the name to this heuristic.

The key point of an efficient implementation of the minimum degree ordering is the representation of fill-ins as the elimination proceeds [19, 26]. The algorithm keeps track of the elimination process storing only pivotal cliques. A clique denotes here a set of row numbers of all, say, $p$ rows that are active in a given pivotal step. The storage of a clique needs only $p$ row numbers, while the symmetric matrix represented by it has $p(p+1)/2$ elements. The sparsity pattern of the Schur complement at the $k$-th step of Gaussian elimination (needed to determine the next minimum degree column) is thus represented implicitly by the sparsity pattern of the decomposed matrix and the pivotal cliques from previous steps of the elimination.

Modern implementations of minimum degree ordering use several enhancements to the basic algorithm [26] and are extremely powerful. Their detailed discussion is beyond the scope of this chapter. However, we would like to focus on one of them that plays a particularly important role. It is a technique that takes advantage of the presence of *indistinguishable* nodes called also *supernodes* in the elimination graph (a set of columns with identical sparsity patterns). Instead of storing several identical cliques, this technique handles all supernodes with only one clique, which leads to obvious savings in a time consuming degree update step of the reordering algorithm. Note that the form of the normal equations matrix

$$AA^T = \sum_{j=1}^{n} a_j a_j^T ,$$

where $a_j$ denotes the $j$-th column of matrix $A$, explains the natural tendency to create many supernodes, as every column $a_j$ of $A$ creates (subject to a symmetric row and column permutation) a dense window in $AA^T$ and larger windows produced by denser columns often cover smaller ones.

### Minimum fill-in ordering

Let us observe that, in general, the function (4.11) considerably overestimates the expected number of fill-ins in a given iteration of the Gaussian elimination because it does not take into account the fact that in many positions of the predicted fill-in, non-zero entries already exist. It is possible that another node, although more expensive in terms of (4.11), would produce less fill-in as the elimination step would mainly update already existing non-zero entries of the Schur complement. An analysis that exactly predicts fill-in and chooses the pivot producing the minimum number (minimum fill-in ordering) is much more involved than the minimum degree ordering. To count predicted fill-in, one has to simulate the elimination step which is quite an expensive operation. Surprisingly, in general, this technique does not offer sufficient advantage over the minimum degree ordering to justify its use.

Our discussion has concentrated up to now on the analysis phase of the decomposition. However, the dominating term in the computational effort of an interior point algorithm is the numerical factorization (repeated several times if necessary). From the mathematical point of view, once the pivot order has been found, the effort of the numerical factorization is uniquely determined:

$$FLOPS = \sum_{i=1}^{m} l_i^2, \tag{4.12}$$

where $l_i$ denotes the number of non-zero elements in column $i$ of $L$. However, in practice, its efficiency still depends very much on how the computations are organized and how well they exploit the specific computer architecture like parallelism, vectorization, cache memory, the use of Basic Linear Algebra System (BLAS) routines, etc. Lower-level BLAS routines offer, for example, a loop unrolling technique that vectorizes very well. Higher-level BLAS routines contain block versions of decomposition algorithms that take advantage of the more efficient organization of the matrix–vector and matrix–matrix products. The gain that results from the use of BLAS may vary significantly on different computers.

Detailed discussion of these issues is beyond the scope of this chapter. Undoubtedly, specializations of the interior point implementations to different computer architectures will draw much attention from the LP community in the near future. The interested reader may consult the preliminary results of [10, 35] and the references therein. A variant of the Cholesky decomposition that is particularly well suited for parallelization comes from [21].

Instead, we shall present a technique that gives considerable time savings on all computer architectures. Of course the savings still vary for different computers. This technique consists in a switch to dense code near the end of decomposition [19]. It exploits the fact that triangular factors become practically completely dense in the last steps of Gaussian elimination. Loops over non-zero entries of the pivot column can then avoid indirect addressing (needed to handle sparse columns).

Table 4 compares the efficiency of our implementation in the case when a switch to dense mode is done (default) and the case when it is disabled. Additionally, it reports the number of decomposition flops required to compute Cholesky factorization of $AA^T$, that part of this effort which is done in dense mode and the size of the dense window. Even on a SUN SPARC-10 computer (without any vectorization facilities) considerable savings are obtained if a dense window is large. Obviously, the savings would be remarkably larger on other computer architectures that take advantage of the specific processor features (e.g. on an IBM RISC 6000 workstation).

Our discussion in this section has naturally concentrated on *direct* approaches for solving reduced KKT systems that for general large-scale LPs are definitely the methods of choice. Saunders' remark [66]: "Major Cholesky would feel proud" seems thus to well summarize the current state of the art of IPM implementations.

**Table 4** Advantages of switch to dense mode

| | No window | | With dense window | | | | |
|---|---|---|---|---|---|---|---|
| Problem | iters | time | iters | time | total flops | dense flops | window |
| 25fv47 | 28 | 26.78 | 28 | 24.05 | 2.55e6 | 1.25e6 | 155 |
| 80bau3b | 43 | 69.43 | 43 | 65.55 | 2.13e6 | 7.41e5 | 130 |
| bnl2 | 30 | 93.79 | 30 | 78.49 | 1.22e7 | 7.92e6 | 287 |
| cycle | 41 | 60.39 | 41 | 57.81 | 3.54e6 | 9.05e5 | 139 |
| d2q06c | 41 | 300.06 | 41 | 254.37 | 2.54e7 | 1.49e7 | 354 |
| ganges | 20 | 7.02 | 20 | 6.90 | 2.43e5 | 5.10e4 | 53 |
| pilot | 44 | 575.02 | 44 | 501.63 | 5.18e7 | 2.27e7 | 408 |
| pilot87 | 43 | 2172.71 | 43 | 1696.19 | 2.18e8 | 1.51e8 | 768 |
| pilotnov | 19 | 25.75 | 19 | 23.89 | 3.64e6 | 1.25e6 | 155 |
| sctap3 | 13 | 6.82 | 13 | 6.54 | 3.89e5 | 1.12e5 | 69 |
| ship12s | 14 | 2.96 | 14 | 2.96 | 1.71e4 | – | – |
| sierra | 18 | 7.39 | 18 | 7.05 | 2.07e5 | 8.53e4 | 63 |
| stocfor2 | 19 | 9.07 | 19 | 8.70 | 3.30e5 | 1.62e4 | 36 |
| woodw | 33 | 40.11 | 33 | 38.28 | 1.90e6 | 4.62e5 | 111 |

To complete the discussion of different methods of computing projections, let us also mention *iterative* approaches. A classical method from this family (i.e. the conjugate gradients algorithm) has been applied to solve reduced KKT systems by several researchers, but to the best of our knowledge it has never proved competitive when applied to general large-scale linear programs. The reason for this is an always present ill-conditioning of KKT systems. The only way to overcome extremely slow convergence of an iterative method is the use of a good preconditioner, but for general linear programs there is no way, to date, to find such a preconditioner at an acceptable computational cost. Instead, for specially structured problems a good preconditioner derived from the interpretation of the problem can sometimes be found [61, 64].

### 4.4 Stepsize

Most interior point methods require that all iterates belong to a neighbourhood of the central path [32]. One way to keep this condition satisfied is to allow only a small reduction in the barrier parameter with a factor

$$\beta = \frac{1}{1 + \gamma/\sqrt{n}}, \tag{4.13}$$

where $\gamma \in (0, 0.1)$ that, in consequence, leads to very short steps (*short-step methods*). Although such an approach allows us to prove nice theoretical properties of the algorithm (all iterates are very close to the central path), it leads to hopelessly slow convergence as the worst case behaviour becomes the practice.

In *medium-step methods* (when $\gamma = \Omega(1)$) and *long-step methods* (when $\gamma = \Omega(\sqrt{n})$), a more optimistic target is chosen and consequently the iterates are allowed to move in a much larger neighbourhood of the central path but more damped Newton steps are required to get close to the new target. The complexity of the long-step methods becomes $O(n)$, which is worse than that of the short-step methods. In practice, these methods offer very fast convergence, significantly faster than is given by the worst case complexity analysis.

In current implementations we do not bother about getting really close to the target. The target is updated after each Newton step. Even worse, different stepsizes are used in primal and dual spaces and the only concern is preserving the non-negativity of the variables. More precisely, maximum possible stepsizes are determined by the formulae

$$\begin{aligned} \alpha_P &:= \max \alpha > 0 : \ (x, s) + \alpha(\Delta x, \Delta s) \geq 0, \\ \alpha_D &:= \max \alpha > 0 : \ (z, t) + \alpha(\Delta z, \Delta t) \geq 0, \end{aligned} \tag{4.14}$$

and these stepsizes are slightly reduced by a factor $\alpha_0 = 0.99995$ to prevent hitting the boundary. The use of such a stepsize rule saves about 40% of the number of iterations compared with the case when the theory of the long-step methods is strictly followed.

The use of different stepsizes in different spaces is due to Kojima *et al.* [43]. However, to preserve polynomial complexity, some additional safeguards were needed.

In another paper, Kojima, Megiddo and Mizuno [42] prove the global convergence of the infeasible primal-dual method. In their stepsize selection, only non-negativity of variables is concerned as in (4.14), but additional conditions are imposed that ensure more uniform progress in reducing both primal and dual feasibility and approaching optimality.

The results presented in Table 5 illustrate the efficiency of our primal-dual implementation measured with the number of iterations to reach optimality in two cases: 1) always identical stepsizes are chosen in both spaces and 2) different stepsizes are allowed (default). These results clearly show that restricting primal and dual stepsizes to be the same causes a remarkable loss in efficiency (218 iterations over 90 problems).

### 4.5 Centering and higher-order methods

Almost all interior point algorithms compose the direction step ($\Delta x$, $\Delta s$, $\Delta y$, $\Delta z$, $\Delta t$) (denoted by $\Delta$ for short) from two parts

$$\Delta = \Delta_a + \Delta_c. \tag{4.15}$$

i.e. combine affine-scaling, $\Delta_a$, and centering, $\Delta_c$, components. The term $\Delta_a$ is obtained by solving (3.7) for $\mu = 0$ and $\Delta_c$ is the solution of an equation like (3.7) with the right-hand side being

$$(0,\ 0,\ 0,\ \mu e - XZe,\ \mu e - STe)^T,$$

where $\mu > 0$ is some centering parameter ($\mu = (x^T z + s^T t)/(2n)$, for example, refers to centering that does not change the current complementarity gap). The

**Table 5** Different stepsizes in different spaces

| Problem | $\alpha_P = \alpha_D$ | $\alpha_P \neq \alpha_P$ |
|---|---|---|
| 25fv47 | 31 | 28 |
| 80bau3b | 44 | 43 |
| bnl2 | 33 | 30 |
| cycle | 49 | 41 |
| d2q06c | 45 | 41 |
| ganges | 21 | 20 |
| pilot | 47 | 44 |
| pilot87 | 45 | 43 |
| pilotnov | 19 | 19 |
| sctap3 | 13 | 13 |
| ship12s | 15 | 14 |
| sierra | 20 | 18 |
| stocfor2 | 21 | 19 |
| woodw | 35 | 33 |
| whole NETLIB | 1996 | 1778 |

term $\Delta_a$ is responsible for "optimization", while $\Delta_c$ keeps the current iterate away from the boundary.

All path-following methods can benefit from the use of the predictor–corrector technique. This technique takes into account higher-order terms in the Newton direction and estimates targets that are more likely to be achieved. Undoubtedly, one of the reasons why it proved so successful in the primal-dual method is Mehrotra's very good heuristic defining the new target that takes advantage of knowledge about the current and predicted complementarity gap. We shall now address this technique in more detail.

The introduction of higher-order terms is combined in it with the adaptive choice of the barrier parameter $\mu$ based on an analysis of the reduction of the complementarity gap achievable when moving in a (predictor, $\mu = 0$) primal-dual affine scaling direction.

The affine scaling (predictor) direction $\Delta_a$ solves the linear system (3.7) for the right-hand side equal to the current violation of the first-order optimality conditions for (3.1)–(3.2), i.e. with $\mu = 0$. This direction is usually "too optimistic" — if a full step of length one could be made in it, the LP problem would be solved in one step. Predictor–corrector makes a hypothetical step in this direction. The maximum stepsizes in the primal, $\alpha_{Pa}$, and in the dual, $\alpha_{Da}$, spaces preserving non-negativity of $(x, s)$ and $(z, t)$ respectively are determined and the predicted complementarity gap

$$g_a = (x + \alpha_{Pa}\Delta x)^T(z + \alpha_{Da}\Delta z) + (s + \alpha_{Pa}\Delta s)^T(t + \alpha_{Da}\Delta t)$$

is computed. It is then used to determine the barrier parameter

$$\mu = \left(\frac{g_a}{g}\right)^2 \frac{g_a}{n}, \tag{4.16}$$

where $g = x^T z + s^T t$ denotes the current complementarity gap. The term $g_a/g$ measures the achievable progress in the affine scaling direction. If only a small step in the affine scaling direction can be made, then $g_a/g$ is close to one and $\mu = g/n$, which means that we only want to improve centrality. If the affine scaling direction offers considerable progress in the reduction of the complementarity gap, then a more optimistic target (closer to the optimum) is chosen.

For such a $\mu$, the corrector direction $\Delta_c$ is computed

$$\begin{bmatrix} A & 0 & 0 & 0 & 0 \\ I & 0 & I & 0 & 0 \\ 0 & A^T & 0 & I & -I \\ Z & 0 & 0 & X & 0 \\ 0 & 0 & T & 0 & S \end{bmatrix} \begin{bmatrix} \Delta x_c \\ \Delta y_c \\ \Delta s_c \\ \Delta z_c \\ \Delta t_c \end{bmatrix} = \begin{bmatrix} 0 \\ 0 \\ 0 \\ \mu e - \Delta X_a \Delta Z_a e \\ \mu e - \Delta S_a \Delta T_a e \end{bmatrix}, \tag{4.17}$$

and finally the direction $\Delta$ of (4.15) is determined.

A single iteration of the (second-order) predictor-corrector primal-dual method thus needs two solves of the same large, sparse linear system for two different right-hand sides: first to solve (3.7) with $\mu = 0$, and then to solve (4.17).

We should note here that the above presentation of the predictor–corrector technique follows computational practice. It abuses mathematics in the sense that the stepsizes $\alpha_P$ and $\alpha_D$ are not taken into account in it (one would expect this when using higher-order Taylor approximations). Readers who are interested in seeing a detailed rigorous presentation of this approach are encouraged to consult [54].

Let us observe that the predictor–corrector mechanism can be applied repeatedly, thus leading to a higher (than two) order method. The direction $\Delta$ of (4.15) can be used as a predictor, and a new corrector term $\Delta_c'$ can later be computed and added to it. A method of order $k$ would compute $k-1$ corrector terms and would require solving the same linear system for $k$ different right-hand sides.

Although the use of higher-order terms usually results in the reduction of the number of iterations to reach optimality, it does not necessarily produce time savings due to the increased effort in a single iteration. To date, second-order methods seem, on average, to be the most efficient. (They are used in most implementations [23, 31, 49, 55].) No one yet knows how to benefit practically from higher-order information. A possible way to take advantage of it is to promote higher-order terms, if the Cholesky factorization is very expensive compared with

**Table 6** Efficiency of higher-order methods

| | Order 1 | | Order 2 | | Order 3 | | Order 4 | |
|---|---|---|---|---|---|---|---|---|
| Problem | iters | time | iters | time | iters | time | iters | time |
| 25fv47 | 44 | 33.26 | 28 | 24.05 | 25 | 24.98 | 23 | 24.44 |
| 80bau3b | 65 | 74.00 | 43 | 65.55 | 35 | 66.19 | 40 | 82.75 |
| bnl2 | 58 | 125.54 | 30 | 78.49 | 25 | 73.20 | 27 | 80.42 |
| cycle | 63 | 74.15 | 41 | 57.81 | 33 | 51.88 | 33 | 55.10 |
| d2q06c | 65 | 364.35 | 41 | 254.37 | 36 | 243.84 | 33 | 236.53 |
| ganges | 32 | 7.50 | 20 | 6.90 | 19 | 8.31 | 19 | 8.45 |
| pilot | 56 | 617.06 | 44 | 501.63 | 34 | 405.90 | 32 | 389.20 |
| pilot87 | 63 | 2419.63 | 43 | 1696.19 | 35 | 1431.03 | 33 | 1384.22 |
| pilotnov | 33 | 33.24 | 19 | 23.89 | 17 | 24.39 | 15 | 23.68 |
| sctap3 | 24 | 8.64 | 13 | 6.54 | 12 | 7.94 | 11 | 8.69 |
| ship12s | 26 | 3.63 | 14 | 2.96 | 14 | 4.46 | 13 | 4.75 |
| sierra | 30 | 8.87 | 18 | 7.05 | 16 | 8.65 | 16 | 9.88 |
| stocfor2 | 31 | 12.90 | 19 | 8.70 | 17 | 10.48 | 17 | 11.80 |
| woodw | 54 | 49.07 | 33 | 38.28 | 27 | 41.31 | 26 | 43.36 |

a single solve (see, e.g. [4] for details).

We end this section with a demonstration of the use of higher-order techniques. Table 6 reports the number of iterations to reach 8-digit precision in the duality gap and the CPU time for several NETLIB problems for different order methods. The method of order 1 is a pure primal-dual one in which the direction comes from the solution of (3.7). The method of order 2 is a classical predictor–corrector one in which one corrector term has been computed. The method of order $k$ computes $k-1$ corrector terms.

The analysis of results presented in Table 6 shows that the use of higher-order information may reduce the number of iterations to reach optimality. However, this translates into computation time savings only in the case when Cholesky factorizations are very expensive (compare to the data of Table 4).

### 4.6 Stopping criteria

Interior point algorithms terminate when the relative primal and dual feasibility and the relative duality gap are reduced to a predetermined tolerance. In case of the primal-dual method, the following conditions are checked to state $p$-digit accurate solutions

$$\frac{||Ax-b||}{1+||x||_\infty} \le 10^{-p} \quad \text{and} \quad \frac{||x+s-u||}{1+||x||_\infty+||s||_\infty} \le 10^{-p}, \tag{4.18}$$

$$\frac{||A^Ty+z-t-c||}{1+||z||_\infty+||t||_\infty} \le 10^{-p}, \tag{4.19}$$

$$\frac{|c^Tx-(b^Ty-u^Tt)|}{1+|b^Ty-u^Tt|} \le 10^{-p}. \tag{4.20}$$

It is worth noting that scaling may change the left-hand sides of equations (4.18)–(4.20) and consequently change the measures of the quality of solution significantly. The denominators of these left-hand sides usually decrease after scaling the problem. An 8-digit exact solution ($p = 8$) is typically required in the literature.

In practice, it is extremely rare that condition (4.20) is satisfied and at the same time one of the conditions (4.18) or (4.19) does not hold. The explanation of this phenomenon comes from the analysis of the first-order optimality conditions (3.5). Observe that the first three equations, which impose primal and dual feasibility, are linear. They are thus "easier" to satisfy for Newton's method than the last two equations which are non-linear, and additionally, change in subsequent interior point iterations. Consequently, the most important and perhaps the only condition that really has to be checked is (4.20).

Results presented in Table 7 report the numbers of primal-dual iterations at which the primal and the dual relative feasibilities and the relative optimality gap were reduced to $10^{-8}$ for a subset of NETLIB problems. Its last three columns give the relative infeasibilities and the relative optimality gap for the 8-digit optimal solution.

Note that for infeasible problems or for problems with unbounded level sets, the scaling of infeasibilities is not appropriate. In these cases the iterates diverge, hence the denominator goes to infinity which might lead to false results. The proper normalization factor, in these cases, would be $1 + ||b||_\infty$ and $1 + ||c||_\infty$ respectively.

### 4.7 Theoretical versus practical worst-case complexity

The theoretical worst-case complexity of the simplex method is not known to be polynomial, but this has never caused problems in practice. In fact, modern simplex codes rarely do more than $m + n$ iterations to reach optimality.

Although the worst-case complexity of interior point methods is polynomial, the situation is somehow similar. A theoretical bound of $O(\sqrt{n}\ \log(1/\epsilon))$ iterations to obtain an $\epsilon$–exact solution to LP is still too pessimistic. In practice, the number of iterations is something like $O(\log n)$. It is very rare, for example, that the predictor–corrector primal-dual infeasible IPM makes more that 50 iterations to reach 8-digit exact solutions. This is never the case when the HOPDM code [31] is applied to solve 90 LPs from the NETLIB suite (only 7 problems need to be solved with more than 40 iterations).

## 5 Remarks on numerical difficulties

### 5.1 Degeneracy and IPMs

The main computational step in all IPMs is solving the system of normal equations

$$AD^2A^Tp = q, \tag{5.1}$$

for some $q$, where $D$ is a diagonal matrix (3.11). Güler and Ye [34] showed that in all central path-following methods, any limit point $(x^*, z^*)$ of the iterates

**Table 7** Attaining feasibilities and optimality

| | Iterations to reach | | | Accuracy of solution | | |
|---|---|---|---|---|---|---|
| Problem | Pr. Feas. | Dl. Feas. | Opt. | Pr. Inf. | Dl. Inf. | Opt. Gap |
| 25fv47 | 25 | 13 | 28 | 3.3e-12 | 5.8e-9 | 1.9e-9 |
| 80bau3b | 23 | 20 | 43 | 1.2e-12 | 9.7e-9 | 4.3e-9 |
| bnl2 | 10 | 22 | 30 | 1.1e-11 | 4.4e-10 | 6.9e-9 |
| cycle | 25 | 41 | 41 | 1.3e-14 | 7.0e-9 | 2.7e-9 |
| d2q06c | 20 | 13 | 41 | 7.1e-15 | 5.7e-12 | 2.5e-10 |
| ganges | 15 | 14 | 20 | 3.2e-11 | 4.1e-9 | 8.3e-11 |
| pilot | 44 | 11 | 44 | 4.6e-9 | 1.6e-13 | 6.6e-9 |
| pilot87 | 39 | 43 | 43 | 2.4e-11 | 8.8e-9 | 3.2e-9 |
| pilotnov | 15 | 7 | 19 | 1.0e-12 | 1.8e-10 | 2.0e-10 |
| sctap3 | 13 | 13 | 13 | 2.9e-12 | 2.4e-12 | 1.0e-9 |
| ship12s | 12 | 13 | 14 | 2.6e-10 | 1.1e-12 | 6.9e-11 |
| sierra | 10 | 16 | 18 | 7.9e-13 | 1.8e-9 | 4.8e-9 |
| stocfor2 | 13 | 18 | 19 | 3.4e-12 | 2.3e-13 | 1.6e-10 |
| woodw | 32 | 10 | 33 | 1.8e-12 | 1.2e-10 | 3.9e-10 |

$(x^k, z^k)$ is in the relative interior of the optimal faces. In fact, they prove that there exists a constant $\gamma$, where $0 < \gamma < 1$, such that the relations

$$\gamma \leq x_j^k \leq 1/\gamma, \quad \forall j \in P_x, \tag{5.2}$$

$$\gamma \leq z_j^k \leq 1/\gamma, \quad \forall j \in P_z, \tag{5.3}$$

are satisfied for all $k \geq 0$. Here $(P_x, P_z)$ is the optimal partition of the LP problem.

Note that $x_{P_z}^k \to 0$ and $z_{P_x}^k \to 0$ in any convergent IPM. Therefore, the limiting behaviour of $A_{P_x}(D_{P_x}^k)^2 A_{P_x}^T$ determines the asymptotic behaviour of the linear systems (5.1). For primal-dual path-following methods we can obtain more information about the matrices $D_{P_x}^k$ as we shall now explain.

For this class of methods, $(D^k)^2 = X^k(Z^k)^{-1}$. So

$$\frac{x_i}{z_i} = \frac{x_i^2}{x^T z} \cdot \frac{x^T z}{x_i z_i} \quad \text{for any } i. \tag{5.4}$$

If $i \in P_x$, it follows from relations (5.2), (5.3), and (5.4) that

$$\frac{\gamma^2}{x^T z} \leq \frac{x_i}{z_i} \leq \frac{1}{\xi \gamma^2 (x^T z)}, \tag{5.5}$$

where $\xi$ is an appropriate constant depending on the specific IPM. This shows that the condition numbers of the matrices $D_{P_x}^k$ are uniformly bounded and bounded away from zero. Thus, when matrix $A_{P_x}$ has full rank, the matrices $A_{P_x}(D^k_{P_x})^2 A^T_{P_x}$ also have full rank, and the condition numbers of the latter matrices are uniformly bounded. The rank of $A_{P_x}$ depends on the degeneracy of the problem, which we discuss below in more detail.

(1) If $(P)$ and $(D)$ are both non-degenerate on their respective optimal faces, then both programs have unique solutions and $A_{P_x}$ and $A_{P_x}(D^k_{P_x})^2 A^T_{P_x}$ are non-singular. The linear systems (5.1) are well conditioned, at least when $\xi$ and $\gamma$ are not too small.

(2) If $(P)$ is degenerate and $(D)$ is non-degenerate, matrix $A_{P_x}$ has less than $m$ columns and so the rank of $A_{P_x}$ is less than $m$ and $A_{P_x}(D^k_{P_x})^2 A^T_{P_x}$ is singular. This means that the linear system (5.1) is ill-conditioned. Numerical problems caused by this ill-conditioning may arise and in fact they were recognized early [27]. However, as we shall discuss in Section 5.2 the method itself has an embedded safeguard against them.

(3) If $(P)$ is non-degenerate and $(D)$ is degenerate, then the rank of $A_{P_x}$ is equal to $m$, hence $A_{P_x}(D^k_{P_x})^2 A^T_{P_x}$ is non-singular. This implies that the linear system (5.1) is well-conditioned, again as long as $\xi$ and $\gamma$ are not too small. Observe that if $(P)$ is non-degenerate, the degeneracy status of $(D)$ matters little from the numerical point of view.

(4) If both $(P)$ and $(D)$ are degenerate, then not much can be said about the partition $(A_{P_x}, A_{P_z})$ and the rank of $A_{P_x}$. Consequently, we can also say nothing about the rank of the $A(D^k)^2A^T$ matrices.

### 5.2 Ill-conditioned normal equations matrix

It became folklore that every iteration of an IPM needs inversion of the matrix that, as the optimum approaches, becomes extremely ill-conditioned. Although this is generally true, its impact on the accuracy of solutions is not as dramatic as one would expect. Ill-conditioning of the coefficient matrix is not the only factor that determines the difficulty of solving the system of linear equations and the accuracy of solutions. An equally important factor is the relation between the matrix and the right-hand side. The following important result [69] theoretically justifies the common experience that numerical problems are rare even if the normal equations matrix is ill-conditioned.

Let us have a closer look at the right-hand side of the normal equations system (3.11). Since it has the form $q = AD^2r + h$, we may split (3.11) into two systems of equations

$$AD^2A^T p_r = AD^2 r \tag{5.6}$$

and

$$AD^2A^T p_h = h. \tag{5.7}$$

Stewart [69] proved that with given $A$ and $r$, for any positive diagonal matrix $D$, the solution of equation (5.6) belongs to a bounded compact set. In other words, even if the normal matrix becomes ill-conditioned, this bad property has limited influence on the solution vector $p_r$.

Unfortunately, in the infeasible primal-dual method, $h$ in (5.7) is non-zero and has no clear dependence on $A$ and $D$. Furthermore, $r$ changes (see (3.10)). If the LP is both primal and dual feasible, then all infeasibilities $\xi_b, \xi_u$ and $\xi_c$ go rapidly to zero. Consequently, $h$ (and $p_h$) vanish while $r$ converges to a fixed point, so Stewart's result ensures the stability of $p_r$.

If the LP is dual infeasible, then the dual iterates diverge, $\xi_c$ does not converge to zero and $r$ diverges. The above stability result does not apply and in practice serious numerical difficulties are observed.

If the LP is primal infeasible, then the primal iterates diverge, primal infeasibilities $(\xi_b, \xi_u)$ do not converge to zero and $r$ diverges. Similarly to the dual infeasible case, the stability result does not apply to (5.6) and eventually, as $h = \xi_b$ might not converge to zero, the solution of (5.7) may cause additional numerical problems.

## 6 Optimal basis identification

From now on in the rest of the chapter, we consider the LP problem and its dual in the standard form, e.g. the primal variables are non-negative and have no upper bounds. This form simplifies the notation and makes it easier to explain and understand the underlying ideas. Hence the primal problem that we consider is

$$\begin{array}{ll} \text{minimize} & c^T x, \\ \text{subject to} & Ax = b, \\ & x \geq 0, \end{array} \tag{6.1}$$

and the dual problem is

$$\begin{array}{ll} \text{maximize} & b^T y, \\ \text{subject to} & A^T y + z = c, \\ & z \geq 0, \end{array} \tag{6.2}$$

where the vectors and matrices have the same dimensions as before.

It is known that a pair of primal and dual feasible solutions $x^*$ and $(y^*, z^*)$ are optimal if $X^* z^* = 0$. The optimal pair is *strictly complementary* if $x^* + z^* > 0$. In this case, we have $P_x = \{i \mid x_i^* > 0\}$ and $P_z = \{i \mid z_i^* > 0\}$. Due to the complementarity property $P_x \cap P_z = \emptyset$, the strict complementarity property is equivalent to $P_x \cup P_z = \{1, \ldots, n\}$. The partition $(P_x, P_z)$ is uniquely determined by the problem data and is called the *optimal partition* of the LP problem.

### 6.1 Do we need an optimal basis?

It is well known that if the LP problem is solved by the simplex method, then an optimal basis (solution) is finally produced as the solution of the problem. Of course, if the problem is degenerate and multiple optimal solutions are present, then one of the possible optimal basis solutions is generated. In this case IPMs (except Iri and Imai's [36] method), contrary to the simplex algorithm, do not produce an optimal basis solution but a primal-dual optimal pair where the primal optimal solution is a solution from the relative interior of the primal optimal face and the dual optimal solution is a solution from the relative interior of the dual optimal face. Hence, IPMs provide an optimal solution with a maximal number of non-zero coordinates in both the primal and dual problems (see e.g. Güler and Ye [34]). It is also proved that these solutions are strictly complementary, and they define the optimal partition of the LP problem.

The existence of strictly complementary primal-dual optimal solutions was first proved by Goldman and Tucker [28]. Balinski and Tucker [9] propose a pivot algorithm to generate such a strictly complementary pair. In contrast, as we will see later on, an optimal basis can be obtained from any primal-dual optimal solution pair in strongly polynomial time.

To have an optimal basis might be important for several reasons. For example, basic solutions have a minimal number of non-zero coordinates, which is advantageous when a solution with a small number of positive coordinates is required for some practical applications. Of course, there are situations when a strictly complementary optimal solution and the knowledge of the optimal partition is needed (see e.g. Greenberg [33]).

Currently, sensitivity and postoptimal analysis are based on the knowledge of optimal bases in all commercial packages. People are used to this approach, although several papers recognized the potential mathematical errors and negative economical consequences of this approach [33, 37, 74]. Correct postoptimal analysis is possible based on solutions found by IPMs, or in fact, by the simplex method. The decision on which approach is preferable depends on the problem instance and is discussed in Section 7.1.

To the best of our knowledge, a basic solution is necessary for cutting plane methods in mixed-integer programming. In branch and bound methods, the problem has to be re-optimized after some small modification. To date, simplex-based solvers are more efficient than IPMs if an "almost optimal" solution is available.

For numerical reasons, or if we are facing a large mixed-integer problem, sometimes a crossover from the IPMs to the simplex method is needed. The efficient techniques for crossover are similar to the basis identification techniques.

These advantages provide sufficient motivation to examine how one can generate an optimal basic solution from an optimal or near-optimal solution obtained by an IPM. It is evident that this question occurs only in the case of degeneracy, since otherwise the primal and dual optimal solutions are unique and are also basic solutions.

## 6.2 How to get an optimal basis

Several attempts were made to create efficient methods that produce an optimal basis if a (strictly complementary) optimal solution is available. The best theoretical algorithm, which also turned out to be very efficient in practice (see e.g. [11]) is due to Megiddo [53]. It finds an optimal basis in strongly polynomial time, provided that optimal solutions are available to both (6.1) and (6.2). Due to its theoretical and practical importance, we discuss it in more detail.

Principally, IPM ends up with a strictly complementary pair of optimal solutions but the scheme is applicable to a slightly more general case. Suppose that $x$ (primal) and $y$, $z$ (dual) optimal solutions are available. Let $A = [A_1, A_2, A_3]$, $x = (x_1, x_2, x_3)$, $z = (z_1, z_2, z_3)$, $c = (c_1, c_2, c_3)$ where index 1 refers to the positive coordinates of $x$, index 2 refers to the zero coordinates of both $x$ and $z$, and index 3 refers to the positive coordinates of $z$. Then, we have $A_1x_1 = b$, $x_1 > 0$, $x_2 = 0$, $x_3 = 0$, and $A_1^Ty = c_1$, $A_2^Ty = c_2$, $A_3^Ty < c_3$.

Starting from such a solution and partition, we are looking for an optimal basis, i.e. a disjoint partition $(\mathcal{B}, \mathcal{N})$ of the indices such that the submatrix of $A$ built of columns from $\mathcal{B}$ is non-singular, $x_{\mathcal{B}} \geq 0$, $z_{\mathcal{B}} = 0$, $x_{\mathcal{N}} = 0$, and $z_{\mathcal{N}} \geq 0$.

The process will subsequently build $\mathcal{B}$ up and, if necessary, modify $x$ and $z$ to satisfy the above sign requirements. Naturally, all columns whose indices enter $\mathcal{B}$ must be linearly independent to meet the non-singularity requirement. As the process starts from an optimal solution, it will not alter the objective values.

The algorithm is split into three phases. In the first one, a maximal linearly independent subset of the columns of $A_1$ is built up; at the same time some coordinates of $x_1$ are purified and so moved to $A_2$. This continues until the resulting $A_1$ contains only linearly independent columns. The indices of $A_1$ then form an initial $\mathcal{B}$. If $\mathcal{B}$ is already a basis, we are done. Otherwise, in the second phase it is extended to a maximal independent subset of $[A_1, A_2]$. If $\mathcal{B}$ is still not a basis, then we proceed with the third phase. Every step of it purifies some indices of $z_3$, i.e. drives the corresponding coordinates to zero. All the corresponding columns are independent of the current $\mathcal{B}$ and moved to $A_2$. They are immediately examined if the current $\mathcal{B}$ can be extended by them (at least one of them has to be eligible). These steps are repeated until $\mathcal{B}$ extends to a basis. The complete algorithm is presented in Fig. 2.

Note that only Gaussian elimination steps (pivoting) are necessary to perform this algorithm. The amount of work involved depends on the degree of degeneracy of the LP problem. In the worst case, not more than $n$ pivots (the dimension of the space) are necessary to identify an optimal basis. The algorithm uses both primal and dual information and generates optimal basic solutions to both the primal and the dual problems.

Megiddo also proves the surprising result that the complexity of identifying an optimal basis, if just a primal optimal solution or just a dual optimal solution is available, is equivalent to solving an LP problem. This result also demonstrates that the primal-dual approach has an important advantage compared to pure primal or dual methods.

Although the above algorithm is clear and elegant, there are several problems to be solved in its practical implementation. One has, for example, only approximate optimal and approximate complementary solutions, hence tolerances and some safeguards are needed in implementations. Another problem consists in finding the partition $P_x$, $P_z$. This is why, in applying Megiddo's algorithm initialized from an IPM optimal pair, the $(x_2, z_2)$ part usually refers to those variables for which no clear decision can be made if the variable index belongs

---

**Initialization**

suppose $x$ primal and $y, z$ dual optimal solutions are available and $x$ and $z$ are initially partitioned as explained above;

**Reduce the positive part of $x$:**

**while** the columns of $A_1$ are dependent **do**

find (e.g. by pivoting) a vector $t$ such that $A_1 t = 0$ (this implies $c_1^T t = 0$);

using $t$, eliminate a positive coordinate (say $j$) from $x_1$, while preserving the non-negativity of $x_1$ (ratio test);

remove column $a_j$ from $A_1$ and add it to $A_2$;

**end**;

Let $B = A_1$; (Note that the columns of $B$ are independent at this stage)

**Extend $B$ to a basis:**

**Extend $B$ using $A_2$:**

**while** $\text{rank}(B) < \text{rank}[A_1, A_2]$ **do**

**if** $\text{rank}[A_1, A_2] > \text{rank}(B)$ **and** column $a_j$ of $[A_1, A_2]$ is independent from $B$

**then** add $a_j$ to $B$;

**end**;

**Extend $B$ using $A_3$:**

**while** $\text{rank}(B) < m$ **do**

find (e.g. by pivoting) a vector $u$ such that $B^T u = 0$ (this implies $A_1^T u = 0$, $A_2^T u = 0$) and $A_3^T u \neq 0$; (Note that $u$ satisfies $b^T u = 0$ since $u^T A_1^T x_1 = b^T u$)

using $u$, eliminate a positive coordinate (say $j$) from $z_3$, while preserving the dual feasibility of $z$ (ratio test);

remove $a_j$ from $A_3$ and add it to $A_2$ and $B$;

**end**.

(We now have an optimal complementary pair $(x, z)$, where $\text{rank}(B) = m$. Using the formulae $Bx_B = b$ and $B^T y = c_B$, we see that the basis $B$ is optimal).

---

FIG. 2. Optimal basis identification algorithm

to $P_x$ or $P_z$. Special care must also be taken to select a stable basis. Megiddo's algorithm has been efficiently implemented [5, 11, 46]. In [5] it is theoretically clarified when a guaranteed crossover can take place.

## 7 Problems to be solved

Although great progress has been made concerning the efficiency of IPM implementations, some relevant questions still remain open. We shall concentrate on only three of them which seem the most important, namely

- implementing postoptimal analysis in a correct way,
- handling problems which have unbounded optimal faces and detecting primal/dual infeasibility, and
- warm start.

### 7.1 Correct postoptimal analysis

In implementing algorithms for LP, one has to consider methods to produce shadow prices and ranges. Currently, implementations of sensitivity and postoptimal analysis in all commercial codes are based on the knowledge of an optimal basis. People are used to this approach, although several papers recognized the potential mathematical errors and negative economical consequences of it ([33, 37, 74]). The ranges and shadow prices obtained from analysing an optimal basis are usually incorrect as one would like to know the linearity intervals of the optimal value function and its left/right derivatives which are the correct shadow prices. This correct information is obtainable at a different cost. It is proven [3, 37] that there is a one to one correspondence between the optimal partitions and the linearity intervals (breakpoints of the value function). These values can be obtained at the cost of the solution of some smaller LPs. To define these LPs, one needs a description of the optimal face. Once a strictly complementary solution is found, the optimal partition $(P_x, P_z)$ is also known. The primal optimal face is then given by

$$\mathcal{P}^* = \{x \mid Ax = b,\ \ x \geq 0,\ x_{P_z} = 0\},$$

while the dual optimal face is

$$\mathcal{D}^* = \{(y, z) \mid A^T y + z = c,\ \ z \geq 0,\ z_{P_x} = 0\}.$$

Alternatively, if the optimal partition is not known, just a primal optimal solution $x^*$ and a dual optimal solution $(y^*, z^*)$ is available, then the optimal faces are characterizable as follows.

$$\begin{aligned} \mathcal{P}^* &= \{x \mid Ax = b,\ \ c^T x = c^T x^*,\ \ x \geq 0\}, \\ \mathcal{D}^* &= \{(y, z) \mid A^T y + z = c,\ \ b^T y = b^T y^*,\ \ z \geq 0\}. \end{aligned}$$

As mentioned above, to get the linearity intervals and correct shadow prices, the solution of some auxiliary LPs is needed. As an illustration, first we define the two LPs to find the left and right end of the actual linearity interval $[\epsilon_{left}, \epsilon_{right}]$ when a single objective coefficient $c_j$ changes. If the optimal partition $(P_x, P_z)$ is known, one needs to solve the following two linear programs:

$$\epsilon_{left} = \min\{\epsilon \mid A^T y + z = c + \epsilon e_j,\ z \geq 0,\ z_{P_x} = 0\}$$

$$\epsilon_{right} = \max\{\epsilon \mid A^T y + z = c + \epsilon e_j,\ z \geq 0,\ z_{P_x} = 0\}.$$

If the optimal partition is not available, just optimal solutions are known, the solution of the two LPs

$$\epsilon_{left} = \min\{\epsilon \mid A^T y + z = c + \epsilon e_j,\ b^T y = c^T x^* + \epsilon x_j^*,\ z \geq 0\},$$

and

$$\epsilon_{right} = \max\{\epsilon \mid A^T y + z = c + \epsilon e_j,\ b^T y = c^T x^* + \epsilon x_j^*,\ z \geq 0\}$$

is necessary. Note that the validity of the resulting ranges follows from the concavity or convexity of the optimal value functions respectively.

Similarly, the correct left and right shadow prices can be obtained by solving some auxiliary LP problems. It is obvious that if we are in the interior of a linearity interval of the value function as $c_j$ varies, then the left and right shadow prices are equal and both equal to $x_j$. If $c_j$ is a breakpoint of the optimal value function, then the left and right shadow prices $(x_{j-left}, x_{j-right})$ can be computed by solving the two LP problems

$$x_{j-left} = \max\{x_j \mid Ax = b,\ x \geq 0,\ x_{P_z} = 0\},$$

$$x_{j-right} = \min\{x_j \mid Ax = b,\ x \geq 0,\ x_{P_z} = 0\}$$

in the optimal partition approach. If the optimal partition is not available the following two LPs are to be solved:

$$x_{j-left} = \max\{x_j \mid Ax = b,\ c^T x = c^T x^*,\ x \geq 0\},$$

$$x_{j-right} = \min\{x_j \mid Ax = b,\ c^T x = c^T x^*,\ x \geq 0\}.$$

If $b_i$ varies then the corresponding values can be obtained similarly on the cost of the solution of some analogous LPs. We will not go into further details here. For a complete analysis based on the optimal partition see [3, 17, 37]. The alternative approaches when the optimal partition is not available are discussed in [17, 56, 74].

To close this section we point out that in order to get the correct ranges and shadow prices for all the primal and dual variables, substantially more time than in the incomplete (incorrect) optimal basis approach is needed. Hence, it is advisable to select a smaller set of "important" variables and to perform the necessary calculations just for this set.

### 7.2 Unbounded optimal faces, infeasible problems

As discussed in Section 3, the current implementations are based on infeasible IPMs. One of the theoretical disadvantages of the implemented infeasible IPMs is that their theoretical complexity is $O(nL)$, instead of the best complexity to date: $O(\sqrt{n}L)$. If the optimal face(s) is unbounded, then the duality gap converges to zero while the sequence(s) of solutions (primal, dual or both according to the unboundedness of the optimal faces respectively) diverges, hence it is difficult to identify optimal solution(s). Similar difficulties occur if the primal or dual or both problems are infeasible. This case always manifests itself with the divergence of the iteration sequence in infeasible IPMs, which in practice might not be easy to identify. Resolving these difficulties is one of the most interesting problems nowadays.

The so-called skew-symmetric self-dual embedding (SSSD) might be a remedy. The SSSD was first introduced by Ye *et al.* [78] using the standard form problems (6.1) and (6.2). They discussed most of the advantages of this embedding and showed that the predictor–corrector algorithms of Mizuno *et al.* [58] solve the LP problem in $O(\sqrt{n}L)$ iterations, yielding the first infeasible IPM with this complexity. Somewhat later, Jansen *et al.* [38] presented the SSSD problem for the symmetric form primal-dual LP pair in a concise introduction to the theory of LP based on IPMs. Before presenting the SSSD embedding, the surprisingly nice properties of it are summarized.

(1) Self-duality, the dual problem is identical to the primal one. One can solve it as an LP or as an LCP.
(2) SSSD is always feasible. Furthermore, the interior of the feasible sets is also non-empty, hence the optimal faces are bounded. IPMs applied to SSSD always converge to an optimal solution.
(3) Optimality of the original problem is detected by convergence, independently from the boundedness/unboundedness of the optimal faces of the original problem.
(4) Infeasibility of the original problem is detected by convergence. Primal, dual, or primal and dual rays for the original problems are identified to prove dual, primal, or dual and primal infeasibility.
(5) A perfectly centered initial pair can always be constructed for SSSD.
(6) Polynomial convergence with the best-known complexity and local quadratic convergence [75, 78] is guaranteed.

**Self-dual embedding**

To exploit fully the symmetry of the SSSD embedding, the problems (6.1) and (6.2) are first transformed to symmetric form. This can be done without increasing the number of variables or the number of constraints. We may assume that in problem (6.1), $\text{rank}(A) = m$; otherwise the redundant constraints can be eliminated. Let $B$ be any basis of $A$. Let $A = [B, N]$, $c^T = [c_B^T, c_N^T]$ and $x^T = [x_B^T, x_N^T]$. Then $Ax = b$, $x \geq 0$ can be written as $x_B + B^{-1}Nx_N = B^{-1}b$, $x \geq 0$ or equivalently $-B^{-1}Nx_N \geq -B^{-1}b$, $x_N \geq 0$. Likewise, $c^Tx = c_B^Tx_B + c_N^Tx_N = c_B^TB^{-1}b + (c_N^T - c_B^TB^{-1}N)x_N$. Hence $(\overline{P})$ can be written equivalently in the symmetric form as

$$\min\left\{(c_N^T - c_B^TB^{-1}N)x_N \mid -B^{-1}Nx_N \geq -B^{-1}b,\ x_N \geq 0\right\}.$$

Note that these transformations need a modified preprocessing of the problem that chooses a sparse $B$ and hopefully produces a sparse $B^{-1}N$.

It is not obvious how such a basis can be constructed. To the best of our knowledge, this has never been implemented yet. The problem of finding a basis that is optimal, in the sense that $B^{-1}N$ is the sparsest possible, imposes some additional requirements on the sparsity problem (see equation (4.5)). Hence, it seems practically more difficult to find efficient heuristics to solve it. Those heuristics should naturally generalize the techniques used to solve the sparsity problem.

Having done the above transformations, we redefine the objective vector, the coefficient matrix and the right-hand side vector, and denote them again by $c, A$ and $b$ respectively. This produces the LP problem in the symmetric form

$$(P) \qquad \min\left\{c^T x \mid Ax \geq b,\ x \geq 0\right\},$$

where $A$ is an $m \times n$ matrix, $c, x \in \mathbb{R}^n$ and $b \in \mathbb{R}^m$. The linear program (P) has associated with it the dual program

$$(D) \qquad \max\left\{b^T y \mid A^T y \leq c,\ y \geq 0\right\}.$$

As usual, a pair $(x^*, y^*)$ will be called *strictly complementary* if $Ax^* \geq b$, $x^* \geq 0$, $A^T y^* \leq c$, $y^* \geq 0$, $(Ax^* - b)^T y^* = (c - A^T y^*)^T x^* = 0$ and moreover, $y^* + (Ax^* - b) > 0$ and $x^* + (c - A^T y^*) > 0$.

To formulate the SSSD embedding we need some further vectors. Let $x^0, z^0, \bar{c} \in \mathbb{R}^n$, $y^0, r^0, \bar{b} \in \mathbb{R}^m$, $\vartheta^0, \tau^0, \kappa^0, \nu^0 \in \mathbb{R}$ and $\alpha, \beta \in \mathbb{R}$ be given as follows.

$$x^0 > 0,\ \ r^0 > 0,\ \ y^0 > 0,\ \ z^0 > 0,\ \ \vartheta^0 > 0,\ \ \tau^0 > 0,\ \ \kappa^0 > 0,\ \ \nu^0 > 0,$$

$$\begin{aligned}
\bar{b} &= \frac{1}{\vartheta^0}(b\tau^0 - Ax^0 + r^0),\\
\bar{c} &= \frac{1}{\vartheta^0}(c\tau^0 - A^T y^0 - z^0),\\
\alpha &= \frac{1}{\vartheta^0}(c^T x^0 - b^T y^0 + \kappa^0),\\
\beta &= \alpha\tau^0 + \bar{b}^T y^0 - \bar{c}^T x^0 + \nu^0\\
&= \frac{1}{\vartheta^0}[(y^0)^T r^0 + (x^0)^T z^0 + \tau^0\kappa^0] + \nu^0 > 0.
\end{aligned}$$

It is worthwhile noting that if $x^0$ is feasible for $(P)$, $\tau^0 = 1$ and $r^0 = Ax^0 - b$, then $\bar{b} = 0$. Also if $y^0$ is feasible for $(D)$, $\tau^0 = 1$ and $z^0 = c - A^T y$, then $\bar{c} = 0$. With some abuse of mathematics, the vectors $\bar{b}$ and $\bar{c}$ measure the amount of scaled infeasibility of the given vectors $x^0$, $r^0$, $y^0$ and $z^0$.

Now consider the following skew symmetric self-dual LP problem:

$$\begin{array}{llccccl}
\text{(SSSD)} & \min & & & \beta\vartheta, & & \\
& \text{s.t.} & & Ax & +\bar{b}\vartheta & -b\tau & \geq 0,\\
& & -A^T y & & -\bar{c}\vartheta & +c\tau & \geq 0,\\
& & -\bar{b}^T y & +\bar{c}^T x & & -\alpha\tau & \geq -\beta,\\
& & b^T y & -c^T x & +\alpha\vartheta & & \geq 0,\\
& & y \geq 0, & x \geq 0, & \vartheta \geq 0, & \tau \geq 0. &
\end{array}$$

Due to the selection of the parameters the positive solution $x = x^0$, $y = y^0$, $\vartheta = \vartheta^0$, $\tau = \tau^0$ is interior feasible for the SSSD problem. Let us denote the slack variables for the problem SSSD by $r, z, \nu$ and $\kappa$ respectively. Also note that if one chooses $x = x^0 = e,\ r = r^0 = e,\ y = y^0 = e,\ z = z^0 = e,\ \kappa = \kappa^0 = 1,\ \nu = \nu^0 = 1,\ \vartheta = \vartheta^0 = 1,\ \tau = \tau^0 = 1$, then this solution is a perfectly centered initial solution for the SSSD problem. The following theorem holds (see [38, 78]).

**Theorem 7.1** *For the given problems (P) and (D) the SSSD embedding is made. Then one has:*

*(i) The SSSD problem is feasible, hence both primal and dual feasible and has an optimal solution.*

*(ii) For any optimal solution of SSSD* $\vartheta^* = 0$.

*(iii) SSSD always has a strictly complementary optimal solution* $(x^*, y^*, \vartheta^*, \tau^*)$.

*(iv) If* $\tau^* > 0$, *then* $\frac{x^*}{\tau^*}$ *and* $\frac{y^*}{\tau^*}$ *are strictly complementary optimal solutions of (P) and (D) respectively.*

*(v) If* $\tau^* = 0$, *then either (P) or (D) or both are infeasible.*

Skew-symmetric self-dual embedding has been implemented [76]. Impressive numerical results are reported that show this approach to be only slightly less efficient than the primal-dual method on feasible problems. Additionally, as the experience of the above-mentioned paper shows, this method detects problem infeasibility really by a convergence (not by a divergence). Although it needs slightly different preprocessing and it has to deal with two dense rows and columns bordered to the matrix of the augmented system (which require special care when solving (3.11)), the SSSD approach may become a computationally attractive alternative to the infeasible primal-dual method.

### 7.3 Warm start

Many practical problems need the solution of a sequence of similar linear programs where small perturbations are made to $b$ and/or $c$. As long as these perturbations are small we naturally expect that the optimal solutions are not far from each other and restarting the optimization from the solution of the old problem (warm start) should be very efficient. This is the case in practice when the simplex method is used. In contrast, we still do not have an efficient implementation of warm start in IPMs that exploits the information contained in the old optimal solution.

Several attempts have been made to solve this problem like using shifted barriers to allow infeasibility of the original variables [25]; applying modified barrier functions [60]; and perturbing the problem to throw it away from the boundary of the positive orthant [49]. However, we may say that, in such a general context, IPM warm start is still far from working satisfactorily. On the other hand, computational experience of [8] shows that applying the (feasible) projective algorithm [65] to restart from a well centered, almost optimal solution that is sufficiently far away from the boundary, works well even if problems that have to subsequently be solved differ considerably.

The difficulty with IPM warm start comes from the fact that the old optimal solution is very close to the boundary and well centered. This point in the perturbed problem still remains close to the boundary but is very badly centered.

Consequently, IPM makes a long sequence of short steps due to the fact that the iterates cannot get rid of the boundary (boundary behaviour). Therefore, one needs a well centered point close to the old optimal one, or an efficient centering method (leaving the boundary) to overcome these difficulties. These two possibilities are discussed below.

Independently of the approach chosen, it would be wise to save a well centered almost optimal solution (say, with $10^{-2}$ relative duality gap) that is still sufficiently far away from the boundary.

- **Efficient centering**. The so-called target-following method offers much flexibility in choosing achievable targets. Path-following methods define targets on the central path that, due to boundary behaviour, are too optimistic in the case of warm start. Using a sequence of traceable targets that improves centrality allows larger steps, therefore speeds up the centering and, finally, the optimization process. Practical methods of defining such a sequence of targets are not well studied yet although the theory is well established (see, e.g. Jansen *et al.* [39] and Roos and Vial [65] in this book).
- **Centered solutions for warm start in SSSD embedding**. Among the spectacular properties of the SSSD embedding listed in the previous section, the ability to always construct a perfectly centered initial point was mentioned. The old well centered optimal or almost optimal solution $x^*$, $r^*$, $y^*$, $z^*$, $\vartheta^*$, $\tau^*$, $\nu^*$, $\kappa^*$, can be used as $x^0$, $r^0$, $y^0$, $z^0$, $\vartheta^0$, $\tau^0$, $\nu^0$, $\kappa^0$, the initial point to embed the perturbed problem. As we have seen in Section 7.2, $\bar{b}$, $\bar{c}$, $\alpha$ and $\beta$ can always be redefined so that the above solution stays well centered. The construction allows simultaneous perturbations of $b$ and $c$. Additionally, it extends to handling new constraints or variables added to the problem (e.g. in build-up or cutting plane schemes). In these cases, we can keep the solution unchanged for the old coordinates (let $\mu$ be the actual barrier parameter) and define the initial value of the new complementary variables to be $\sqrt{\mu}$ as $\mu$ was the barrier parameter of the old centered solution. This results in a perfectly centered initial solution.

## Bibliography


1. I. Adler, N. Karmarkar, M.G.C. Resende and G. Veiga. Data structures and programming techniques for the implementation of Karmarkar's algorithm. *ORSA Journal on Computing*, 1:84–106, 1989.
2. I. Adler, N. Karmarkar, M.G.C. Resende and G. Veiga. An implementation of Karmarkar's algorithm for linear programming. *Mathematical Programming*, 44:297–336, 1989.
3. I. Adler and R.D.C. Monteiro. A geometric view of parametric linear programming. *Algorithmica*, 8:161–176, 1992.
4. A. Altman and J. Gondzio. An efficient implementation of a higher order primal–dual interior point method for large sparse linear programs. *Archives*

*of Control Sciences*, 2:23–40, 1993.

5. E.D. Andersen and Y. Ye. Combining interior–point and pivoting algorithms for linear programming. Research report, Department of Management Studies, The University of Iowa, Iowa City IA, USA, May 1994.
6. M. Arioli, J.W. Demmel and I.S. Duff. Solving sparse linear systems with sparse backward error. *SIAM Journal on Matrix Analysis and Applications*, 10:165–190, 1989.
7. M. Arioli, I.S. Duff and P.P.M. de Rijk. On the augmented system approach to sparse least–squares problems. *Numerische Mathematik*, 55:667–684, 1989.
8. O. Bahn, J.-L. Goffin, J.P. Vial and O. du Merle. Experimental behaviour of an interior point cutting plane algorithm for convex programming: an application to geometric programming. *Discrete Applied Mathematics*, 49:3–23, 1994.
9. M.L. Balinski and A.W. Tucker. Duality theory of linear programs: A constructive approach with applications. *SIAM Review*, 11:499–581, 1969.
10. R.H. Bisseling, T.M. Doup and L.D.J.C. Loyens. A parallel interior point algorithm for linear programs on a network of transputers. *Annals of Operations Research*, 43:51–86, 1993.
11. R.E. Bixby. Progress in linear programming. *ORSA Journal on Computing*, 6:15–22, 1994.
12. A.L. Brearley, G. Mitra and H.P. Williams. Analysis of mathematical programming problems prior to applying the simplex algorithm. *Mathematical Programming*, 15:54–83, 1975.
13. J.R. Bunch and B.N. Parlett. Direct methods for solving symmetric indefinite systems of linear equations. *SIAM Journal on Numerical Analysis*, 8:639–655, 1971.
14. S.F. Chang and S.T. McCormick. A hierarchical algorithm for making sparse matrices sparser. *Mathematical Programming*, 56:1–30, 1992.
15. I.C. Choi, C.L. Monma and D.F. Shanno. Further development of a primal–dual interior point method. *ORSA Journal on Computing*, 2:304–311, 1990.
16. G.B. Dantzig. *Linear Programming and Extensions.* Princeton University Press, Princeton N.J., 1963.
17. J.J. de Jong, B. Jansen, C. Roos and T. Terlaky. Sensitivity analysis in linear programming: Just be careful! Technical Report AMER.93.022, KSLA, Amsterdam, The Netherlands, 1994.
18. I.I. Dikin. Iterative solution of problems of linear and quadratic programming. *Doklady Akademii Nauk SSSR*, 174:747–748, 1967. Translated into English in *Soviet Mathematics Doklady*, 8:674–675.
19. I.S. Duff, A.M. Erisman and J.K. Reid. *Direct Methods for Sparse Matrices.* Oxford University Press, New York, 1989.
20. I.S. Duff, N.I.M. Gould, J.K. Reid, J.A. Scott and K. Turner. The factor-

ization of sparse symmetric indefinite matrices. *IMA Journal of Numerical Analysis*, 11:181–204, 1991.

21. I.S. Duff and J.K. Reid. The multifrontal solution of indefinite sparse symmetric linear equations. *ACM Transactions on Mathematical Software*, 9:302–325, 1983.
22. A.V. Fiacco and G.P. McCormick. *Non-linear Programming: Sequential Unconstrained Minimization Techniques.* Wiley, New York, 1968.
23. J.J.H. Forrest and J.A. Tomlin. Implementing interior point linear programming methods in the optimization subroutine library. *IBM Systems Journal*, 31:26–38, 1992.
24. R. Fourer and S. Mehrotra. Solving symmetric indefinite systems in an interior point method for linear programming. *Mathematical Programming*, 62:15–39, 1993.
25. R.M. Freund. Theoretical efficiency of a shifted barrier function algorithm for linear programming. *Linear Algebra and Its Applications*, 152:19–41, 1991.
26. A. George and J.W.H. Liu. The evolution of the minimum degree ordering algorithm. *SIAM Review*, 31:1–19, 1989.
27. P.E. Gill, W. Murray, M.A. Saunders, J.A. Tomlin and M.H. Wright. On projected Newton barrier methods for linear programming and an equivalence to Karmarkar's projective method. *Mathematical Programming*, 36:183–209, 1986.
28. A.J. Goldman and A.W. Tucker. Theory of linear programming. In H.W. Kuhn and A.W. Tucker, (eds.), *Linear Inequalities and Related Systems*, 53–97. Princeton University Press, Princeton, New Jersey, 1956.
29. G.H. Golub and C. Van Loan. *Matrix Computations.* Johns Hopkins University Press, Baltimore and London, second edition, 1989.
30. J. Gondzio. Splitting dense columns of constraint matrix in interior point methods for large scale linear programming. *Optimization*, 24:285–297, 1992.
31. J. Gondzio. Presolve analysis of linear programs prior to applying the interior point method. Technical Report 1994.3, Department of Management Studies, University of Geneva, Switzerland, 1994. To appear in *ORSA Journal on Computing.*
32. C.C. Gonzaga. Path–following methods for linear programming. *SIAM Review*, 34:167–224, 1992.
33. H.J. Greenberg. The use of the optimal partition in a linear programming solution for postoptimal analysis. *Operations Research Letters*, 15:179–185, 1994.
34. O. Güler and Y. Ye. Convergence behaviour of some interior–point algorithms. *Mathematical Programming*, 60:215–228, 1993.
35. H. Hafsteinsson, R. Levkovitz and G. Mitra. Solving large scale linear pro-

gramming problems using an interior point method on a massively parallel SIMD computer. In D.J. Evans, (ed.), *Applied Parallel Computing.* Gordon and Breach Publishers, 1993.

36. M. Iri and H. Imai. A multiplicative barrier function method for linear programming. *Algorithmica*, 1:455–482, 1986.
37. B. Jansen, C. Roos and T. Terlaky. An interior point approach to postoptimal and parametric analysis in linear programming. Technical Report 92-21, Faculty of Technical Mathematics and Informatics, Technical University Delft, Delft, The Netherlands, 1992.
38. B. Jansen, C. Roos and T. Terlaky. The theory of linear programming: Skew–symmetric self–dual problems and the central path. *Optimization*, 29:225–233, 1993.
39. B. Jansen, C. Roos, T. Terlaky and J.P. Vial. Primal-dual target following algorithm for linear programming. Technical Report 93-107, Faculty of Technical Mathematics and Informatics, Technical University Delft, Delft, The Netherlands, 1993.
40. N.K. Karmarkar. A new polynomial-time algorithm for linear programming. *Combinatorica*, 4:373–395, 1984.
41. N.K. Karmarkar, J.C. Lagarias, L. Slutsman and P. Wang. Power series variants of Karmarkar-type algorithms. *AT&T Technical Journal*, 68:20–36, 1989.
42. M. Kojima, N. Megiddo and S. Mizuno. A primal-dual infeasible interior point algorithm for linear programming. *Mathematical Programming*, 61:263–280, 1993.
43. M. Kojima, N. Megiddo and S. Mizuno. Theoretical convergence of large-step-primal-dual interior point algorithms for linear programming. *Mathematical Programming*, 59:1–21, 1993.
44. M. Kojima, S. Mizuno and A. Yoshise. A primal-dual interior point algorithm for linear programming. In N. Megiddo, (ed.), *Progress in Mathematical Programming: Interior Point and Related Methods*, 29–48. Springer-Verlag, New York, 1989.
45. E. Kranich. Interior point methods for mathematical programming: A bibliography. Discussion Paper 171, Institute of Economy and Operations Research, Fern Universitat Hagen, P.O.Box 940, D-5800 Hagen 1, Germany, 1991.
46. R. Levkovitz and G. Mitra. Experimental investigations in combining primal-dual interior point method and simplex based LP solvers. *Annals of Operations Research*, 58:19–38, 1995.
47. I.J. Lustig, R.E. Marsten and D.F. Shanno. Computational experience with a primal–dual interior point method for linear programming. *Linear Algebra and its Applications*, 152:191–222, 1991.
48. I.J. Lustig, R.E. Marsten and D.F. Shanno. On implementing Mehrotra's

predictor–corrector interior point method for linear programming. *SIAM Journal on Optimization*, 2:435–449, 1992.

49. I.J. Lustig, R.E. Marsten and D.F. Shanno. Interior point methods for linear programming: Computational state of the art. *ORSA Journal on Computing*, 6:1–14, 1994.
50. H.M. Markowitz. The elimination form of the inverse and its application to linear programming. *Management Science*, 3:255–269, 1957.
51. K.A. McShane, C.L. Monma and D.F. Shanno. An implementation of a primal-dual interior point method for linear programming. *ORSA Journal on Computing*, 1:70–89, 1989.
52. N. Megiddo. Pathways to the optimal set in linear programming. In N. Megiddo, (ed.), *Progress in Mathematical Programming: Interior Point and Related Methods*, 131–158. Springer-Verlag, New York, 1989.
53. N. Megiddo. On finding primal- and dual-optimal bases. *ORSA Journal on Computing*, 3:63–65, 1991.
54. S. Mehrotra. Higher order methods and their performance. Technical Report 90-16R1, Department of Industrial Engineering and Management Sciences, Northwestern University, Evanston, USA, 1991.
55. S. Mehrotra. On the implementation of a primal–dual interior point method. *SIAM Journal on Optimization*, 2:575–601, 1992.
56. S. Mehrotra and R.D.C. Monteiro. Parametric and range analysis for interior point methods. Technical report, Dept. of Systems and Industrial Engineering, University of Arizona, Tucson, AZ 85721, USA, April 1992.
57. S. Mizuno. Polynomiality of Kojima–Megiddo–Mizuno infeasible interior point algorithm for linear programming. *Mathematical Programming*, 67:109–119, 1994.
58. S. Mizuno, M.J. Todd and Y. Ye. On adaptive-step primal-dual interior point algorithms for linear programming. *Mathematics of Operations Research*, 18:964–981, 1992.
59. C.L. Monma and A.J. Morton. Computational experience with a dual affine variant of Karmarkar's method for linear programming. *Operations Research Letters*, 6:261–267, 1987.
60. R. Polyak. Modified barrier functions (theory and methods). *Mathematical Programming*, 54:177–222, 1992.
61. L. Portugal, F. Bastos, J. Judice, J. Paixõ and T. Terlaky. An investigation of interior point algorithms for the linear transportation problems. Report 93-100, Faculteit der Technische Wiskunde en Informatica, Technische Universiteit Delft, Nederlands, 1993.
62. F.A. Potra. An infeasible interior-point predictor–corrector algorithm for linear programming. Technical Report 26, Department of Mathematics, University of Iowa, Iowa City IA, USA, 1992.
63. J.K. Reid. A sparsity-exploiting variant of the Bartels–Golub decomposition

for linear programming bases. *Mathematical Programming*, 24:55–69, 1982.

64. M.G.C. Resende and G. Veiga. An efficient implementation of a network interior point method. Technical report, AT&T Bell Laboratories, Murray Hill, NJ, USA, February 1992.
65. C. Roos and J.P. Vial. Interior point methods. In J.E. Beasley, (ed.), *Advances in Linear and Integer Programming*. Oxford University Press, Oxford, England, 1996.
66. M.A. Saunders. Major Cholesky would feel proud. *ORSA Journal on Computing*, 6:23–27, 1994.
67. D.F. Shanno and A. Bagchi. A unified view of interior point methods for linear programming. *Annals of Operations Research*, 22:55–70, 1990.
68. G. Sonnevend. An "analytic center" for polyhedrons and new classes of global algorithms for linear (smooth, convex) programming, in *System modelling and optimization: Proceedings of the 12th IFIP–conference*. In A. Prékopa, J. Szelezsán and B. Strazicky, (eds.), *Lecture Notes in Control and Information Sciences*, 84:866–876. Springer Verlag, Berlin, Germany, 1986.
69. G.W. Stewart. On scaled projections and pseudoinverses. *Linear Algebra and Its Applications*, 112:189–193, 1989.
70. W.F. Tinney and J.W. Walker. Direct solution of sparse network equations by optimally ordered triangular factorization. In *Proceedings of IEEE*, 55:1801–1809, 1967.
71. K. Turner. Computing projections for the Karmarkar algorithm. *Linear Algebra and its Applications*, 152:141–154, 1991.
72. R.J. Vanderbei. Splitting dense columns in sparse linear systems. *Linear Algebra and its Applications*, 152:107–117, 1991.
73. R. Vanderbei and T.J. Carpenter. Symmetric indefinite systems for interior point methods. Technical Report SOR 91-7, Department of Civil Engineering and Operations Research, Princeton University, Princeton, New Jersey, 1991.
74. J.E. Ward and R.E. Wendell. Approaches to sensitivity analysis in linear programming. *Annals of Operations Research*, 27:3–38, 1990.
75. F. Wu, S. Wu and Y. Ye. On quadratic convergence of the $O(\sqrt{n}L)$-iteration homogeneous and self-dual linear programming algorithm. Technical report, Department of Management Studies, The University of Iowa, Iowa City IA, USA, 1992.
76. X. Xu, P.F. Hung and Y. Ye. A simplified homogeneous and self-dual linear programming algorithm and its implementation. Technical report, Department of Management Sciences, The University of Iowa, Iowa City IA, USA, 1993.
77. M. Yannakakis. Computing the minimum fill–in is NP–complete. *SIAM Journal on Algebraic Discrete Methods*, 2:77–79, 1981.

78. Y. Ye, M.J. Todd and S. Mizuno. An $O(\sqrt{n}L)$-iteration homogeneous and self-dual linear programming algorithm. *Mathematics of Operations Research*, 19:53–67, 1994.
79. Y. Zhang. On the convergence of an infeasible interior-point algorithm for linear programming and other problems. Research Report 92-07, Dept. of Mathematics and Statistics, University of Maryland, Baltimore, MD 21228, USA, 1992.

# 4

# Interior point algorithms for network flow problems

**Mauricio G.C. Resende**

*AT&T Bell Laboratories, Murray Hill, NJ 07974-2070 USA*

**Panos M. Pardalos**

*The University of Florida, Gainesville, FL 32611-6595 USA*

## 1 Introduction

A large number of problems in transportation, communications and manufacturing can be modelled as network flow problems. In these problems one seeks to find the most efficient, or optimal, way to move flow (e.g. materials, information, buses, electrical currents) on a network (e.g. postal network, computer network, transportation grid, power grid). Many of these optimization problems are special classes of linear programming problems, with combinatorial properties that allow the development of efficient solution techniques. In this chapter, we limit our discussion to linear network flow problems. For a treatment of non-linear network flow problems, the reader is referred to [17, 28, 29, 48].

Given a directed graph $G = (\mathcal{N}, \mathcal{A})$, where $\mathcal{N}$ is a set of $m$ nodes and $\mathcal{A}$ a set of $n$ arcs, let $(i, j)$ denote a directed arc from node $i$ to node $j$. Every node is classified in one of the following three categories. *Source* nodes produce more flow than they consume. *Sink* nodes consume more flow than they produce. *Transshipment* nodes produce as much flow as they consume. Without loss of generality, one can assume that the total flow produced in the network equals the total flow consumed. Each arc has associated with it an origin node and a destination node, implying a direction for flow to follow. Arcs have limitations (often called capacities or bounds) on how much flow can move through them. The flow on arc $(i, j)$ must be no less than $l_{ij}$ and can be no greater than $u_{ij}$. To set up the problem in the framework of an optimization problem, a unit flow cost $c_{ij}$, incurred by each unit of flow moving through arc $(i, j)$, must be defined. Besides being restricted by lower and upper bounds at each arc, flows must satisfy another important condition, known as Kirchhoff's Law (conservation of flow), which states that for every node in the network, the sum of all incoming flow plus the flow produced at the node must equal the sum of all outgoing flow plus the flow consumed at the node. The objective of the *minimum cost network flow problem* is to determine the flow on each arc of the network, such that all of the flow produced in the network is moved from the source nodes to the sink nodes in the most cost-effective way, whilst not violating Kirchhoff's Law and the flow limitations on the arcs. The minimum cost network flow problem can be formulated as the following linear program:

$$\begin{array}{lll} \text{minimize} & \sum_{(i,j)\in\mathcal{A}} c_{ij}x_{ij}, & (1.1) \\ \text{subject to} & \sum_{(j,k)\in\mathcal{A}} x_{jk} - \sum_{(k,j)\in\mathcal{A}} x_{kj} = b_j, \;\; j \in \mathcal{N}, & (1.2) \\ & l_{ij} \le x_{ij} \le u_{ij}, \;\; (i,j) \in \mathcal{A}. & (1.3) \end{array}$$

In this formulation, $x_{ij}$ denotes the flow on arc $(i,j)$ and $c_{ij}$ is the cost of transporting one unit of flow on arc $(i,j)$. For each node $j \in \mathcal{N}$, let $b_j$ denote a quantity associated with node $j$ that indicates how much flow is produced or consumed at the node. If $b_j > 0$, node $j$ is a source. If $b_j < 0$, node $j$ is a sink. Otherwise ($b_j = 0$), node $j$ is a transshipment node. For each arc $(i,j) \in \mathcal{A}$, as before, let $l_{ij}$ and $u_{ij}$ denote, respectively, the lower and upper bounds on flow on arc $(i,j)$. The case where $u_{ij} = \infty$, for all $(i,j) \in \mathcal{A}$, gives rise to the *uncapacitated* network flow problem. Without loss of generality, $l_{ij}$ can be set to zero. Most often, the problem data (i.e. $c_{ij}, u_{ij}, l_{ij}$, for $(i,j) \in \mathcal{A}$ and $b_j$, for $j \in \mathcal{N}$) are assumed to be integer, and many software codes adopt this assumption. However, there can exist applications where the data are real numbers, and algorithms should be capable of handling problems with real data.

Constraints of type (1.2) are referred to as the flow conservation equations, while constraints of type (1.3) are called the flow capacity constraints. In matrix notation, the above network flow problem can be formulated as a linear program of the special form

$$\min \{c^\top x \mid Ax = b, \; l \le x \le u\},$$

where $A$ is the $m \times n$ *node-arc incidence matrix* of the graph $G = (\mathcal{N}, \mathcal{A})$, i.e. for each arc $(i,j)$ in $\mathcal{A}$ there is an associated column in matrix $A$ with exactly two non-zero entries: an entry 1 in row $i$ and an entry $-1$ in row $j$. Note that of the $mn$ entries of $A$, only $2n$ are non-zero and because of this the node-arc incidence matrix is not a space-efficient representation of the network. There are many other ways to represent a network. A popular representation is the *node-node adjacency* matrix $B$. This is an $m \times m$ matrix with an entry 1 in position $(i,j)$ if arc $(i,j) \in \mathcal{A}$ and 0 otherwise. Such a representation is efficient for dense networks, but is inefficient for sparse networks. A more efficient representation for sparse networks is the *adjacency list*, where for each node $i \in \mathcal{N}$ there exists a list of arcs emanating from node $i$, i.e. a list of nodes $j$ such that $(i,j) \in \mathcal{A}$. The *forward star* representation is a multi-array implementation of the adjacency list data structure. The adjacency list allows easy access to the arcs emanating from a given node, but no easy access to the incoming arcs. The *reverse star* representation allows easy access to the list of arcs incoming into $i$. Another representation that is much used in interior point network flow implementations is a simple *arc list*, where the arcs are stored in a linear array. The complexity of an algorithm for solving network flow problems depends greatly on the network representation and the data structures used for maintaining and updating intermediate computations.

We denote the $i$-th column of $A$ by $A_i$, the $i$-th row of $A$ by $A_{.i}$ and a submatrix of $A$ formed by columns with indices in set $S$ by $A_S$. If graph $G$ is disconnected and has $p$ connected components, there are exactly $p$ redundant flow conservation constraints, which are sometimes removed from the problem formulation. We rule out a trivially infeasible problem by assuming

$$\sum_{j \in \mathcal{N}^k} b_j = 0, \quad k = 1, \ldots, p, \tag{1.4}$$

where $\mathcal{N}^k$ is the set of nodes for the $k$-th component of $G$.

Often it is further required that the flow $x_{ij}$ be integer, i.e. we replace (1.3) by

$$l_{ij} \leq x_{ij} \leq u_{ij}, \; x_{ij} \text{ integer}, \; (i,j) \in \mathcal{A}. \tag{1.5}$$

Since the node-arc incidence matrix $A$ is totally unimodular, when the data is integer all vertex solutions of the linear program are integer. An algorithm that finds a vertex solution, such as the simplex method, will necessarily produce an integer optimal flow. In certain types of network flow problems, such as the assignment problem, one may only be interested in solutions having integer flows, since fractional flows do not have a logical interpretation.

In the remainder of this chapter we assume, without loss of generality, that $l_{ij} = 0$ for all $(i,j) \in \mathcal{A}$ and that $c \neq 0$. A simple change of variables can transform the original problem into an equivalent one with $l_{ij} = 0$ for all $(i,j) \in \mathcal{A}$. The case where $c = 0$ is a simple feasibility problem, and can be handled by solving a maximum flow problem [3].

Many important combinatorial optimization problems are special cases of the minimum cost network flow problem. Such problems include the linear assignment and transportation problems, and the maximum flow and shortest path problems. In the transportation problem, the underlying graph is bipartite, i.e. there exist two sets $\mathcal{S}$ and $\mathcal{T}$ such that $\mathcal{S} \cup \mathcal{T} = \mathcal{N}$ and $\mathcal{S} \cap \mathcal{T} = \emptyset$ and arcs occur only from nodes of $\mathcal{S}$ to nodes of $\mathcal{T}$. The set $\mathcal{S}$ is usually called the set of source nodes and the set $\mathcal{T}$ the set of sink nodes. For the transportation problem, the right-hand side vector in (1.2) is given by

$$b_j = \begin{cases} s_j, & \text{if } j \in \mathcal{S}, \\ -t_j, & \text{if } j \in \mathcal{T}, \end{cases}$$

where $s_j$ is the supply at node $j \in \mathcal{S}$ and $t_j$ is the demand at node $j \in \mathcal{T}$. The assignment problem is a special case of the transportation problem, in which $s_j = 1$ for all $j \in \mathcal{S}$ and $t_j = 1$ for all $j \in \mathcal{T}$.

The computation of the maximum flow from node $s$ to node $t$ in $G = (\mathcal{N}, \mathcal{A})$ can be done by computing a minimum cost flow in $G' = (\mathcal{N}', \mathcal{A}')$, where $\mathcal{N}' = \mathcal{N}$

and $\mathcal{A}' = \mathcal{A} \cup (t, s)$, where

$$c_{ij} = \begin{cases} 0, & \text{if } (i,j) \in \mathcal{A}, \\ -1, & \text{if } (i,j) = (t,s), \end{cases}$$

and

$$u_{ij} = \begin{cases} \text{cap}(i,j), & \text{if } (i,j) \in \mathcal{A}, \\ \infty, & \text{if } (i,j) = (t,s), \end{cases}$$

where $\text{cap}(i,j)$ is the capacity of arc $(i,j)$ in the maximum flow problem.

The shortest paths from node $s$ to all nodes in $\mathcal{N} \setminus \{s\}$ can be computed by solving an uncapacitated minimum cost network flow problem in which $c_{ij}$ is the length of arc $(i,j)$ and the right-hand side vector in (1.2) is given by

$$b_j = \begin{cases} m-1, & \text{if } j = s, \\ -1, & \text{if } j \in \mathcal{N} \setminus \{s\}. \end{cases}$$

Although all of the above combinatorial optimization problems are formulated as minimum cost network flow problems, several specialized algorithms have been devised for solving them efficiently.

In many practical applications, flows in networks with more than one commodity need to be optimized. In the multicommodity network flow problem, $k$ commodities are to be moved in the network. The set of commodities is denoted by $\mathcal{K}$. Let $x_{ij}^k$ denote the flow of commodity $k$ in arc $(i,j)$. The multicommodity network flow problem can be formulated as the following linear program:

$$\text{minimize} \quad \textstyle\sum_{k \in \mathcal{K}} \sum_{(i,j) \in \mathcal{A}} c_{ij}^k x_{ij}^k, \tag{1.6}$$

$$\text{subject to} \quad \textstyle\sum_{(j,l) \in \mathcal{A}} x_{jl}^k - \sum_{(l,j) \in \mathcal{A}} x_{lj}^k = b_j^k, \quad j \in \mathcal{N}, \quad k \in \mathcal{K}, \tag{1.7}$$

$$\textstyle\sum_{k \in \mathcal{K}} x_{ij}^k \leq u_{ij}, \quad (i,j) \in \mathcal{A}, \tag{1.8}$$

$$x_{ij}^k \geq 0, \quad (i,j) \in \mathcal{A}, \quad k \in \mathcal{K}. \tag{1.9}$$

The minimum cost network flow problem is a special case of the multicommodity network flow problem, in which there is only one commodity.

In the 1940s, Hitchcock [30] proposed an algorithm for solving the transportation problem and later Dantzig [14] developed the simplex method for linear programming problems. In the 1950s, Kruskal [41] developed a minimum spanning tree algorithm and Prim [51] devised an algorithm for the shortest path problem. During that decade, commercial digital computers were introduced widely. The first book on network flows was published by Ford and Fulkerson [19] in 1962. Since then, active research has produced a variety of algorithms, data structures and software for solving network flow problems. For an introduction to network flow problems and applications, see the books [3, 7, 17, 19, 39, 42, 58, 62].

The focus of this chapter is on recent computational approaches, based on interior point methods, for solving large-scale network flow problems. In the last two decades, many other approaches have been developed. A history of computational approaches up to 1977 is summarized in [11]. In addition, several computational studies had established the fact that, for network flow problems, specialized network simplex algorithms were orders of magnitude faster than simply applying a general simplex linear programming code. (See, e.g. the studies in [23, 27, 39, 43].) A collection of FORTRAN codes of efficient algorithms of that period can be found in [59]. Another important class of network optimization algorithms and codes are the relaxation methods described in [8]. More recent computational research is included in [25, 31].

In 1984, Karmarkar introduced a new polynomial time algorithm for solving linear programming problems [37]. This algorithm and many of its variants, known as interior point methods, have been used to efficiently solve network flow problems. This is the main topic of the remainder of this chapter.

The chapter is organized as follows. In Section 2 we provide a brief survey of the literature in the field of interior point network flow methods. A discussion of complexity issues is made. In Section 3 we focus on several important components of interior point network flow implementations, including a discussion on iterative methods for solving the large linear systems that occur in interior point methods, preconditioners, identification of the optimal partition, and recovery of the optimal flow. These points are illustrated on a particular interior point method, the dual affine scaling algorithm. In Section 4 we present some computational results, comparing an interior point network flow method with an efficient, commercially available, network simplex code. Concluding remarks are made in Section 5.

## 2 Implementation of interior point network flow methods

In this section we present issues related to efficient implementation of interior point methods for solving network flow problems. We present a brief review of the research literature relating to interior point network flow methods and discuss the computational complexity of interior point network flow methods.

### 2.1 Literature review

After the introduction of Karmarkar's algorithm in 1984, many groups of researchers hurried to implement the method. One of the first implementations is described in Adler *et al.* [1, 2]. They implemented what is now known as the dual affine scaling algorithm using direct factorization for solving, at each iteration, a direction-finding system of linear equations, of the form

$$ADA^{\top} u = t, \tag{2.1}$$

where $A$ is an $m \times n$ constraint matrix, $D$ is a diagonal $n \times n$ scaling matrix, and $u$ and $t$ are $m$-vectors. Though that implementation was shown to compare favourably with an efficient implementation of the simplex method [44] on general linear programming problems from the test problem set NETLIB [20], the code was orders of magnitude slower than the network simplex implementation NETFLO [39] on small assignment problems.

At the same time, Aronson *et al.* [5] implemented PTANET, a variant of Karmarkar's projective transformation algorithm, for solving network flow problems. The code, an implementation of a primal projective algorithm, used the LSQR subroutine of Paige and Saunders [46] to solve (2.1). Their implementation took advantage of the network structure to build the $Q$-factor in the LSQR method. No preconditioners were used, possibly explaining the large number of LSQR iterations taken. For example, a $20 \times 20$ linear assignment problem took 6931 LSQR iterations in 7 interior point iterations before stalling at 93% of the optimum. They limited their computational testing to small dense assignment problems of size up to $80 \times 80$. PTANET was compared with NETFLO and IBM MPSX/370 and was found to be hundreds of times slower than NETFLO and tens of times slower than IBM MPSX/370.

An implementation of the dual affine scaling method with centering, called the approximate dual projective algorithm (ADP) was described in 1988 at the Tokyo Symposium of the Mathematical Programming Society by Karmarkar and Ramakrishnan [38]. Their code had a key ingredient for solving network flow problems. They introduced a preconditioned conjugate gradient algorithm for solving (2.1). ADP obtained a feasible dual solution in an initial phase, an $\epsilon$-optimal dual solution in a second phase, terminating in a third phase, with an $\epsilon$-optimal primal-dual solution. They reported computational results on maximum flow, minimum cost network flow problems on grid graphs and compared their results only with the general linear programming code MINOS [44]. On a $100 \times 100$ grid graph, ADP was 123 times faster than MINOS and on a $200 \times 200$ graph, ADP was 252 times faster.

At the same time, a computational study [52] compared the dual affine scaling implementation in AT&T's KORBX mathematical programming system [12] with the implementation of Bertsekas' relaxation algorithm, RELAXT-3 [10], on a set of problems generated with NETGEN [40]. RELAXT-3 was up to two orders of magnitude faster than KORBX. The KORBX implementation solved the system in (2.1) using direct factorization and took no advantage of the structure of the network.

Armacost and Mehrotra [4] report a comparison of their implementations of a network simplex method and the DAS algorithm using direct factorization for solving (2.1). Their DAS implementation produces an approximate dual optimal solution $y^*$, stopping when $b^\top(y^{k+1} - y^k)/\max(1, |b^\top y^k|) \leq 10^{-6}$. They generate assignment, transportation and minimum cost network flow problems using NETGEN. On problems having 100 to 400 nodes and 150 to 1,600 arcs, the network simplex code was 2 to 20 times faster than the interior point code. On dense 1,000 node problems, the simplex code was up to 80 times faster, whilst on 1,000 node sparse problems, it was at most 3 times faster. They conclude that the results by Karmarkar and Ramakrishnan on grid graphs suggest that the preconditioned conjugate gradient approach for solving (2.1) may be the most promising way to proceed.

Yeh [66] proposed and implemented the reduced dual affine scaling (RDAS) interior point algorithm for solving assignment and transportation problems. This method uses a reduced working submatrix of the constraint matrix to compute the direction. System (2.1) is solved via a preconditioned conjugate gradient method with a diagonal preconditioner $M$ of the form $M = \text{diag}(AGA^\top)$. The method switches to a network simplex algorithm after a feasible primal basis is identified. In a comparative study with Yeh's network simplex implementation on assignment problems of dimension $n = 200$, 300, 500 and 2,500, the simplex-to-RDAS CPU time ratios were 0.53, 1.01, 1.39 and 3.57 respectively, indicating a relative improvement in the interior point method with the dimension of the problem.

Resende and Veiga [55] report computational results on a modification of the implementation of the dual affine scaling algorithm described in [1, 2] where the optimal primal basic solution $\tilde{x}$ is obtained by solving the linear system $A_{\text{MST}}\tilde{x} = b$, where $A_{\text{MST}}$ is the submatrix of $A$ corresponding to the edges of a maximum weighted spanning tree of $G$, with appropriately defined arc weights. The algorithm stops when

$$c^\top \tilde{x} - b^\top y^k < 1 \quad \text{and} \quad \tilde{x} \geq 0,$$

where $y^k$ is the dual iterate at step $k$. They compare two codes that differ only by how (2.1) is solved: one by direct factorization and the other by a conjugate gradient method using diagonal and maximum weighted spanning tree preconditioners. On a 3,000 × 22,000 assignment problem, they show a 116 speedup factor of the iterative solver with respect to the solver using the direct method. This study further indicated that direct methods may not be appropriate for solving systems of type (2.1) that occur in network flow problems.

Prior to the paper by Resende and Veiga [57], most studies that compared interior point methods with network simplex methods used their own implementations of the simplex method. In [57] an implementation of the DAS algorithm for bipartite uncapacitated networks was compared with NETFLO and RELAX on large assignment problems. The interior point code used was the one based on the preconditioned conjugate gradient in [55]. The matrix–vector multiplication of the conjugate gradient method was implemented in parallel on an Alliant FX/80 computer. On a 500 × 500 assignment problem, the CPU time observed with eight processors was 6.5 times faster than on a single processor for the conjugate gradient and 5.5 times faster for the overall DAS algorithm. The parallel interior point code was compared with NETFLO and RELAX, both running on a single processor on the same machine. The relative performance of the DAS code with respect to the other codes improved with the size of the network problem solved. The break-even points were 20,000 node networks for NETFLO and 50,000 node networks for RELAX.

Kaliski and Ye [33] implemented a build-up variant of the dual potential reduction algorithm for solving transportation problems with $n \gg m$. Their implementation used the preconditioned conjugate gradient with the maximum weighted spanning tree preconditioner to solve (2.1). They differ from previous implementations in the way they compute the spanning tree, using a hybrid

QuickSort procedure. To speed up the convergence of the duality gap, they use an indicator function to guess a dual face, and project the dual interior solution on the face and test optimality. On small problems, with 50 to 2,000 nodes and arcs, they report that NETFLO was 2 to 10 times faster than their code.

In 1991, the first year-long DIMACS algorithm implementation challenge [31] was initiated. This challenge focused on network flow and matching algorithms. A large set of network flow test problems was proposed and collected for the challenge. All codes solved a subset of those instances. Three entries in the network flow challenge involved interior point approaches.

The first approach, by Joshi, Goldstein and Vaidya [32], described the implementation and testing of a primal path-following interior point algorithm using a preconditioned conjugate gradient with the maximum weighted spanning tree preconditioner to solve (2.1). Their code was compared with NETFLO and RELAXT-3. Though the performance of this code, relative to NETLFO and RELAXT-3, was not competitive in most instances, it was shown to improve with problem size. The largest speedups with respect to NETFLO and RELAXT-3 were 2.4 and 25, respectively.

The second approach, by Ramakrishnan, Karmarkar and Kamath [53], extended the ADP code by adding an initial solution module and a stopping heuristic module for handling assignment problems. The code, called ADP/A, was used to solve assignment problems having up to 32,768 nodes and was compared with the code AUCTION of Bertsekas and Castañon [9]. Ramakrishnan, Karmarkar and Kamath report that, unlike AUCTION, their code has little CPU-time sensitivity to problem instances. Because of its high CPU-time sensitivity, AUCTION is much faster than ADP/A on some problem classes, whilst being much slower on others. That code was also used to solve large dense assignment problems to verify conjectures on the value of the optimal assignments of randomly distributed cost data [47]. Computational results on dense assignment problems with up to 20,000 nodes are reported.

The third approach, by Resende and Veiga [56], describes DLNET, an implementation of a dual affine scaling algorithm using a diagonal and maximum weighted spanning tree preconditioned conjugate gradient algorithm to solve (2.1). That code was compared with NETFLO, RELAXT-3, CPLEX and CPLEX NETOPT on 250 instances in 13 problem classes, having from 256 to 262,144 nodes. Compared to RELAXT-3 and NETFLO, DLNET had the slowest CPU-time growth rates in 11 of the 13 classes. It was up to 152 times faster than RELAXT-3 and 18 times faster than NETFLO. The study also pointed out that the variance in the CPU times for fixed size problems was smallest for the interior point code. The study also compared CPLEX and CPLEX NETOPT with NETFLO and showed that overall NETOPT was twice as fast as NETFLO, but was slower in 4 of the 13 problem classes. NETOPT was up to 180 times faster than the CPLEX primal simplex algorithm.

A new indicator to identify the optimal face of network linear programs was introduced in [54] and incorporated in DLNET. Using this new indicator DLNET was able to find complementary primal-dual integer optimal solutions for all of the DIMACS instances. The DLNET version used in [56] found primal-optimal only solutions in 4% of the instances.

Using a new preconditioner (incomplete $QR$ decomposition), Portugal *et al.* [49] implemented a preconditioned conjugate gradient based DAS, primal-dual, and predictor–corrector interior point algorithms for solving linear transportation problems. On this class of problems, the new preconditioner outperforms both the diagonal and maximum weighted spanning tree preconditioners. They limit their computational experiments to instances with up to 2,000 nodes, and conclude that the primal-dual and predictor–corrector algorithms are more appropriate for well-scaled transportation problems, while DAS is more appropriate for well-scaled assignment problems.

In [50], the infeasible-primal feasible-dual interior point algorithm for linear programming was presented and its implementation to solve network flow problems described. Using a standard primal-dual indicator function [21, 18], along with the stopping criterion, based on solving a maximum flow problem, the code PDNET was shown to outperform DLNET, CPLEX NETOPT (ver. 3.0) [13] and CS (ver. 1.2) [25] on most classes of problems tested.

### 2.2 Complexity issues

Long before the discovery of polynomial time algorithms for linear programming, strongly polynomial algorithms for many network flow problems were known [3]. The best-known result for minimum cost network flow, due to Orlin [45], is $O(n^2 \log m + mn(\log m)^2)$, where, as before, $m$ and $n$ denote the number of nodes and arcs, respectively.

Recently, Vavasis and Ye [65] proposed a strongly polynomial interior point algorithm for minimum cost network flow problems having complexity $O(n^{6.5} \log n)$. Contrasted with the result of Tardos [61], which requires that all data be integer, the new algorithm generalizes to real-number data.

Prior to the result of Vavasis and Ye, the best-known complexity result for minimum cost network interior point methods, due to Vaidya [63], was $O(m^2 \sqrt{n} \log(m\Gamma))$, where it is assumed that all costs and upper and lower bounds are integers whose magnitudes are bounded above by $\Gamma$.

With regard to the complexity of multicommodity flow problems, the best-known algorithm is the interior point method analysed by Kamath and Palmon [35]. One variant of this interior point algorithm finds an exact solution in $O(k^{2.5}n^{1.5}mL\log(nDU))$, where $L = km + n$, $D$ is the largest demand, $U$ is the largest capacity and $k$ is the number of commodities. This variant uses the conjugate gradient method. A second variant, which has a complexity bound of $O((k^{0.5}n^3 + km^{1.5}n^{1.5})L\log(mDU))$, can be improved to $O((k^{0.5}n^{2.7} + km^{1.2}n^{1.5} + kn^{2.5})L\log(mDU))$, using fast matrix multiplication. This variant uses matrix inversion and rank one updates.

Previously, the fastest known algorithm for multicommodity flow was also an interior point algorithm, due to Kapoor and Vaidya [36, 63]. This algorithm achieves a complexity bound of $O(k^3n^{0.5}m^3L\log(mDU))$ and reduces to $O(k^{2.5}n^{0.5}m^2L\log(mDU))$ using fast matrix multiplication. It should be noted, however, that algorithms using fast matrix multiplication are not of practical interest, unless the dimensions are very large.

## 3 Components of interior point network flow methods

As seen in the literature review, since Karmarkar's breakthrough in 1984, many variants of his algorithm, including dual affine scaling, with and without centering, reduced dual affine scaling, primal path-following (method of centers), primal-dual path following, predictor–corrector primal-dual path-following, and infeasible primal-dual path-following, have been used to solve network flow problems. Though these algorithms are, in some sense, different, they share many of the same computational requirements. The key ingredients for efficient implementation of these algorithms are:

(1) The solution of the linear system $ADA^{\top}u = t$, where $D$ is a diagonal $n \times n$ scaling matrix and $u$ and $t$ are $m$-vectors. This requires an iterative algorithm for computing approximate directions, preconditioners, stopping criteria for the iterative algorithm, etc.

(2) The recovery of the desired optimal solution. This may depend on how the problem is presented (integer data or real data), and what type of solution is required (fractional or integer solution, $\epsilon$-optimal or exact solution, primal optimal or primal-dual optimal solution, etc.).

In this section we present in detail these components, illustrating their implementation in the dual affine scaling network flow algorithm DLNET of Resende and Veiga [56]. To illustrate the use of some of the techniques described we present plots of several quantities obtained by running DLNET on a 16,384 node, 131,072 arc minimum cost network flow problem. We follow closely the presentations in [49, 54, 56].

### 3.1 The dual affine scaling algorithm

The dual affine scaling (DAS) algorithm [6, 15, 60, 64] was one of the first interior point methods to be shown to be able to compete computationally with the simplex method [1, 2]. As before, let $A$ be an $m \times n$ matrix, $c$, $u$, and $x$ be $n$-dimensional vectors and $b$ an $m$-dimensional vector. The DAS algorithm solves the linear program

$$\min \{c^{\top} x \mid Ax = b,\ 0 \leq x \leq u\}$$

indirectly, by solving its dual

$$\max \{b^{\top} y - u^{\top} z \mid A^{\top} y - z + s = c,\ z \geq 0, s \geq 0\}, \tag{3.1}$$

where $z$ and $s$ are $n$-dimensional vectors and $y$ is a $m$-dimensional vector. The algorithm starts with an initial interior solution $\{y^0, z^0, s^0\}$ such that

$$A^{\top} y^0 - z^0 + s^0 = c,\ z^0 > 0,\ s^0 > 0,$$

and iterates according to

$$\{y^{k+1}, z^{k+1}, s^{k+1}\} = \{y^k, z^k, s^k\} + \alpha\, \{\Delta y, \Delta z, \Delta s\},$$

where the search directions $\Delta y, \Delta z$ and $\Delta s$ satisfy

$$\begin{aligned} A(Z_k^2 + S_k^2)^{-1} A^\top \Delta y &= b - A Z_k^2 (Z_k^2 + S_k^2)^{-1} u, \\ \Delta z &= Z_k^2 (Z_k^2 + S_k^2)^{-1} (A^\top \Delta y - S_k^2 u), \\ \Delta s &= \Delta z - A^\top \Delta y, \end{aligned}$$

where

$$Z_k = \mathrm{diag}(z_1^k, \ldots, z_n^k) \text{ and } S_k = \mathrm{diag}(s_1^k, \ldots, s_n^k)$$

and $\alpha$ is such that $z^{k+1} > 0$ and $s^{k+1} > 0$, i.e. $\alpha = \gamma \times \min\{\alpha_z, \alpha_s\}$, where $0 < \gamma < 1$ and

$$\alpha_z = \min\{-z_i^k/(\Delta z)_i \mid (\Delta z)_i < 0,\ i = 1, \ldots, n\},$$

$$\alpha_s = \min\{-s_i^k/(\Delta s)_i \mid (\Delta s)_i < 0,\ i = 1, \ldots, n\}.$$

The dual problem (3.1) has a readily available initial interior point solution

$$\begin{aligned} y_i^0 &= 0, & i &= 1, \ldots, n, \\ s_i^0 &= c_i + \lambda, & i &= 1, \ldots, n, \\ z_i^0 &= \lambda, & i &= 1, \ldots, n, \end{aligned}$$

where $\lambda$ is a scalar such that $\lambda > 0$ and $\lambda > -c_i$, $i = 1, \ldots, n$. The algorithm described above has two important parameters, $\gamma$ and $\lambda$. For example, in DLNET, $\gamma = 0.95$ and $\lambda = 2\, \|c\|_2$.

Figure 1 shows the progress of the interior point dual objective function solving the 16,384 node, 131,072 arc example. The optimal objective function value is 10,631,080,962 and was found at iteration 69. The objective function values prior to iteration 21 are all negative and are not plotted.

### 3.2 Computing the direction

The computational efficiency of interior point network flow methods relies heavily on a preconditioned conjugate gradient algorithm to solve the direction finding system at each iteration. The preconditioned conjugate gradient algorithm is used to solve

$$M^{-1}(A D_k A^\top) \Delta y = M^{-1} \bar{b}, \tag{3.2}$$

where $M$ is a positive definite matrix and, in the case of the DAS algorithm, $\bar{b} = b - A Z_k^2 D_k$, and $D_k = (Z_k^2 + S_k^2)^{-1}$ is a diagonal matrix of positive elements. The objective is to make the preconditioned matrix

$$M^{-1}(A D_k A^\top) \tag{3.3}$$

less ill-conditioned than $A D_k A^\top$, and improve the convergence of the conjugate gradient algorithm.

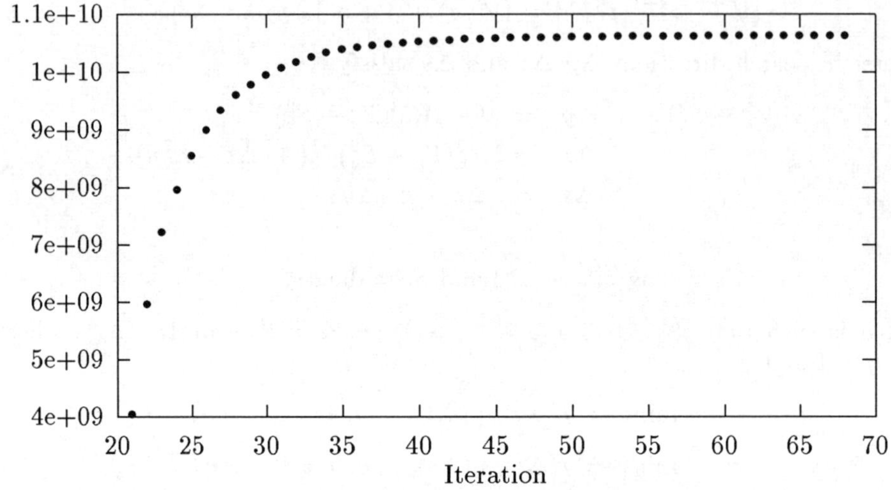


FIG. 1. Interior point dual objective function (positive values)

The pseudocode for the preconditioned conjugate gradient algorithm is presented in Fig. 2. The computationally intensive steps in the preconditioned conjugate gradient algorithm are lines 3, 7 and 11 of the pseudocode. These lines correspond to a matrix–vector multiplication (7) and solving linear systems of equations (3 and 11). Line 3 is computed once and lines 7 and 11 are computed once every conjugate gradient iteration. The matrix–vector multiplications are of the form $AD_kA^\top p_i$, carried out without forming $AD_kA^\top$ explicitly. One way to compute the above matrix–vector multiplication is to decompose it into three sparse matrix–vector multiplications. Let

$$\zeta' = A^\top p_i \quad \text{and} \quad \zeta'' = D_k\zeta'.$$

Then

$$(A\,(D_k\,(A^\top p_i))) = A\zeta''.$$

The complexity of this matrix–vector multiplication is $O(n)$, involving $n$ additions, $2n$ subtractions and $n$ floating point multiplications.

The preconditioned residual is computed in lines 3 and 11 when the system of linear equations

$$Mz_{i+1} = r_{i+1} \tag{3.4}$$

is solved, where $M$ is a positive definite matrix. An efficient implementation requires a preconditioner that can make (3.4) easy to solve. On the other hand, one needs a preconditioner that makes (3.3) well conditioned. In the next subsection, we show several preconditioners that satisfy, to some extent, these two

**procedure pcg**$(A, D_k, \bar{b}, \epsilon_{cg}, \Delta y)$
    $\Delta y_0 := 0;$
    $r_0 := \bar{b};$
    $z_0 := M^{-1} r_0;$
    $p_0 := z_0;$
    $i := 0;$
    **do** stopping criterion not satisfied $\rightarrow$
        $q_i := A D_k A^\top p_i;$
        $\alpha_i := z_i^\top r_i / p_i^\top q_i;$
        $\Delta y_{i+1} := \Delta y_i + \alpha_i p_i;$
        $r_{i+1} := r_i - \alpha_i q_i;$
        $z_{i+1} := M^{-1} r_{i+1};$
        $\beta_i := z_{i+1}^\top r_{i+1} / z_i^\top r_i;$
        $p_{i+1} := z_{i+1} + \beta_i p_i;$
        $i := i + 1$
    **od**;
    $\Delta y := \Delta y_i$
**end pcg**;

FIG. 2. The preconditioned conjugate gradient algorithm

criteria.

To determine when the approximate direction $\Delta y_i$ produced by the conjugate gradient algorithm is satisfactory, one can compute the angle $\theta$ between $(AD_kA^\top)\Delta y_i$ and $\bar{b}$ and stop when $|1 - \cos\theta| < \epsilon_{cos}$, where $\epsilon_{cos}$ is some small tolerance. In practice, one can initially use $\epsilon_{cos} = 10^{-3}$ and tighten the tolerance as the interior point iterations proceed, as $\epsilon_{cos} = \epsilon_{cos} \times 0.95$. The exact computation of

$$\cos\theta = \frac{|\bar{b}^\top (AD_kA^\top)\Delta y_i|}{\|\bar{b}\|_2 \cdot \|(AD_kA^\top)\Delta y_i\|_2}$$

has the complexity of one conjugate gradient iteration and is therefore expensive if computed at each conjugate gradient iteration. One way to proceed is to compute the cosine every $l_{cos}$ conjugate gradient iterations. A more efficient procedure follows from the observation that $(AD_kA^\top)\Delta y_i$ is approximately equal to $\bar{b} - r_i$, where $r_i$ is the estimate of the residual at the $i$-th conjugate gradient iteration. Using this approximation, the cosine can be estimated by

$$\cos\theta = \frac{|\bar{b}^\top (\bar{b} - r_i)|}{\|\bar{b}\|_2 \cdot \|(\bar{b} - r_i)\|_2}.$$

Since, in practice, the conjugate gradient method finds good directions in a few iterations, this estimate has been shown to be effective and can be computed at each conjugate gradient iteration. Figure 3 shows the accuracy of the direction computed by the preconditioned conjugate gradient algorithm on the 16,384

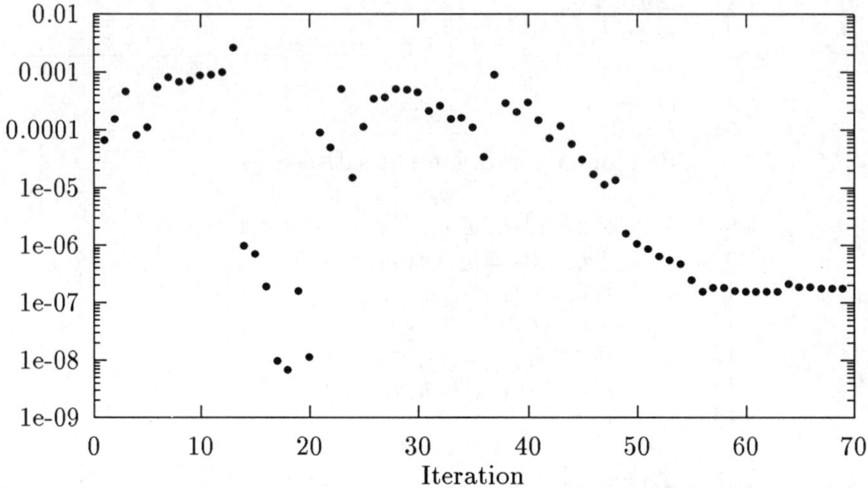


FIG. 3. $|1 - \cos\theta|$ value at termination of conjugate gradient

node, 131,072 arc example. The tolerance $\epsilon_{cos}$ was set to $10^{-3}$ on all interior point iterations, except iteration 13. On that iteration, the conjugate gradient reached the limit of 256 ($= 2 \times \sqrt{16,384}$) iterations without attaining the desired precision. Stopping was tested every 5 conjugate gradient iterations, and because of that the $|1-\cos\theta|$ value was always smaller than $10^{-3}$. The diagonal preconditioner was used from iteration 1 to iteration 13. From iteration 14 on the spanning tree preconditioner took effect.

### 3.3 Network preconditioners for conjugate gradient method

A useful preconditioner for the conjugate gradient algorithm must be one that allows the efficient solution of (3.4), whilst at the same time causing the number of conjugate gradient iterations to be small. Four preconditioners have been found useful in conjugate gradient based interior point network flow methods: diagonal, maximum weighted spanning tree, incomplete $QR$ decomposition and the Karmarkar–Ramakrishnan preconditioner for general linear programming.

A diagonal matrix constitutes the most straightforward preconditioner used in conjunction with the conjugate gradient algorithm [26]. They are simple to compute, taking $O(n)$ double precision operations, and can be effective [55, 57, 66]. In the diagonal preconditioner, $M = \text{diag}\,(AD_kA^\top)$, and the preconditioned residue systems of lines 3 and 11 of the conjugate gradient pseudocode in Fig. 2 can each be solved in $O(m)$ double precision divisions.

A preconditioner that is observed to improve with the number of interior point iterations is the maximum weighted spanning tree preconditioner. Since the underlying graph $G$ is not necessarily connected, one can identify a maximal forest using as weights the diagonal elements of the current scaling matrix,

$$w = D_k e, \tag{3.5}$$

where $e$ is a unit $n$-vector. In practice, Kruskal's and Prim's algorithms have been used to compute the maximal forest. Kruskal's algorithm, implemented with the data structures in [62] has been applied to arcs, ordered approximately with a bucket sort [32, 56], or exactly using a hybrid QuickSort [33]. Prim's algorithm is implemented in [49] using the data structures presented in [3].

At the $k$-th interior point iteration, let $\mathcal{S}_k$ be the submatrix of $A$ with columns corresponding to arcs in the maximal forest, $t_1, \ldots, t_q$. The preconditioner can be written as

$$M = \mathcal{S}_k \mathcal{D}_k \mathcal{S}_k^{\top},$$

where, for example, in the DAS algorithm

$$\mathcal{D}_k = \operatorname{diag}(1/z_{t_1}^2 + 1/s_{t_1}^2, \ldots, 1/z_{t_q}^2 + 1/s_{t_q}^2).$$

For simplicity of notation, we include in $\mathcal{S}_k$ the linearly dependent rows corresponding to the redundant flow conservation constraints. At each conjugate gradient iteration, the preconditioned residue system

$$(\mathcal{S}_k \mathcal{D}_k \mathcal{S}_k^{\top}) z_{i+1} = r_{i+1} \tag{3.6}$$

is solved with the variables corresponding to redundant constraints set to zero. As with the diagonal preconditioner, (3.6) can be solved in $O(m)$ time, as the system coefficient matrix can be ordered into a block triangular form. Figure 4 shows the number of iterations taken by the preconditioned conjugate gradient during each interior point iteration on the 16,384 node, 131,072 arc example. The diagonal preconditioner was used from iteration 1 to iteration 13. From iteration 14 on the spanning tree preconditioner took effect. Besides checking the angle every five iterations the conjugate gradient was also stopped by a residual stopping criterion. This occurred in iterations 16 and 19.

Portugal *et al.* [49] introduced a preconditioner based on an incomplete $QR$ decomposition (IQRD) for use in interior point methods to solve transportation problems. They showed empirically, for that class of problems, that this preconditioner mimics the diagonal preconditioner during the initial iterations of the interior point method, and the spanning tree preconditioner in the final interior point method iterations, while causing the conjugate gradient method to take fewer iterations than either method during the intermediate iterations. In [50], the use of this preconditioner is extended to general minimum cost network flow problems. In the following discussion, we omit the iteration index $k$ from the notation for the sake of simplicity. Let $\bar{\mathcal{T}} = \{1, 2, \ldots, n\} \setminus \mathcal{T}$ be the index set of the arcs not in the computed maximal spanning tree, and let

$$D = \begin{bmatrix} D_{\mathcal{T}} & \\ & D_{\bar{\mathcal{T}}} \end{bmatrix},$$

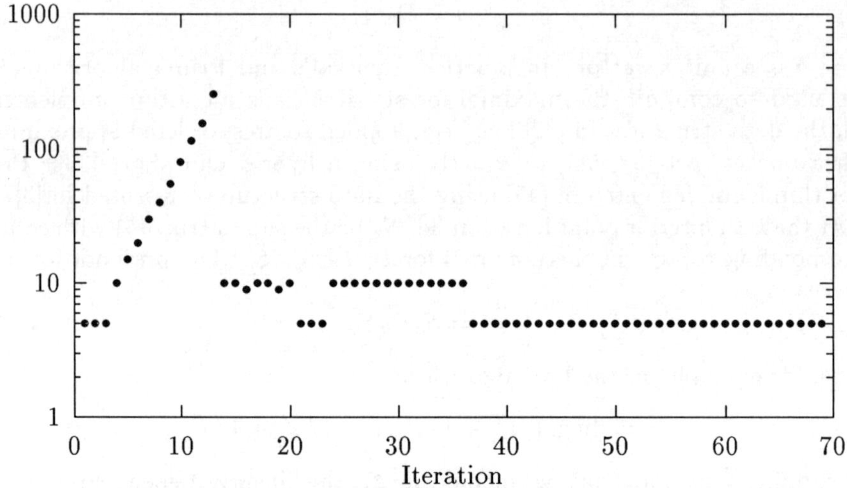


FIG. 4. Conjugate gradient iterations

where $D_{\mathcal{T}} \in \mathbb{R}^{q \times q}$ is the diagonal matrix with the arc weights of the maximal spanning tree and $D_{\bar{\mathcal{T}}} \in \mathbb{R}^{(n-q)\times(n-q)}$ is the diagonal matrix with weights of the arcs not in the maximal spanning tree. Then

$$\begin{aligned} ADA^\top &= \left[\begin{array}{cc} A_{\mathcal{T}} & A_{\bar{\mathcal{T}}} \end{array}\right] \left[\begin{array}{cc} D_{\mathcal{T}} & \\ & D_{\bar{\mathcal{T}}} \end{array}\right] \left[\begin{array}{c} A_{\mathcal{T}}^\top \\ A_{\bar{\mathcal{T}}}^\top \end{array}\right] \\ &= \left[\begin{array}{cc} A_{\mathcal{T}} D_{\mathcal{T}}^{\frac{1}{2}} & A_{\bar{\mathcal{T}}} D_{\bar{\mathcal{T}}}^{\frac{1}{2}} \end{array}\right] \left[\begin{array}{c} D_{\mathcal{T}}^{\frac{1}{2}} A_{\mathcal{T}}^\top \\ D_{\bar{\mathcal{T}}}^{\frac{1}{2}} A_{\bar{\mathcal{T}}}^\top \end{array}\right]. \end{aligned}$$

The Cholesky factorization of $ADA^\top$ can be found by simply computing the $QR$ factorization of

$$\bar{A} = \left[\begin{array}{c} D_{\mathcal{T}}^{\frac{1}{2}} A_{\mathcal{T}}^\top \\ D_{\bar{\mathcal{T}}}^{\frac{1}{2}} A_{\bar{\mathcal{T}}}^\top \end{array}\right]. \tag{3.7}$$

In fact, if $Q\bar{A} = R$, then

$$ADA^\top = \bar{A}^\top \bar{A} = R^\top Q^\top Q R = R^\top R.$$

The computation of the $QR$ factorization is not recommended here, since besides being more expensive than a Cholesky factorization, it also destroys the sparsity of the matrix $\bar{A}$. Instead, Portugal *et al.* [49] propose an *incomplete* $QR$ decomposition of $\bar{A}$. Applying Givens rotations [22] to $\bar{A}$, using the diagonal

```
procedure iqrd(T, T̄, N, D, 𝒟, F)
    do i ∈ N \ {r} →
        j = ρ_i;
        if (i, j) ∈ A_T → 𝒟_ii = D_ij fi;
        if (j, i) ∈ A_T → 𝒟_ii = D_ji fi;
    od;
    do (i, j) ∈ A_T̄ →
        if i ∈ N \ {r} → 𝒟_ii = 𝒟_ii + D_ij fi;
        if j ∈ N \ {r} → 𝒟_jj = 𝒟_jj + D_ij fi;
    od;
    do i ∈ N \ {r} →
        j = ρ_i;
        if j ∈ N \ {r} →
            if (i, j) ∈ A_T → F_ij = D_ij/𝒟_ii fi;
            if (j, i) ∈ A_T → F_ji = D_ji/𝒟_ii fi;
        fi;
    od;
end iqrd;
```

FIG. 5. Computing the $F$ and $\mathcal{D}$ matrices in IQRD

elements of $D^{\frac{1}{2}}_{\mathcal{T}} A^{\top}_{\mathcal{T}}$, the elements of $D^{\frac{1}{2}}_{\bar{\mathcal{T}}} A^{\top}_{\bar{\mathcal{T}}}$ become null. No fill-in is incurred in this factorization. See [49] for an example illustrating this procedure. After the factorization, we have the preconditioner

$$M = F\mathcal{D}F^{\top},$$

where $F$ is a matrix with a diagonal of ones that can be reordered to triangular form, and $\mathcal{D}$ is a diagonal matrix with positive elements.

To avoid square root operations, $\mathcal{D}$ and $F$ are obtained without explicitly computing $\mathcal{D}^{\frac{1}{2}}F^{\top}$. Suppose that the maximum spanning tree is rooted at node $r$, corresponding to the flow conservation equation that has been removed from the formulation. Furthermore, let $\mathcal{A}_{\mathcal{T}}$ denote the subset of arcs belonging to the tree and let $\rho_i$ represent the predecessor of node $i$ in the tree. The procedure used to compute the non-zero elements of $\mathcal{D}$ and the off-diagonal non-zero elements of $F$ is presented in the pseudocode in Fig. 5.

The computation of the preconditioned residual with $F\mathcal{D}F^{\top}$ requires $O(m)$ divisions, multiplications and subtractions, since $\mathcal{D}$ is a diagonal matrix and $F$ can be permuted into a triangular matrix with diagonal elements equal to one. The construction of $F$ and $\mathcal{D}$, which constitute the preconditioner, requires $O(n)$ additions and $O(m)$ divisions.

In practice, the diagonal preconditioner is effective during the initial iterations of the DAS algorithm. As the DAS iterations progress, the spanning tree preconditioner is more effective as it becomes a better approximation of the matrix $AD_kA^{\top}$. Arguments as to why this preconditioner is effective are given in

[32, 49]. The DLNET implementation begins with the diagonal preconditioner and monitors the number of iterations required by the conjugate gradient algorithm. When the conjugate gradient takes more than $\beta\sqrt{m}$ iterations, where $\beta > 0$, DLNET switches to the spanning tree preconditioner. Upper and lower limits to the number of DAS iterations using a diagonal preconditioned conjugate gradient are specified.

Another preconditioner used in an interior point implementation is the one for general linear programming, developed by Karmarkar and Ramakrishnan and used in [38, 53]. This preconditioner is based on a dynamic scheme to drop elements of the original scaled constraint matrix $DA$, as well as from the factors of the matrix $ADA^\top$ of the linear system, and uses the incomplete Cholesky factors as the preconditioner. Because of the way elements are dropped, this preconditioner mimics the diagonal preconditioner in the initial iterations and the tree preconditioner in the final iterations of the interior point algorithm.

### 3.4 Identifying the optimal partition

One way to stop an interior point algorithm before the (fractional) interior point iterates converge is to estimate (or guess) the optimal partition of arcs at each iteration, attempt to recover the flow from the partition and, if a feasible flow is produced, test if that flow is optimal. In the discussion that follows we describe a strategy to partition the set of arcs in the dual affine scaling algorithm. The discussion follows [54] closely, using a dual affine scaling method for uncapacitated networks to illustrate the procedure.

Let $A \in \mathbb{R}^{m\times n}$, $c, x, s \in \mathbb{R}^n$ and $b, y \in \mathbb{R}^m$. Consider the linear programming problem

$$\begin{aligned} \text{minimize} \quad & c^\top x, \\ \text{subject to} \quad & Ax = b, \quad x \geq 0, \end{aligned} \tag{3.8}$$

and its dual

$$\begin{aligned} \text{maximize} \quad & b^\top y, \\ \text{subject to} \quad & A^\top y + s = c, \quad s \geq 0. \end{aligned} \tag{3.9}$$

The dual affine scaling algorithm starts with an initial dual solution $y^0 \in \{y : s = c - A^\top y > 0\}$ and obtains iterate $y^{k+1}$ from $y^k$ according to $y^{k+1} = y^k + \alpha^k d_y^k$, where the search direction $d_y$ is $d_y^k = (AD_k^{-2}A^\top)^{-1}b$ and $D_k = \text{diag}(s_1^k, \ldots, s_n^k)$. A step moving a fraction $\gamma$ of the way to the boundary of the feasible region is taken at each iteration, namely,

$$\alpha^k = \gamma \times \min\{-s_i^k/(d_s^k)_i \ : \ (d_s^k)_i < 0, \ i = 1, \ldots, n\}, \tag{3.10}$$

where $d_s^k = -A^\top d_y^k$ is a unit displacement vector in the space of slack variables. At each iteration, a tentative primal solution is computed by $x^k = D_k^{-2}A^\top(AD_k^{-2}A^\top)^{-1}b$. The set of optimal solutions is referred to as the *optimal face*. We use the index set $N_*$ for the always-active index set on the optimal face of the primal, and $B_*$ for its complement. It is well known that $B_*$ is the always-active index set on the optimal face of the dual, and $N_*$ is its complement. An *indicator* is a quantity to detect whether an index belongs to $N_*$ or $B_*$. We next describe three indicators that can be implemented in the DAS algorithm. For pointers to other indicators see [18].

Under a very weak condition, the iterative sequence of the DAS algorithm converges to a relative interior point of a face on which the objective function is constant, i.e. the sequence $\{y^k\}$ converges to an interior point of a face on which the objective function is constant. Let $B$ be the always-active index set on the face and $N$ be its complement, and let $b^\infty$ be the limiting objective function value. There exists a constant $C_0 > 0$ such that

$$\limsup_{k\to\infty} \frac{s_i^k}{b^\infty - b^\top y^k} \leq C_0 \tag{3.11}$$

for all $i \in B$, while

$$\frac{s_i^k}{b^\infty - b^\top y^k} \tag{3.12}$$

diverges to infinity for all $i \in N$. Denote by $s^\infty$ the limiting slack vector. Then $s_N^\infty > 0$ and $s_B^\infty = 0$. The vector

$$u^k \equiv \frac{(D^k)^{-1} d_s^k}{b^\infty - b^\top y^k} = \frac{D^k x^k}{b^\infty - b^\top y^k} \tag{3.13}$$

plays an important role, since

$$\lim_{k\to\infty} (u^k)^\top e = \lim_{k\to\infty} \frac{(s^k)^\top x^k}{b^\infty - b^\top y^k} = 1. \tag{3.14}$$

Consequently, in the limit $b^\infty - b^\top y^k$ can be estimated by $(s^k)^\top x^k$ asymptotically, and (3.11) can be stated as

$$\limsup_{k\to\infty} \frac{s_i^k}{(s^k)^\top x^k} \leq C_0.$$

Then, if $i \in B$, for any $\beta$ such that $0 < \beta < 1$,

$$\limsup_{k\to\infty} \frac{s_i^k}{((s^k)^\top x^k)^\beta} = 0,$$

since $((s^k)^\top x^k)^\beta$ converges to zero at a slower rate than $((s^k)^\top x^k)$ for any $\beta$ such that $0 < \beta < 1$. Therefore, if $\beta = 1/2$, the following indicator has the property that $\lim_{k\to\infty} N^k = N_*$.

**Indicator 1:** Let $C_1 > 0$ be any constant, and define

$$N^k = \{i \in E \ : \ s_i^k \leq C_1 \sqrt{(s^k)^\top x^k}\}. \tag{3.15}$$

This indicator is available under very weak assumptions, so it can be used to detect $B_*$ and $N_*$ without any substantial restriction on stepsize. On the other hand, it gives the correct partition only if the limit point $y^\infty$ happens to be a relative interior point of the optimal face of the dual and thus lacks a firm theoretical justification. However, since we know by experience that $y^\infty$ usually lies in the relative interior of the optimal face, we may expect that it should work well in practice. Another potential problem with this indicator is that it is not scaling invariant, so that it will behave differently if the scaling of the problem is changed.

Now we assume that the stepsize is asymptotically less than, or equal to, 2/3. Then the limiting point exists in the interior of the optimal face and $b^\infty$ is the optimal value. Specifically, $\{y^k\}$ converges to an interior point of the optimal face of the dual problem, $\{x^k\}$ converges to the analytic center of the optimal face of the primal problem, and $\{b^\top y^k\}$ converges linearly to the optimal value $b^\infty$ asymptotically, where the (asymptotic) reduction rate is exactly $1-\gamma$. Furthermore, one can show that

$$\lim_{k\to\infty} u_i^k = 1/|B_*| \quad \text{for } i \in B_* \tag{3.16}$$

$$\lim_{k\to\infty} u_i^k = 0 \quad \text{otherwise.} \tag{3.17}$$

The vector $u^k$ is not available because the exact optimal value is unknown a priori, but $b^\infty - b^\top y^k$ can be estimated by $(s^k)^\top x^k$ to obtain

$$\lim_{k\to\infty} \frac{s_i^k x_i^k}{(s^k)^\top x^k} = 1/|B_*| \quad \text{for } i \in B_*, \tag{3.18}$$

$$\lim_{k\to\infty} \frac{s_i^k x_i^k}{(s^k)^\top x^k} = 0 \quad \text{otherwise.} \tag{3.19}$$

On the basis of this fact, we have the following procedure to construct $N^k$, which asymptotically coincides with $N_*$:

**Indicator 2:** Let $\delta$ be a constant between 0 and 1. We obtain $N^k$ according to the following procedure:

(1) Sort $g_i^k = s_i^k x_i^k/(s^k)^\top x^k$ according to its order of magnitude. Denote by $i_l$ the index for the $l$-th largest component.

(2) For $p := 1, 2, \ldots$ compare $g_{i_p}$ and $\delta/p$, and let $p^*$ be the first number such that $g_{i_{p^*}} \le \delta/p^*$. Then set

$$N^k = \{i_1, i_2, \ldots, i_{p^*-1}\}. \tag{3.20}$$

To state the third, and most practical indicator, let us turn our attention to the asymptotic behaviour of $s_i^{k+1}/s_i^k$. If $i \in N_*$, then $s_i^k$ converges to a positive value, and hence

$$\lim_{k\to\infty} \frac{s_i^{k+1}}{s_i^k} = 1. \tag{3.21}$$

If $i \in B_*$, $s_i^k$ converges to zero. Since

$$\lim_{k \to \infty} \frac{s_i^k x_i^k}{b^\infty - b^\top y^k} = \frac{1}{|B_*|}, \tag{3.22}$$

$x_i^k$ converges to a positive number, and the objective function reduces with a rate of $1-\gamma$, then

$$\lim_{k \to \infty} \frac{s_i^{k+1}}{s_i^k} = 1 - \gamma, \tag{3.23}$$

which leads to the following indicator:

**Indicator 3:** Take a constant $\eta$ such that $1-\gamma < \eta < 1$. Then let

$$N^k = \{i \; : \; \frac{s_i^{k+1}}{s_i^k} \geq \eta\} \tag{3.24}$$

be defined as the index set. Then $N^k = N_*$ holds asymptotically.

Of the three indicators described here, Indicators 2 and 3 have the firmest theoretical basis. Furthermore, unlike Indicator 1, both are scaling invariant. The above discussion can be easily extended for the case of capacitated network flow problems. DLNET uses Indicator 3 to identify the set of active arcs defining the optimal face by examining the ratio between subsequent iterates of each dual slack. At the optimum, the flow on each arc can be classified as being either at its upper bound, or at its lower bound, or active. From the discussion above, if the flow on arc $i$ converges to its upper bound,

$$\lim_{k \to \infty} s_i^k / s_i^{k-1} = 1 - \gamma \quad \text{and} \quad \lim_{k \to \infty} z_i^k / z_i^{k-1} = 1.$$

If the flow on arc $i$ converges to its lower bound,

$$\lim_{k \to \infty} s_i^k / s_i^{k-1} = 1 \quad \text{and} \quad \lim_{k \to \infty} z_i^k / z_i^{k-1} = 1 - \gamma.$$

If the flow on arc $i$ is active,

$$\lim_{k \to \infty} s_i^k / s_i^{k-1} = 1 - \gamma \text{ and } \lim_{k \to \infty} z_i^k / z_i^{k-1} = 1 - \gamma.$$

From a practical point of view, scale invariance is the most interesting feature of this indicator. An implementable version can use constants which depend only on the stepsize factor $\gamma$. Let $\kappa_0 = 0.7$ and $\kappa_1 = 0.9$. At each iteration of DLNET, the arcs are classified as follows:

- If $s_i^k/s_i^{k-1} < \kappa_0$ and $z_i^k/z_i^{k-1} > \kappa_1$, then arc $i$ is set to its upper bound.
- If $s_i^k/s_i^{k-1} > \kappa_1$ and $z_i^k/z_i^{k-1} < \kappa_0$, then arc $i$ is set to its lower bound.
- Otherwise, arc $i$ is set active, defining the tentative optimal face.

Figures 6, 7 and 8 illustrate the behaviour of the ratios of dual slacks corresponding to arcs converging to a value away from its bounds, to its lower bound, and to its upper bound, respectively. The stepsize fraction used is $\gamma = 2/3$. The

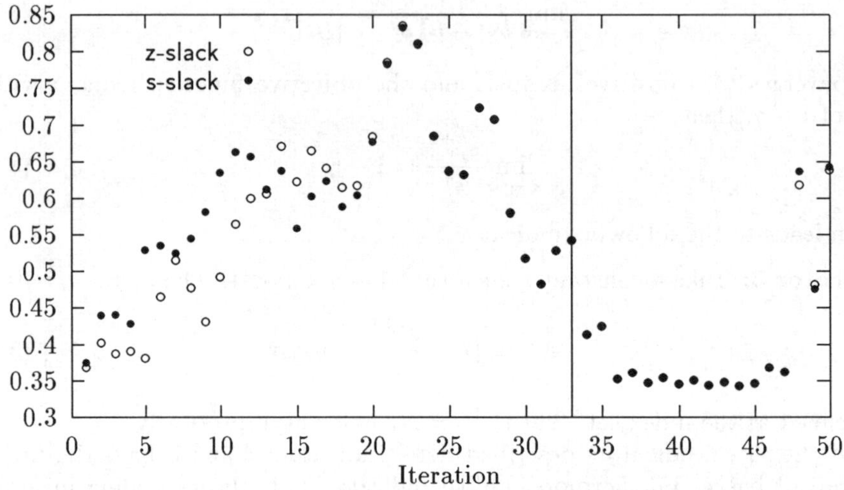


FIG. 6. Dual slack ratio (edge left active)

vertical line in each of the figures shows when the indicator successfully identified the optimal partition. The data points to the right of the vertical line illustrate the behaviour of the indicator as the iterates of the algorithm converge to the optimal solution.

### 3.5 Recovering the optimal flow

The simplex method restricts the sequence of solutions it generates to nodes of the linear programming polytope. Since the matrix $A$ of the network linear program is totally unimodular, when a simplex variant is applied to a network flow problem with integer data, the optimal solution is also integer. On the other hand, an interior point algorithm generates a sequence of interior point (fractional) solutions. Unless the primal optimal solution is unique, the primal solution that an interior point algorithm converges to is not guaranteed to be integer. In an implementation of an interior point network flow method, one would like to be capable of recovering an integer flow even when the problem has multiple optima. We discuss below the stopping strategies implemented in DLNET and used to recover an integer optimal solution.

Besides the indicator described in Section 3.4, DLNET uses the arcs of the spanning forest of the tree preconditioner as an indicator. If there exists a unique optimal flow, this indicator correctly identifies an optimal primal basic sequence, and an integer flow can be easily recovered by solving a triangular system of linear equations. In general, however, the arc indices do not converge to a basic sequence. Let $\mathcal{T} = \{t_1, \ldots, t_q\}$ denote the set of arc indices in the

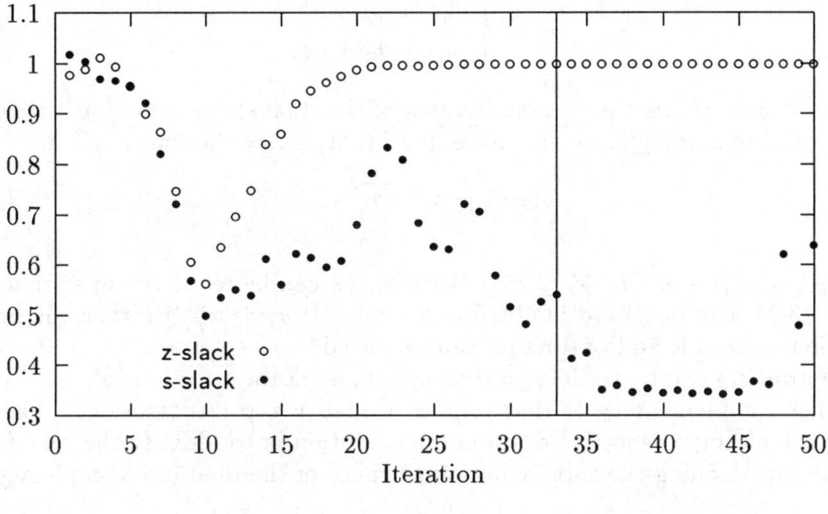


FIG. 7. Dual slack ratio (edge set to upper bound)

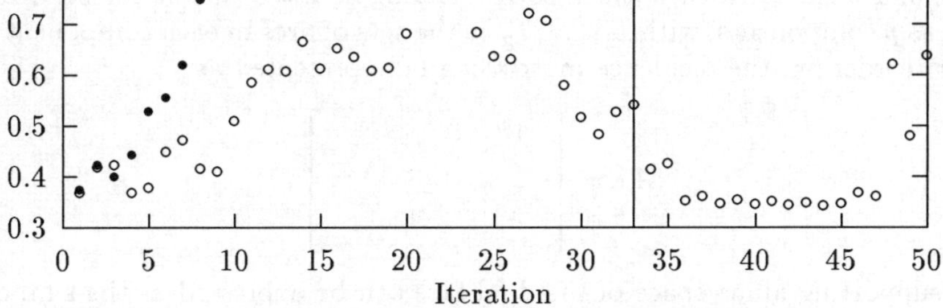


FIG. 8. Dual slack ratio (edge set to lower bound)

spanning forest. To obtain a tentative primal basic solution, first set flow on arcs not in the forest to either their upper or lower bound, i.e. for all $i \in \mathcal{A} \setminus \mathcal{T}$:

$$x_i^* = \begin{cases} 0 & \text{if } s_i^k > z_i^k, \\ u_i & \text{otherwise,} \end{cases}$$

where $s^k$ and $z^k$ are the current iterates of the dual slack vectors as defined in (3.1). The remaining basic arcs have flows that satisfy the linear system

$$A_{\mathcal{T}} x_{\mathcal{T}}^* = b - \sum_{i \in \Omega^-} u_i A_i, \tag{3.25}$$

where $\Omega^- = \{i \in \mathcal{A} \setminus \mathcal{T} : s_i^k \leq z_i^k\}$. Because $A_{\mathcal{T}}$ can be reordered in a triangular form, (3.25) can be solved in $O(m)$ operations. If $u_{\mathcal{T}} \geq x_{\mathcal{T}}^* \geq 0$ then the primal solution is feasible and optimality can be tested.

Optimality can be verified producing a dual feasible solution $(y^*, s^*, z^*)$ that is either complementary or that implies a duality gap less than one. The first step in building a tentative optimal dual solution is to identify the set of dual constraints defining the supporting affine space of the dual face complementary to $x^*$,

$$\mathcal{F} = \{i \in \mathcal{T} \ : \ 0 < x_i^* < u_i\},$$

i.e. the set of arcs with zero dual slacks. Since, in general, $x^*$ is not feasible, $\mathcal{F}$ is usually determined by the indicators of Section 3.4, as the index-set of active arcs. To ensure a complementary primal-dual pair, the current dual interior vector $y^k$ is projected orthogonally onto this affine space. The solution $y^*$ of the least squares problem

$$\min_{y^* \in \mathbb{R}^m} \{\|y^* - y^k\|_2 \ : \ A_{\mathcal{F}}^\top y^* = c_{\mathcal{F}}\} \tag{3.26}$$

is the projected dual iterate.

Let $G_{\mathcal{F}} = (\mathcal{N}, \mathcal{F})$ be the subgraph of $G$ with $\mathcal{F}$ as its set of arcs. Since this subgraph is a forest, its incidence matrix, $A_{\mathcal{F}}$, can be reordered into a block triangular form, with each block corresponding to a tree in the forest. Assume $G_{\mathcal{F}}$ has $p$ components, with $T_1, \ldots, T_p$ as the sets of arcs in each component tree. After reordering, the incidence matrix can be represented as

$$A_{\mathcal{F}} = \begin{bmatrix} A_{T_1} & & \\ & \ddots & \\ & & A_{T_p} \end{bmatrix}.$$

The supporting affine space of the dual face can be expressed as the sum of orthogonal one-dimensional subspaces. The operation in (3.26) can be performed by computing the orthogonal projections onto each individual subspace independently, and therefore can be completed in $O(m)$ time. For $i = 1, \ldots, p$, denote the number of arcs in $T_i$ by $m_i$, and the set of nodes spanned by those arcs by

$\mathcal{N}_i$. $A_{T_i}$ is an $(m_i+1)\times m_i$ matrix and each subspace

$$\Psi_i = \{y_{\mathcal{N}_i} \in \mathbb{R}^{m_i+1} \ : \ A_{T_i}^\top y_{\mathcal{N}_i} = c_{T_i}\}$$

has dimension one. For all $y_{\mathcal{N}_i} \in \Psi_i$,

$$y_{\mathcal{N}_i} = y^0_{\mathcal{N}_i} + \alpha_i y^h_{\mathcal{N}_i}, \tag{3.27}$$

where $y^0_{\mathcal{N}_i}$ is a given solution in $\Psi_i$ and $y^h_{\mathcal{N}_i}$ is a solution of the homogeneous system $A_{T_i}^\top y_{\mathcal{N}_i} = 0$. Since $A_{T_i}$ is the incidence matrix of a tree, the unit vector is a homogeneous solution. The given solution $y^0_{\mathcal{N}_i}$ can be computed by selecting $v \in \mathcal{N}_i$, setting $y^0_v = 0$, removing the row corresponding to node $v$ from matrix $A_{T_i}$ and solving the resulting triangular system

$$\tilde{A}_{T_i}^\top y_{\mathcal{N}_i \setminus \{v\}} = c_{T_i}.$$

With the representation in (3.27), the orthogonal projection of $y_{\mathcal{N}_i}$ onto the subspace $\Psi_i$ is

$$y^*_{\mathcal{N}_i} = y^0_{\mathcal{N}_i} + \frac{e^\top_{\mathcal{N}_i}(y_{\mathcal{N}_i} - y^0_{\mathcal{N}_i})}{(m_i+1)} e_{\mathcal{N}_i}$$

where $e$ is the unit vector. The orthogonal projection, as indicated in (3.26), is obtained by combining the projections onto each subspace,

$$y^* = (y^*_{\mathcal{N}_1}, \ldots, y^*_{\mathcal{N}_q}).$$

A feasible dual solution is built by computing the slacks as

$$z^*_i = \begin{cases} -\delta_i & \text{if } \delta_i < 0, \\ 0 & \text{otherwise}, \end{cases} \qquad s^*_i = \begin{cases} 0 & \text{if } \delta_i < 0, \\ \delta_i & \text{otherwise}, \end{cases}$$

where $\delta_i = c_i - A^\top y^*$.

If the solution of (3.25) is feasible, optimality can be checked at this point, using the projected dual solution as a lower bound on the optimal flow. The primal and dual solutions, $x^*$ and $(y^*, s^*, z^*)$, are optimal if complementary slackness is satisfied, i.e. if for all $i \in \mathcal{A} \setminus \mathcal{T}$ either $s^*_i > 0$ and $x^*_i = 0$ or $z^*_i > 0$ and $x^*_i = u_i$. Otherwise, the primal solution, $x^*$, is still optimal if the duality gap is less than 1, i.e. if $c^\top x^* - b^\top y^* + u^\top z^* < 1$.

However, in general, the method proceeds by attempting to find a feasible flow $x^*$ that is complementary to the projected dual solution $y^*$. Based on the projected dual solution $y^*$, a refined tentative optimal face is selected by redefining the set of active arcs as

$$\tilde{\mathcal{F}} = \{i \in \mathcal{A} \ : \ |c_i - A_{\cdot i}^\top y^*| < \epsilon\}.$$

Next, the method attempts to build a primal feasible solution, $x^*$, complementary to the tentative dual optimal solution by setting the inactive arcs to lower or upper bounds, i.e. for $i \in \mathcal{A} \setminus \tilde{\mathcal{F}}$,

$$x_i^* = \begin{cases} 0, & \text{if } i \in \Omega^+ = \{i \in \mathcal{A} \setminus \tilde{\mathcal{F}} \ : \ c_i - A_{\cdot i}^\top y^* > 0\}, \\ u_i, & \text{if } i \in \Omega^- = \{i \in \mathcal{A} \setminus \tilde{\mathcal{F}} \ : \ c_i - A_{\cdot i}^\top y^* < 0\}. \end{cases}$$

By considering only the active arcs, a *restricted network* is built, represented by the constraint set

$$A_{\tilde{\mathcal{F}}} x_{\tilde{\mathcal{F}}} = \tilde{b} = b - \sum_{i \in \Omega^-} u_i A_i, \tag{3.28}$$

$$0 \le x_i \le u_i, \ \ i \in \tilde{\mathcal{F}}. \tag{3.29}$$

Clearly, from the flow balance constraints (3.28), if a feasible flow $x^*_{\tilde{\mathcal{F}}}$ for the restricted network exists, it defines, along with $x^*_{\Omega^+}$ and $x^*_{\Omega^-}$, a primal feasible solution complementary to $y^*$. A feasible flow for the restricted network can be determined by solving a maximum flow problem on the *augmented network* defined by the underlying graph $\tilde{G} = (\tilde{\mathcal{N}}, \tilde{\mathcal{A}})$, where

$$\tilde{\mathcal{N}} = \{\sigma\} \cup \{\theta\} \cup \mathcal{N}$$

and

$$\tilde{\mathcal{A}} = \Sigma \cup \Theta \cup \tilde{\mathcal{F}}.$$

In addition, for each arc $(i, j) \in \tilde{\mathcal{F}}$ there is an associated capacity $u_{ij}$. The additional arcs are such that

$$\Sigma = \{(\sigma, i) \ : \ i \in \mathcal{N}^+\},$$

with associated capacity $\tilde{b}_i$ for each arc $(\sigma, i)$, and

$$\Theta = \{(i, \theta) \ : \ i \in \mathcal{N}^-\},$$

with associated capacity $-\tilde{b}_i$ for each arc $(i, \theta)$, where $\mathcal{N}^+ = \{i \in \mathcal{N} \ : \ \tilde{b}_i > 0\}$ and $\mathcal{N}^- = \{i \in \mathcal{N} \ : \ \tilde{b}_i < 0\}$. It can be shown that if $\mathcal{M}_{\sigma,\theta}$ is the maximum flow value from $\sigma$ to $\theta$, and $\tilde{x}$ is a maximal flow on the augmented network, then $\mathcal{M}_{\sigma,\theta} = \sum_{i \in \mathcal{N}^+} \tilde{b}_i$ if and only if $\tilde{x}_{\tilde{\mathcal{F}}}$ is a feasible flow for the restricted network. Therefore, finding a feasible flow for the restricted network involves the solution of a maximum flow problem. Furthermore, this feasible flow is integer, as we can select a maximum flow algorithm that provides an integer solution.

Figures 9 and 10 show, respectively, the difference between the flow requested and the maximum flow found in the maximum flow termination criterion and the number of primal infeasibilities in the spanning tree on the 16,384 node, 131,072 arc example. At iteration 69, the requested flow of 18,720,642 units was found and a primal-dual complementary integer solution was recovered. That iteration still had four primal infeasibilities, the smallest number over all interior point iterations.

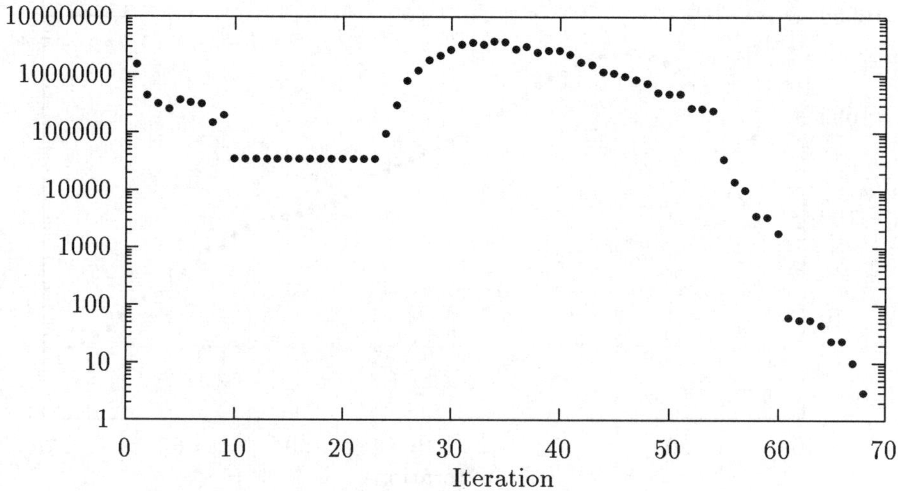


FIG. 9. Difference between the flow requested and the maximum flow found

## 4 Computational challenge to the network simplex

In this section, we illustrate interior point network flow methods by comparing DLNET [56], an implementation of the dual affine scaling algorithm for network flows, with CPLEX NETOPT, the network simplex implementation in the CPLEX mathematical programming system [13]. The experiment was run on an SGI Challenge computer with the configuration summarized in Fig. 11.

Version 2.2a of DLNET was used in the experiments. The stepsize parameter $\gamma^k$ was set to 0.99 for the first 10 dual affine scaling iterations and at each iteration, the relative dual objective function improvement

$$\mathcal{I}^k = \frac{(b^\top y^k - u^\top z^k) - (b^\top y^{k-1} - u^\top z^{k-1})}{|b^\top y^k - u^\top z^k|}$$

is computed. After iteration 10, the stepsize is determined as follows:

$$\gamma^k = \begin{cases} 0.95, & \text{if } \mathcal{I}^k \geq 0.0001, \\ 0.66, & \text{if } 0.00001 \leq \mathcal{I}^k < 0.0001, \\ 0.50, & \text{if } \mathcal{I}^k < 0.00001. \end{cases}$$

Indicator 3, described in Section 3.4, with parameters $\kappa_0 = 0.7$ and $\kappa_1 = 0.9$ is used. DLNET uses diagonal and spanning tree preconditioners and the switching heuristic described in Section 3.3. The PB stopping criterion was checked every iteration on which the spanning tree preconditioner was used. The MF criterion was never checked during the first 5 iterations. Once $\mathcal{I}^k < 10^{-10}$ the MF was checked every 10 iterations, and when $\mathcal{I}^k < 10^{-12}$ it was checked every 5

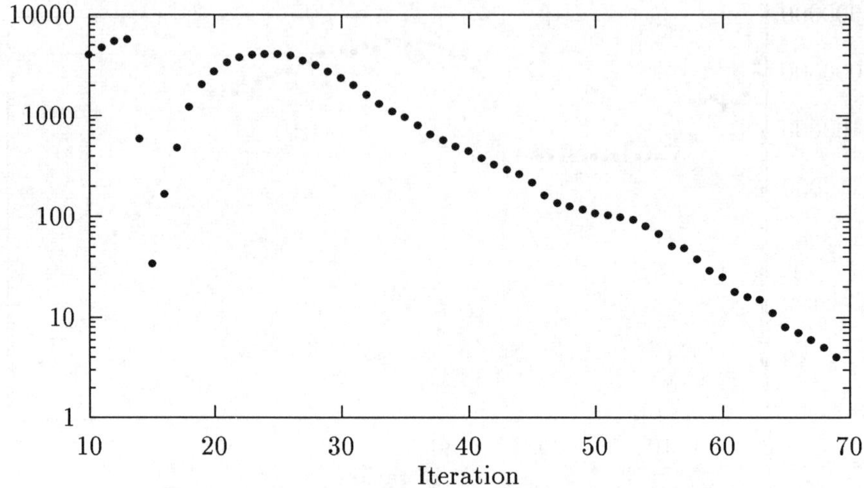


FIG. 10. Number of primal infeasibilities in spanning tree

```
16 150 MHZ IP19 Processors
CPU: MIPS R4400 Processor Chip Revision:  5.0
FPU: MIPS R4010 Floating Point Chip Revision:  0.0
Data cache size:  16 Kbytes
Instruction cache size:  16 Kbytes
Secondary unified instruction/data cache size:  1 Mbyte
Main memory size:  1536 Mbytes, 8-way interleaved
```

FIG. 11. Hardware configuration (partial output of system command `hinv`)

iterations. The conjugate gradient cosine parameter $\epsilon_{cos}$ was set fixed to $10^{-3}$ throughout all iterations and the exact cosine rule was checked every 5 conjugate gradient iterations.

Version 3.0 (1994) of CPLEX NETOPT was used with default parameter settings.

All problems considered in this section are from the set of network flow benchmarks collected for the First DIMACS Algorithm Implementation Challenge [31]. The generators can be retrieved from the DIMACS FTP site: **dimacs.rutgers.edu**.

For each problem class, instances of increasing sizes were generated and run on both codes. For each class, the runs are summarized with a table and a figure. The tables list network dimensions, number of interior point iterations, total CPU time for the interior point algorithm, total number of conjugate gradient iterations, number of calls to the maximum flow (MF) stopping criterion, total

**Table 1** Problem statistics: test problem class Grid-Density-8

| network size | | DLNET | | | | | | CPLEX | |
|---|---|---|---|---|---|---|---|---|---|
| | | int point alg | | CG | max flow | | stopping | | |
| $\|\mathcal{N}\|$ | $\|\mathcal{A}\|$ | itrs | time | itrs | calls | time | criterion | itr | time |
| 256 | 2048 | 21 | 1.16 | 181 | 0 | - | PB | 4137 | 0.36 |
| 1024 | 8192 | 29 | 6.46 | 234 | 0 | - | PB | 34172 | 8.02 |
| 4096 | 32768 | 32 | 40.47 | 376 | 0 | - | PB | 184044 | 139.83 |
| 16384 | 131072 | 43 | 419.32 | 615 | 0 | - | PB | 875318 | 4162.20 |
| 65536 | 524288 | 65 | 6336.61 | 1114 | 1 | 255.56 | MF | 4244165 | 106956.76 |

**Table 2** Problem statistics: test problem class Grid-Density-16

| network size | | DLNET | | | | | | CPLEX | |
|---|---|---|---|---|---|---|---|---|---|
| | | int point alg | | CG | max flow | | stopping | | |
| $\|\mathcal{N}\|$ | $\|\mathcal{A}\|$ | itrs | time | itrs | calls | time | criterion | itr | time |
| 512 | 8192 | 60 | 12.30 | 426 | 1 | 0.05 | MF | 24725 | 2.59 |
| 2048 | 32768 | 79 | 76.45 | 661 | 1 | 0.44 | MF | 148281 | 63.34 |
| 8192 | 131072 | 99 | 814.52 | 1052 | 1 | 4.06 | MF | 884825 | 1673.75 |
| 32768 | 524288 | 94 | 3939.72 | 1223 | 1 | 58.57 | MF | 5399631 | 46479.68 |

CPU to run the MF stopping criterion, number of simplex iterations, and total CPU time for the network simplex code. The figures show, in a log-log plot, CPU time ratios between the simplex code and the interior point code.

The first class of problem is Grid-Density. The networks in this class of problems are obtained by removing the minimum number of arcs from a grid graph embedded on a torus such that all vertical paths wrap around and no horizontal path wraps around, and adding a source and sink node with arcs going from the source to all nodes on one side of the resulting tube network and from all nodes on the other side to the sink. Network density is controlled by adding extra arcs in both the horizontal and vertical directions. All nodes other than the source and sink nodes are transshipment nodes. Costs and capacities are uniformly distributed in the intervals $[0, 4096]$ and $[0, 16384]$, respectively.

These instances are generated with the minimum cost network flow test problem generator `goto.c` by Goldberg [16]. Two subclasses of problems are generated: Grid-Density-8 and Grid-Density-16. Let $m$ and $n$ denote the number of nodes and arcs of the network, respectively. For Grid-Density-8, $n = 8 \;:\; m$; and for Grid-Density-16, $n = 16 \;:\; m$. For Grid-Density-8, instances having $256, 1024, \ldots, 65536$ nodes are generated. For Grid-Density-16, instances with 512, 2048, ..., 32768 nodes are generated. The random number generator seed 270001 was used to generate all instances for each problem size.

Table 1 and Fig. 12 summarize the runs for Grid-Density-8. Table 2 and Fig. 13 summarize the runs for Grid-Density-16.

The next class is Grid-Graph. The networks in this class are formed on a grid of nodes of height $h$ and width $w$. Two additional nodes complete the node

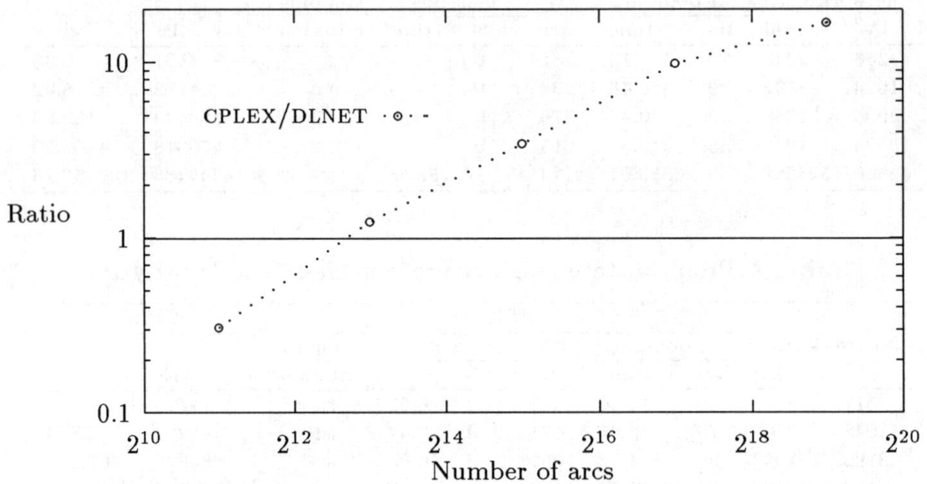


FIG. 12. CPU time ratios for problem class Grid-Density-8

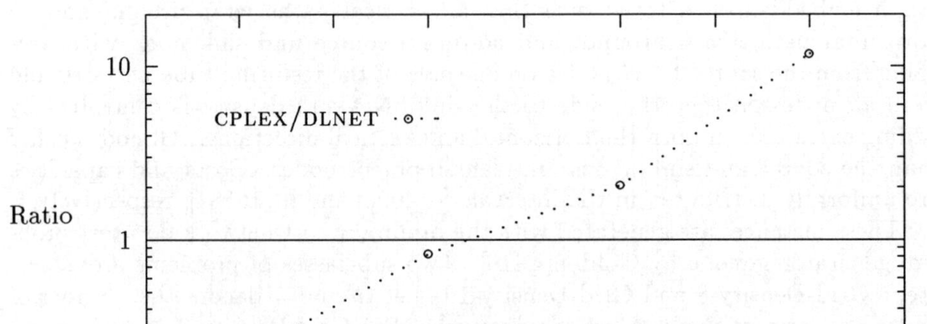


FIG. 13. CPU time ratios for problem class Grid-Density-16

**Table 3** Problem statistics: test problem class Grid-Long

| network size | | DLNET | | | | | | CPLEX | |
|---|---|---|---|---|---|---|---|---|---|
| | | int point alg | | CG | max flow | | stopping | | |
| $\lvert\mathcal{N}\rvert$ | $\lvert\mathcal{A}\rvert$ | itrs | time | itrs | calls | time | criterion | itr | time |
| 1026 | 2000 | 29 | 2.56 | 402 | 0 | - | PB | 2171 | 0.29 |
| 4098 | 7952 | 73 | 29.24 | 1112 | 1 | 0.33 | MF | 7776 | 4.59 |
| 16386 | 31760 | 84 | 284.88 | 2517 | 0 | - | PB | 29729 | 85.77 |
| 65538 | 126992 | 169 | 4925.54 | 4228 | 0 | - | PB | 92599 | 1471.92 |

**Table 4** Problem statistics: test problem class Grid-Wide

| network size | | DLNET | | | | | | CPLEX | |
|---|---|---|---|---|---|---|---|---|---|
| | | int point alg | | CG | max flow | | stopping | | |
| $\lvert\mathcal{N}\rvert$ | $\lvert\mathcal{A}\rvert$ | itrs | time | itrs | calls | time | criterion | itr | time |
| 1026 | 2096 | 30 | 2.48 | 391 | 0 | - | PB | 2449 | 0.23 |
| 4098 | 8432 | 38 | 13.90 | 536 | 0 | - | PB | 10685 | 1.28 |
| 16386 | 33776 | 49 | 77.05 | 458 | 0 | - | PB | 37044 | 8.63 |
| 65538 | 135152 | 74 | 702.05 | 377 | 0 | - | PB | 140428 | 55.84 |

set: a source node $S$ and a sink node $T$. Arcs go from the source to each node in the first column of the grid and from each node in the last column of the grid to the sink node. On the grid, arcs go from node to nearest neighbour node oriented left to right and top to bottom. Grid arc costs and capacities are generated uniformly in the interval $[1, 10000]$. Arcs from the source and into the sink have cost zero and are uncapacitated. All nodes, except for source and sink, are transshipment nodes. The source has a supply of $M_{ST}$, the maximum flow from $S$ to $T$, and the sink has a demand of $M_{ST}$.

These instances are generated with the minimum cost network flow test problem generator `ggraph1.f` of Resende [16]. Two subclasses of problems are generated: Grid-Wide and Grid-Long. Grid-Wide uses parameters $w = 16$ and $h = 32, 64, \ldots$, while Grid-Long uses $h = 16$ and $w = 32, 64, \ldots$.

The random number generator seed 270001 was used to generate the instances. For both Grid-Long and Grid-Wide, instances having $1026, 4098, \ldots, 65538$ nodes were generated. Table 3 and Fig. 14 summarize the runs for Grid-Long. Table 4 and Fig. 15 summarize the runs for Grid-Wide.

Most computational studies of network optimization codes in the past have used the minimum cost network flow test generator NETGEN [40] of Klingman, Napier and Stutz. We tested the codes on two classes of networks generated with NETGEN: Netgen-Lo and Netgen-Hi. Figure 16 lists the NETGEN parameters used to generate the class Netgen-Lo. Networks in class Netgen-Hi are generated with the same parameters except for `maxcap`, which is set to 16,384. For both subclasses, an instance of each size was generated. Sizes correspond to the settings $x = 8, 9, \ldots, 16$.

Table 5 and Fig. 17 summarize the runs for Netgen-Hi. Table 6 and Fig. 18

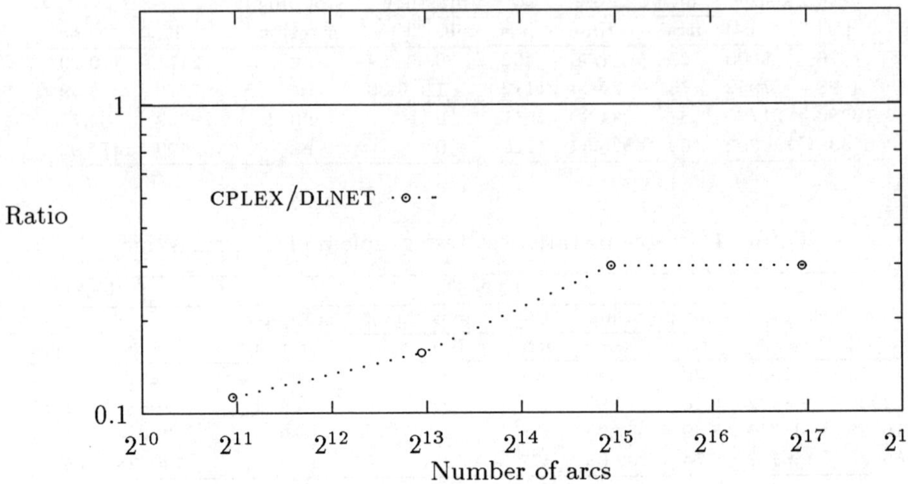


FIG. 14. CPU time ratios for problem class Grid-Long

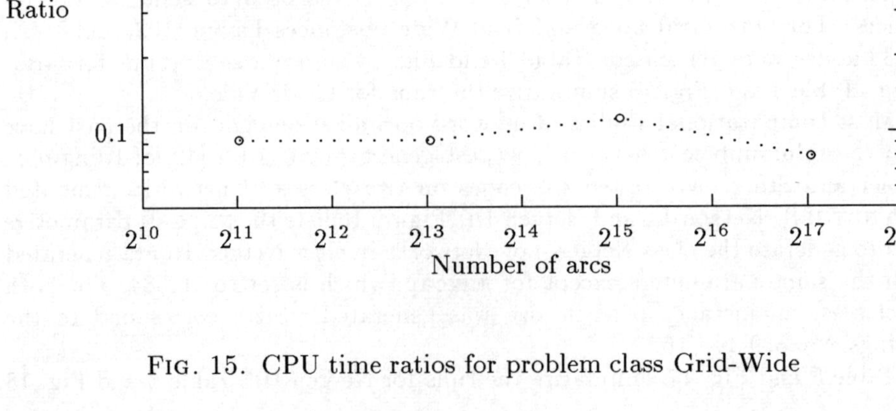


FIG. 15. CPU time ratios for problem class Grid-Wide

| | | |
|---|---|---|
| `seed` | Random number seed: | 270001 |
| `problem` | Problem number (for output): | 1 |
| `nodes` | Number of nodes: | $m = 2^x$ |
| `sources` | Number of sources: | $2^{x-2}$ |
| `sinks` | Number of sinks: | $2^{x-2}$ |
| `density` | Number of (requested) arcs: | $2^{x+3}$ |
| `mincost` | Minimum arc cost: | 0 |
| `maxcost` | Maximum arc cost: | 4096 |
| `supply` | Total supply: | $2^{2\,:\,(x-2)}$ |
| `tsources` | Transshipment sources: | 0 |
| `tsinks` | Transshipment sinks: | 0 |
| `hicost` | Skeleton arcs with max cost: | 100% |
| `capacitated` | Capacitated arcs: | 100% |
| `mincap` | Minimum arc capacity: | 1 |
| `maxcap` | Maximum arc capacity: | 16 |

FIG. 16. NETGEN specification file for Netgen-Lo

**Table 5** Problem statistics: test problem class Netgen-Hi

| network size | | DLNET | | | | | | CPLEX | |
|---|---|---|---|---|---|---|---|---|---|
| | | int point alg | | CG | max flow | | stopping | | |
| $\|\mathcal{N}\|$ | $\|\mathcal{A}\|$ | itrs | time | itrs | calls | time | criterion | itr | time |
| 256 | 2048 | 29 | 1.55 | 255 | 0 | - | PB | 1033 | 0.18 |
| 1024 | 8214 | 41 | 10.43 | 537 | 0 | - | PB | 6679 | 1.22 |
| 4096 | 32877 | 61 | 82.97 | 989 | 0 | - | PB | 29330 | 13.77 |
| 16384 | 131409 | 74 | 835.42 | 2012 | 0 | - | PB | 137023 | 248.34 |
| 65536 | 525803 | 107 | 14861.27 | 5153 | 1 | 162.33 | MF | 1077281 | 21046.19 |

summarize the runs for Netgen-Lo.

The last class of problems considered is Mesh-4. Networks in this class are formed on a grid of nodes embedded on a torus. Arcs connect nodes in the same row or column of the grid. All horizontal arcs have the same orientation. Similarly, all vertical arcs are oriented in the same direction. Each node has arcs going to its 8 nearest neighbours (four in the horizontal direction, the other

**Table 6** Problem statistics: test problem class Netgen-Lo

| network size | | DLNET | | | | | | CPLEX | |
|---|---|---|---|---|---|---|---|---|---|
| | | int point alg | | CG | max flow | | stopping | | |
| $\|\mathcal{N}\|$ | $\|\mathcal{A}\|$ | itrs | time | itrs | calls | time | criterion | itr | time |
| 256 | 2048 | 19 | 1.11 | 233 | 0 | - | PB | 2876 | 0.24 |
| 1024 | 8214 | 36 | 9.72 | 532 | 0 | - | PB | 16956 | 2.08 |
| 4096 | 32858 | 42 | 60.75 | 644 | 0 | - | PB | 63856 | 19.48 |
| 16384 | 131409 | 73 | 584.65 | 1220 | 1 | 10.54 | MF | 218163 | 231.59 |
| 65536 | 525803 | 102 | 5703.28 | 1685 | 1 | 190.26 | MF | 1212399 | 9459.77 |

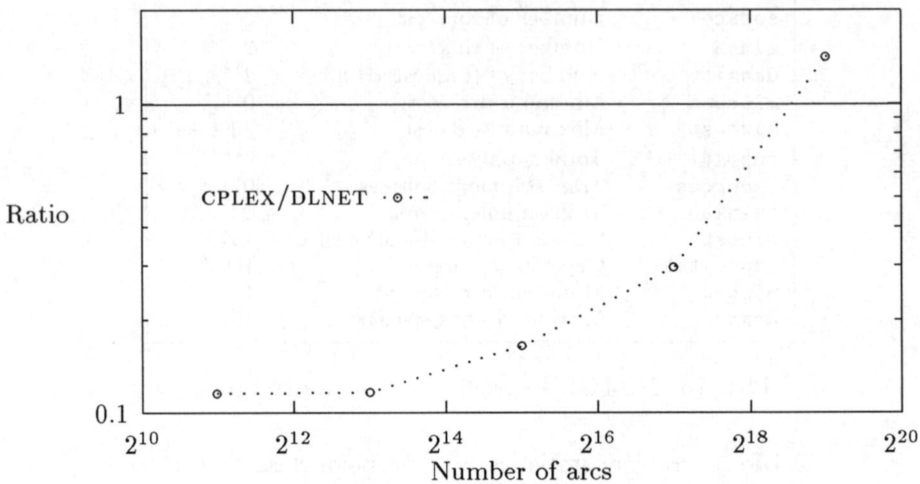


FIG. 17. CPU time ratios for problem class Netgen-Hi

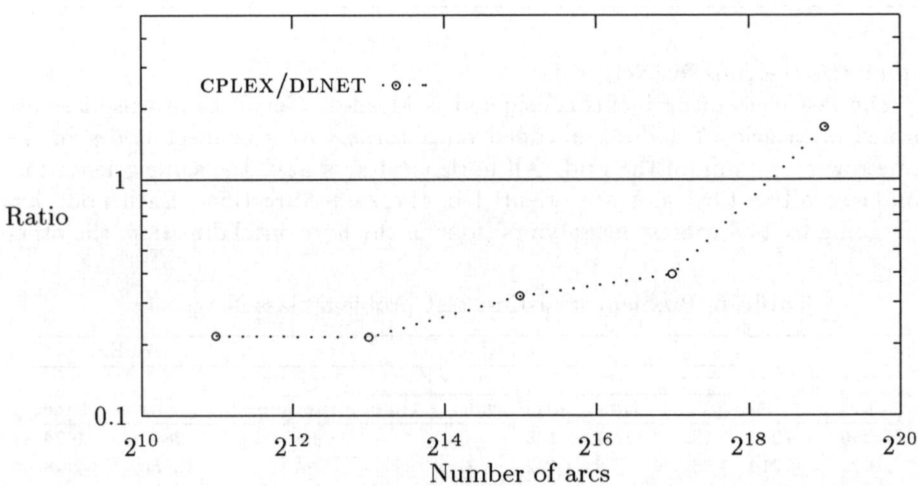


FIG. 18. CPU time ratios for problem class Netgen-Lo

**Table 7** Problem statistics: test problem class Mesh-4

| network size | | DLNET | | | | | | CPLEX | |
|---|---|---|---|---|---|---|---|---|---|
| | | int point alg | | CG | max flow | | stopping | | |
| $\lvert\mathcal{N}\rvert$ | $\lvert\mathcal{A}\rvert$ | itrs | time | itrs | calls | time | criterion | itr | time |
| 256 | 2048 | 21 | 1.18 | 245 | 0 | - | PB | 4790 | 0.36 |
| 1024 | 8192 | 56 | 11.67 | 475 | 1 | 0.10 | MF | 25459 | 4.13 |
| 4096 | 32768 | 88 | 90.02 | 779 | 1 | 0.68 | MF | 125315 | 68.34 |
| 16384 | 131072 | 60 | 520.57 | 662 | 1 | 7.82 | MF | 712187 | 1406.08 |
| 65536 | 524288 | 83 | 5972.02 | 1035 | 1 | 104.39 | MF | 3231833 | 28801.18 |

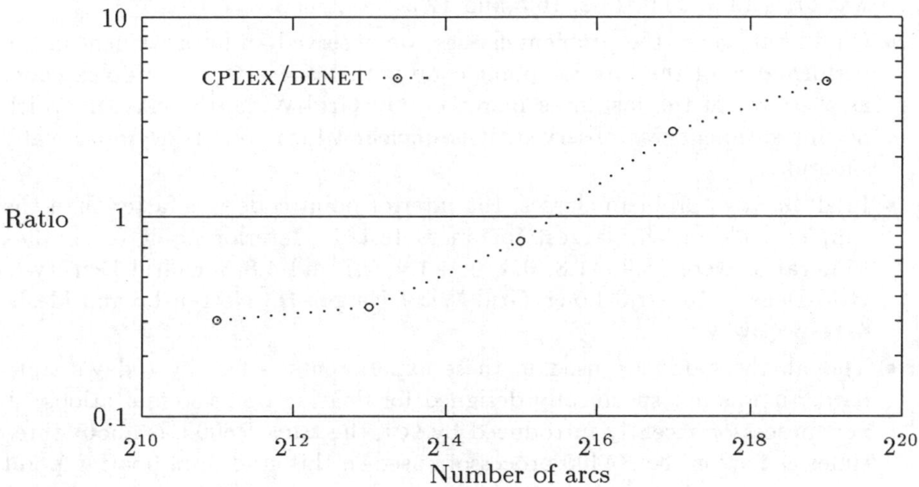


FIG. 19. CPU time ratios for problem class Mesh-4

four in the vertical direction). Costs are generated uniformly in the interval $[-1000, 1000]$. Capacities are generated in the interval $[-1000, 1000]$ with a bias that makes longer arcs have smaller capacities.

The instances are generated with the minimum cost network flow test generator `mesh.c` by Goldberg [16]. All node sets are $h \times w$ grids. We generated all instances with $h = w$. Instances are generated with parameters $h = w = 16, 32, \ldots, 256$, having $256, 1024, \ldots, 65536$ nodes. The random number generator seed 270001 was used to generate the instances.

Table 7 and Fig. 19 summarize the runs for Mesh-4.

We conclude this section with some remarks:

- Both codes found optimal solutions to all the instances tested.

- Both codes produced integer flows as output.
- Of the 32 instances optimized, the interior point code stopped with the maximum flow (MF) criterion on 13 instances and with the primal basic (PB) criterion on 19 instances.
- On all 32 instances, the interior point code found optimal primal-dual complementary solutions.
- The triggering mechanism for the maximum flow stopping criterion, based on relative dual improvement, worked well, since on every instance MF was never called more than once.
- The preconditioners used resulted in a small average number of conjugate gradient iterations. For the largest instance in each problem class, the average number of conjugate gradient iterations per interior point iteration was 17.1, 13.0, 25.0, 48.2, 16.5 and 12.5.
- On all but one of the problem classes, we observed an improvement in the performance of the interior point code with respect to the simplex code, as the sizes of the instances increase. On Grid-Wide, the class in which no improvement was observed, it is unclear which code is asymptotically superior.
- In all but two problem classes, the interior point code was faster than the simplex code on the largest instances tested. Interior point to simplex time ratios were 16.9, 11.8, 0.3, 0.1, 1.4, 1.7 and 4.8, for Grid-Density-8, Grid-Density-16, Grid-Long, Grid-Wide, Netgen-Hi, Netgen-Lo and Mesh-4, respectively.
- Though the hardware used in these experiments is fast by today's standards, it was not specifically designed for floating point computations. A new processor recently introduced by SGI, the MIPS R8000, is about three times as fast as the R4400 processor (used in this study) on floating point computations. Because DLENT does a large amount of floating point computations, it is expected to benefit from the new hardware, whilst NETOPT, which does mainly integer arithmetic is not expected to benefit at all.
- Parallel implementation of the conjugate gradient algorithm has been shown to improve the performance of conjugate gradient based interior point codes more than parallel implementations speed up the simplex method. One should expect larger time ratios in favour of interior point codes on parallel architectures.

## 5 Concluding remarks

In this chapter, we discussed different issues related to the implementation of interior point methods for solving linear network flow problems. These methods provide, in addition to the plethora of existing solution techniques, new ways to solve large-scale network flow problems. Though interior point methods have not, so far, been shown to outperform all other methods in several classes of large-scale problems, there exist many classes where interior point codes are currently the most efficient computational solution approaches. In this way, interior point methods should not be thought of as a substitute for existing algorithms, but rather they are an addition to the class of computationally efficient network flow algorithms. It is expected that in the near future robust interior point codes for solving network flow problems will be widely available.

In addition to the codes developed for the single commodity network flow problem, initial attempts to use interior point methods to solve the more difficult multicommodity network flow problem can be found in [24, 34].

## Bibliography


1. I. Adler, N. Karmarkar, M.G.C. Resende and G. Veiga. Data structures and programming techniques for the implementation of Karmarkar's algorithm. *ORSA Journal on Computing*, 1:84–106, 1989.
2. I. Adler, N. Karmarkar, M.G.C. Resende and G. Veiga. An implementation of Karmarkar's algorithm for linear programming. *Mathematical Programming*, 44:297–335, 1989.
3. N.K. Ahuja, T.L. Magnanti and J.B. Orlin. *Network Flows.* Prentice Hall, Englewood Cliffs, NJ, 1993.
4. A. Armacost and S. Mehrotra. Computational comparison of the network simplex method with the affine scaling method. *Opsearch*, 28:26–43, 1991.
5. J. Aronson, R. Barr, R. Helgason, J. Kennington, A. Loh and H. Zaki. The projective transformation algorithm of Karmarkar: A computational experiment with assignment problems. Technical Report 85-OR-3, Department of Operations Research, Southern Methodist University, Dallas, TX, August 1985.
6. E.R. Barnes. A variation on Karmarkar's algorithm for solving linear programming problems. *Mathematical Programming*, 36:174–182, 1986.
7. M.S. Bazaraa, J.J. Jarvis and H.D. Sherali. *Linear Programming and Network Flows.* Wiley, New York, NY, 1990.
8. D. Bertsekas. *Linear Network Optimization: Algorithms and Codes.* MIT Press, Cambridge, MA, 1991.
9. D.P. Bertsekas and D.A. Castañon. The auction algorithm for transportation problems. *Annals of Operations Research*, 20:67–96, 1989.
10. D.P. Bertsekas and P. Tseng. Relaxation methods for minimum cost ordinary and generalized network flow problems. *Operations Research*, 36:93–114, 1988.
11. G.H. Bradley, G.G. Brown and G.W. Graves. Design and implementation of large scale primal transshipment algorithms. *Management Science*, 24:1–34, 1977.
12. Y.C. Cheng, D.J. Houck, J.M. Liu, M.S. Meketon, L. Slutsman, R.J. Vanderbei and P. Wang. The AT&T KORBX System. *AT&T Technical Journal*, 68:7–19, 1989.

13. CPLEX Optimization, Inc. *Using the CPLEX$^{\text{TM}}$callable library and CPLEX$^{\text{TM}}$mixed integer library including the CPLEX$^{\text{TM}}$linear optimizer and CPLEX$^{\text{TM}}$mixed-integer optimizer – Version 3.0.* Incline Village, NV, 1994.
14. G.B. Dantzig. Application of the simplex method to a transportation problem. In T.C. Koopsmans, (ed.), *Activity Analysis of Production and Allocation.* John Wiley and Sons, 1951.
15. I.I. Dikin. Iterative solution of problems of linear and quadratic programming. *Soviet Mathematics Doklady,* 8:674–675, 1967.
16. DIMACS. The first DIMACS international algorithm implementation challenge: The benchmark experiments. Technical report, DIMACS, New Brunswick, NJ, 1991.
17. D.-Z. Du and P.M. Pardalos, (eds.). *Network Optimization Problems: Algorithms, Applications and Complexity.* World Scientific, 1993.
18. A.S. El-Bakry, R.A. Tapia and Y. Zhang. A study of indicators for identifying zero variables in interior-point methods. Technical Report TR91-15, Department of Mathematical Sciences, Rice University, Houston, TX 77251, 1991.
19. L.R. Ford and D.R. Fulkerson. *Flows in Networks.* Princeton University Press, Princeton, NJ, 1962.
20. D. Gay. Electronic mail distribution of linear programming test problems. *COAL Newsletter,* 13:10–12, 1985.
21. D.M. Gay. Stopping tests that compute optimal solutions for interior-point linear programming algorithms. Technical report, AT&T Bell Laboratories, Murray Hill, NJ, 1989.
22. A. George and M. Heath. Solution of sparse linear least squares problems using Givens rotations. *Linear Algebra and Its Applications,* 34:69–83, 1980.
23. F. Glover and D. Klingman. The simplex SON algorithm for LP/embedded network problems. *Mathematical Programming Study,* 15:148–176, 1981.
24. J.-L. Goffin, J. Gondzio, R. Sarkissian and J.P. Vial. Solving non-linear multicommodity flow problems by the analytic center cutting plane method. Technical report, GERARD, McGill University, Montreal, 1994.
25. A.V. Goldberg. An efficient implementation of a scaling minimum-cost flow algorithm. Technical report, Department of Computer Science, Stanford University, Stanford, CA 94305, 1992.
26. G.H. Golub and C.F. van Loan. *Matrix Computations.* The Johns Hopkins University Press, Baltimore, MD, 1983.
27. M.D. Grigoriadis. An efficient implementation of the network simplex method. *Mathematical Programming Study,* 26:83–111, 1986.
28. G.M. Guisewite. Network problems. In R. Horst and P.M. Pardalos, (eds.), *Handbook of Global Optimization.* Kluwer Academic Publishers, 1995.
29. G.M. Guisewite and P.M. Pardalos. Minimum concave cost network flow problems: Applications, complexity, and algorithms. *Annals of Operations*

*Research*, 25:75–100, 1990.

30. F.L. Hitchcock. The distribution of product from several sources to numerous facilities. *Journal of Mathematical Physics*, 20:224–230, 1941.
31. D.S. Johnson and C.C. McGeoch, (eds.). *Network flows and matching: First DIMACS implementation challenge*, volume 12 of *DIMACS Series in Discrete Mathematics and Theoretical Computer Science.* American Mathematical Society, 1993.
32. A. Joshi, A.S. Goldstein and P.M. Vaidya. A fast implementation of a path-following algorithm for maximizing a linear function over a network polytope. In D.S. Johnson and C.C. McGeoch, (eds.), *Network Flows and Matching: First DIMACS Implementation Challenge*, *DIMACS Series in Discrete Mathematics and Theoretical Computer Science*, 12:267–298. American Mathematical Society, 1993.
33. J.A. Kaliski and Y. Ye. A decomposition variant of the potential reduction algorithm for linear programming. *Management Science*, 39:757–776, 1993.
34. A.P. Kamath, N.K. Karmarkar and K.G. Ramakrishnan. Computational and complexity results for an interior point algorithm for multicommodity flow problems. Technical Report TR-21/93, Dipartimento di Informatica, Università di Pisa, 1993. Extended abstracts of *NETFLOW'93*.
35. A. Kamath and O. Palmon. Improved interior point algorithms for exact and approximate solution of multicommodity flow problems. Technical report, Department of Computer Science, Stanford University, Stanford, CA 94305, 1994.
36. S. Kapoor and P.M. Vaidya. Fast algorithms for convex quadratic programming and multicommodity flows. In *Proceedings of 18th Annual ACM Symposium on the Theory of Computing*, 147–159, 1986.
37. N. Karmarkar. A new polynomial-time algorithm for linear programming. *Combinatorica*, 4:373–395, 1984.
38. N.K. Karmarkar and K.G. Ramakrishnan. Computational results of an interior point algorithm for large scale linear programming. *Mathematical Programming*, 52:555–586, 1991.
39. J.L. Kennington and R.V. Helgason. *Algorithms for Network Programming.* John Wiley and Sons, New York, NY, 1980.
40. D. Klingman, A. Napier and J. Stutz. Netgen: A program for generating large scale capacitated assignment, transportation, and minimum cost flow network problems. *Management Science*, 20:814–821, 1974.
41. J.B. Kruskal. On the shortest spanning tree of a graph and the traveling salesman problem. *Proceedings of the American Mathematical Society*, 7:48–50, 1956.
42. E. Lawler. *Combinatorial Optimization: Networks and Matroids.* Holt, Rinehart and Winston, 1976.
43. J.M. Mulvey. Testing of a large scale network optimization program. *Math-*

*ematical Programming*, 15:291–315, 1978.

44. B.A. Murtagh and M.A. Saunders. MINOS 5.1 user's guide. Technical Report SOL 83-20R, Systems Optimization Laboratory, Department of Operations Research, Stanford University, Stanford, CA 94305, 1983. Revised 1987.
45. J.B. Orlin. A faster strongly polynomial minimum cost flow algorithm. In *Proceedings of 20th Annual ACM Symposium on the Theory of Computing*, 377–387, 1988.
46. C.C. Paige and M.A. Saunders. Algorithm 583 – LSQR: Sparse linear equations and least square problems. *ACM Transactions on Mathematical Software*, 8:195–209, 1982.
47. P.M. Pardalos and K.G. Ramakrishnan. On the expected optimal value of random assignment problems: Experimental results and open questions. *Computational Optimization and Applications*, 2:261–271, 1993.
48. P.M. Pardalos and H. Wolkowicz, (eds.). *Quadratic assignment and related problems*, volume 16 of *DIMACS Series in Discrete Mathematics and Theoretical Computer Science*. American Mathematical Society, 1994.
49. L. Portugal, F. Bastos, J. Júdice, J. Paixão and T. Terlaky. An investigation of interior point algorithms for the linear transportation problem. Technical report, Department of Mathematics, University of Coimbra, Coimbra, Portugal, 1993. To appear in *SIAM J. Sci. Computing*.
50. L. Portugal, M.G.C. Resende, G. Veiga and J. Júdice. A truncated primal-infeasible dual-feasible network interior point method. Technical report, AT&T Bell Laboratories, Murray Hill, New Jersey, 1994.
51. R.C. Prim. Shortest connection networks and some generalizations. *Bell System Technical Journal*, 36:1389–1401, 1957.
52. A. Rajan. An empirical comparison of KORBX against RELAXT, a special code for network flow problems. Technical report, AT&T Bell Laboratories, Holmdel, NJ, 1989.
53. K.G. Ramakrishnan, N.K. Karmarkar and A.P. Kamath. An approximate dual projective algorithm for solving assignment problems. In D.S. Johnson and C.C. McGeoch, (eds.), *Network Flows and Matching: First DIMACS Implementation Challenge*, *DIMACS Series in Discrete Mathematics and Theoretical Computer Science*, 12:431–451. American Mathematical Society, 1993.
54. M.G.C. Resende, T. Tsuchiya and G. Veiga. Identifying the optimal face of a network linear program with a globally convergent interior point method. In W.W. Hager, D.W. Hearn and P.M. Pardalos, (eds.), *Large Scale Optimization: State of the Art*, 362–387. Kluwer Academic Publishers, 1994.
55. M.G.C. Resende and G. Veiga. Computing the projection in an interior point algorithm: An experimental comparison. *Investigación Operativa*, 3:81–92, 1993.

56. M.G.C. Resende and G. Veiga. An efficient implementation of a network interior point method. In D.S. Johnson and C.C. McGeoch, (eds.), *Network Flows and Matching: First DIMACS Implementation Challenge, DIMACS Series in Discrete Mathematics and Theoretical Computer Science*, 12:299–348. American Mathematical Society, 1993.
57. M.G.C. Resende and G. Veiga. An implementation of the dual affine scaling algorithm for minimum cost flow on bipartite uncapacitated networks. *SIAM Journal on Optimization*, 3:516–537, 1993.
58. Günther Ruhe. *Algorithmic Aspects of Flows in Networks*. Kluwer Academic Publishers, Boston, MA, 1991.
59. B. Simeone, P. Toth, G. Gallo, F. Maffioli and S. Pallottino, (eds.). *Fortran codes for network optimization*, volume 13 of *Annals of Operations Research*. J.C. Baltzer, Basel, Switzerland, 1988.
60. L.P. Sinha, B.A. Freedman, N.K. Karmarkar, A. Putcha and K.G. Ramakrishnan. Overseas network planning. In *Proceedings of the Third International Network Planning Symposium - NETWORKS'86*, 8.2.1–8.2.4, June 1986.
61. É. Tardos. A strongly polynomial algorithm to solve combinatorial linear programs. *Operations Research*, 34:250–256, 1986.
62. R.E. Tarjan. *Data Structures and Network Algorithms*. Society for Industrial and Applied Mathematics, Philadelphia, PA, 1983.
63. P.M. Vaidya. Speeding up linear programming using fast matrix multiplication. In *Proceedings of 30th IEEE Annual Symposium on Foundations of Computer Science*, 332–337, 1989.
64. R.J. Vanderbei, M.S. Meketon and B.A. Freedman. A modification of Karmarkar's linear programming algorithm. *Algorithmica*, 1:395–407, 1986.
65. S.A. Vavasis and Y. Ye. Accelerated interior point method whose running time depends only on $A$ (extended abstract). In *Proceedings of 26th Annual ACM Symposium on the Theory of Computing*, 512–521, 1994.
66. Q.-J. Yeh. *A reduced dual affine scaling algorithm for solving assignment and transportation problems*. PhD thesis, Columbia University, New York, NY, 1989.

# 5

# Branch and cut algorithms

**Abilio Lucena**

*Laboratório Nacional de Computação Científica,*
*Rio de Janeiro, Brazil 22290*

**John E. Beasley**

*The Management School, Imperial College, London SW7 2AZ, England*

## 1 Introduction

Let $\mathbb{R}_+^p$ and $\mathbb{Z}_+^n$ be, respectively, the set of non-negative real-valued $p$-dimensional vectors and the set of non-negative integral valued $n$-dimensional vectors. For variables $x \in \mathbb{R}_+^p$ and $y \in \mathbb{Z}_+^n$, a *linear mixed-integer programming* (MIP) problem is given by

$$\min\{cx + hy : \ Ax + Gy \geq b, \ x \in \mathbb{R}_+^p, \ y \in \mathbb{Z}_+^n\}, \tag{1.1}$$

where $c$ is a $p$-vector, $h$ is an $n$-vector, $b$ is an $m$-vector, $A$ is an $m \times p$ matrix, and $G$ is an $m \times n$ matrix. The *feasible region* associated with (1.1) is

$$P = \{Ax + Gy \geq b, \ x \in \mathbb{R}_+^p, \ y \in \mathbb{Z}_+^n\}, \tag{1.2}$$

while

$$z = cx + hy \tag{1.3}$$

is the *objective function* to be optimized over $P$. Whenever all variables in (1.1) are restricted to be integral, the problem is known as a *linear integer programming* (IP) problem. Conversely, a pure *linear programming* (LP) problem results when all variables in (1.1) are restricted to be real valued.

As is frequently the case for MIP, instead of attempting to optimize (1.3) directly over $P$, it may be advantageous to divide that region into a finite number of smaller regions and optimize the objective function over each smaller region individually. In the process one must ensure that an optimal solution to (1.1) is contained in at least one of the smaller regions that were generated. This basic solution strategy can be further extended to include the newly defined MIP problems. In particular, the idea would be attractive when the resulting

The authors wish to thank Dr. C.E. Ferreira of the University of São Paulo for some helpful comments on this chapter.

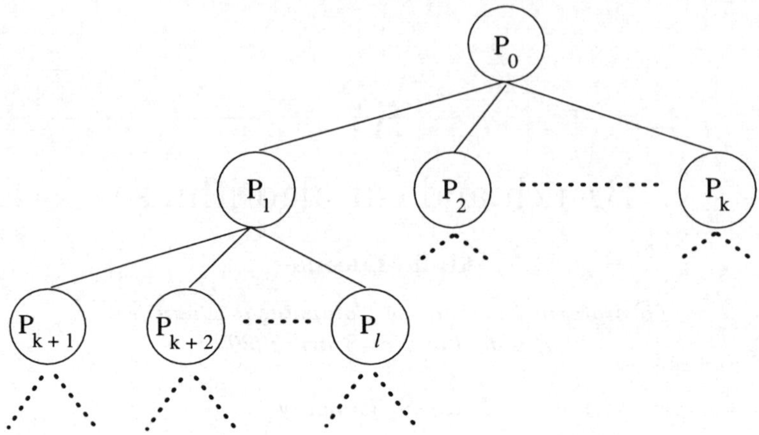


FIG. 1. Search tree

problems cannot be easily solved with the tools to hand. For most MIP problems, taken to the limit, the idea would imply almost total enumeration. Therefore it would only be of use for extremely small problems. A convenient representation for the process just described is in terms of a tree (see Fig. 1). A node in the *search tree* of Fig. 1, say node $k$, represents the MIP problem given by

$$\min\{cx + hy : (x, y) \in P_k\}, \tag{1.4}$$

where $P_k \subseteq P$. Notice that problem (1.1) would be represented, in this case, at the root node ($k = 0$) of the tree (i.e. $P_0 = P$). As one moves down the search tree, the regions $P_k$ become smaller and smaller. Eventually, one may be able, for a given $P_k$, to either solve (1.4) to optimality, or else determine that no optimal solution to (1.1) is contained in $P_k$. For such a scheme to be applicable in practice, one tends to work with *relaxations* of (1.4) instead of (1.4) proper. For instance, a relaxation of (1.4) is

$$\min\{w(x, y) : (x, y) \in R_k\}, \tag{1.5}$$

where $P_k \subseteq R_k$ and $w(x, y) \leq cx + hy, (x, y) \in R_k$. Clearly, (1.5) gives a lower bound on (1.4). To be fair, nevertheless, (1.5) must be a lot easier to solve than (1.4). An example of a relaxation that is going to be extensively used in this chapter is the *LP relaxation.* A valid LP relaxation of (1.1), for instance, has the objective function $w(x, y) = cx + hy$ and the feasible region $R = \{Ax + Gy \geq b, \; (x, y) \in \mathbb{R}_+^{p+n}\}$. One should notice that the solution strategy outlined above is considerably enhanced with the introduction of MIP relaxations. With that modification in place, at a given (search) tree node, whenever the associated MIP problem cannot be easily solved, a relaxation of it is solved instead. As a result, a better judgement can be made of the subsequent actions to follow. The solution strategy can be enhanced even further if one is also capable of generating a good upper bound (corresponding to a feasible solution of

1.1), $v(ub)$ say. This would result in smaller search trees since a tree node (i.e. the optimization problem defined over, say $P_k$) is *fathomed* (i.e. the corresponding problem can be discarded) whenever it can be established that $P_k$ does not contain a solution that improves on $v(ub)$. The overall scheme just described is generically called *branch and bound* (see Chapter 4 in Garfinkel and Nemhauser [18], Garfinkel [17] or Chapter II.4 in Nemhauser and Wolsey [40]). For branch and bound algorithms branching deals with the process of conveniently dividing feasible regions, at every tree node, into smaller regions. Bounding deals with the generation of upper and lower bounds for the (sub)problems involved.

A *branch and cut* algorithm (see Padberg and Rinaldi [42]) can be viewed as branch and bound where one systematically attempts to obtain stronger LP relaxations (at every node of the search tree). Stronger is used here in the sense of implying larger objective function values. In particular, this is attempted with the aid of *cutting planes* (i.e. linear inequalities which, depending on their type, could be valid only for the feasible regions implied or else globally valid across the search tree). Cutting planes in branch and cut algorithms are polyhedral in nature (see Chapter I.4 in Nemhauser and Wolsey [40] for the theory of polyhedral cutting planes) and are globally valid across the search tree. This represents a major departure from traditional cutting planes such as Gomory cuts (see Gomory [23, 24]), for instance. These usually can be guaranteed to be valid only for the subproblems for which they were generated. This would imply, amongst other things, massive computer memory requirements for storing the cuts generated during the search tree. Furthermore, in general, polyhedral cutting planes do not possess the (widely reported) poor convergence properties associated with the use of general cutting planes (Gomory cuts in particular). Polyhedral cutting planes are easier to explain if one concentrates on IPs. The basic ideas conveyed, nevertheless, can be readily extended to MIPs.

Consider a polyhedral region

$$R = \{x \in \mathbb{R}^n_+ : \ Ax \geq b\}, \tag{1.6}$$

where $(A, b)$ is an $m \times (n+1)$ matrix. Then the IP problem we consider is:

$$\min\{cx : x \in P\}, \tag{1.7}$$

where $P = R \cap \mathbb{Z}^n_+$ is the feasible region associated with this problem. Defining $conv(P)$ as the convex hull of the elements of $P$, it is possible to show (see page 106 in [40]) that the IP can be solved to optimality by simply solving the LP program

$$\min\{cx : x \in conv(P)\}. \tag{1.8}$$

Therefore, the key message conveyed is that the more an LP relaxation of (1.7) approximates $conv(P)$, the better. As much as is practicable, therefore, one should aim at obtaining the strongest possible approximations of $conv(P)$ as the LP relaxation of choice. For most IP problems, an explicit description of *conv*$(P)$ by linear inequalities is out of the question. Nevertheless, experience shows that dramatic computational gains can be realized by working with "good partial descriptions" of *conv*$(P)$. In essence, at a tree node of a branch and cut algorithm, an LP relaxation of the problem (subproblem) is made stronger with the introduction of cutting planes that induce *facets* (or, if these are not available, hopefully, *faces* of high dimension) of the polyhedron *conv*$(P)$.

We will describe the key elements in a branch and cut algorithm by concentrating on two combinatorial optimization problems: the Minimum Spanning Tree (MST) problem and the Steiner Problem in Graphs (SPG). For a given connected graph, MSTs can be obtained in polynomial time while the SPG is NP-hard even for planar graphs (see Garey and Johnson [16]). In spite of this substantial difference, as will become apparent in the following sections, the two problems have quite a lot in common.

The term "branch and cut" was coined by Padberg and Rinaldi [42]. Most of the basic methodology associated with branch and cut algorithms can be traced to the seminal work of Dantzig *et al.* [11] for the travelling salesman problem (TSP). A considerable amount of experimentation has occurred before arriving at the fairly consensual setting for branch and cut algorithms that exists today. An incomplete list of references that contributed to this state of affairs follows.

Grötschel [26] solved a 120-city TSP using polyhedral cutting planes (which were generated visually). Padberg and Hong [41] used an automatic procedure to solve to optimality TSPs with up to 120 cities. Crowder and Padberg [10] used the technique to solve TSPs of sizes up to 318 cities. They used polyhedral cutting planes (which were generated automatically) and a commercially available MIP solver. For [10] cutting planes were only used to obtain an initial LP relaxation bound. From that point on the MIP solver was used to perform branch and bound whenever optimality could not be achieved at the root node of the search tree. Grötschel *et al.* [27] described an exact solution algorithm for the Linear Ordering Problem that could be considered as the very first true branch and cut algorithm. In [27] polyhedral cutting planes could be generated at any node of the search tree. Padberg and Rinaldi [42] pushed the methodology to new heights with their branch and cut algorithm for the TSP. They were capable of solving to proven optimality TSPs with 2,392 cities. Finally, Applegate, Bixby, Chvátal and Cook [2] developed a branch and cut algorithm that has been able to solve to proven optimality a TSP with 7,397 cities.

## 2 Solving the Minimum Spanning Tree problem by LP

Let $G = (V, E)$ be a finite, connected, undirected graph with a set of vertices $V$ and a set of edges $E$. Sets $V$ and $E$ have cardinalities $m = |V|$ and $n = |E|$ respectively. A tree is a connected acyclic partial subgraph of $G$. The tree is said to be spanning if it contains exactly $|V| - 1$ edges. Given costs $c_e$ for all edges $e \in E$ the cost of a tree equals the sum of the costs of the edges it contains. The MST problem is to find a spanning tree of minimum cost. For a connected graph $G$, the greedy algorithm of Kruskal [31]:

- **Input:** A connected graph $G = (V, E)$.
- **Algorithm:** "Select, in increasing value of their costs, edges that do not form a cycle with edges already chosen".
- **Output:** An MST of $G$.

finds an MST of $G$ in $O(|E|\log|V|)$ time. For the purposes of this chapter, nevertheless, we are also going to generate MSTs from solutions to linear programs. This is in spite of the fact that LP should not be expected to be competitive with the greedy algorithm given above. Rather the objective here is to allow a smooth introduction to branch and cut algorithms for NP-hard problems. Some of the key ingredients found in those algorithms are to be highlighted in this process.

Specifically, we are going to solve an LP formulation of the MST problem with exponentially many inequalities. This is made possible by explicitly using an inequality only when it becomes violated in a relaxation of the original formulation. As violated inequalities are introduced into the relaxation and the resulting strengthened relaxation reoptimized, we will eventually reach a point where no more violated inequalities exist. At that stage an MST of $G$ will have been found.

In the presence of an excessively large number of variables the algorithm may incorporate the use of "column generation" (see Padberg and Rinaldi [42], for instance). This involves initially restricting attention, in our linear program, to a small, "attractive", subset of all variables. From time to time the remaining variables are checked to discover suitable candidates to join this subset. The selection criteria used in deciding whether a variable is suitable or not is the LP reduced cost. For a minimization problem a negative reduced cost indicates the possibility of further reducing the objective function value. One drawback of this scheme is that, in principle, one can only attempt to fix, a priori, the value of variables when no variable with a negative reduced cost exists (i.e. when the LP relaxation to the original IP problem has been obtained). In this chapter, we propose a scheme which avoids that restriction and allows for the early fixing of variables.

Given a set $S \subseteq V$ of vertices, let $E(S) \subseteq E$ be the set of edges with both endpoints in $S$. Associate a real-valued variable $x_e$ with every edge $e \in E$ and denote by $x(E(S))$, $S \subseteq V$, $E(S) \subseteq E$, the sum $\sum_{e\in E(S)} x_e$. Consider the polyhedral region $P_0$ defined by

$$x(E) = |V| - 1, \tag{2.1}$$

$$x(E(S)) \leq |S| - 1, \quad S \subseteq V, \quad 2 \leq |S| \leq |V| - 1 \tag{2.2}$$

$$x_e \geq 0, \quad e \in E. \tag{2.3}$$

The MST problem can be formulated as

$$\min\{\sum_{e\in E} c_e x_e : x_e \in P_0\}, \tag{2.4}$$

where inequalities (2.2) are known as subtour elimination constraints (SECs). It can be shown that the extreme points of the constraint set (2.1)–(2.3) correspond exactly to incidence vectors of spanning trees in $G$. (For a discussion on a similar description for a closely related polyhedron, the 1-tree polytope, see page 262 of Grötschel and Padberg [28].) Therefore, for every spanning tree of $G$, the incidence vector $x \in \mathbb{R}^{|E|}$, of components $x_e = 1$ if edge $e$ is in the tree and $x_e = 0$ if edge $e$ is not in the tree, has a one-to-one correspondence with the extreme points of (2.1)–(2.3). The convex hull of incidence vectors of spanning trees in $G$ (i.e. the spanning tree polytope) is therefore completely described by the linear inequalities in (2.1)–(2.3). This is certainly a very special situation which arises from the association of the MST problem with matroids (see Lawler [32] for applications of matroids to combinatorial optimization in general and Edmonds [13] for the more specific case discussed here). In general, obtaining complete linear descriptions for polytopes associated with more complex (NP-hard) combinatorial optimization problems is no easy task. In those situations, one would have to content oneself with partial descriptions of the polytopes involved.

An optimal solution to (2.4) would imply an optimal MST for $G$. Nevertheless, such a linear program would be enormous, even for fairly small graphs. For instance, it would involve 380 variables and in excess of one million constraints for a complete graph $G$ of only 20 vertices. It would therefore be unwise, in general, to attempt to solve (2.4) by making explicit use of all inequalities in (2.2). This is further supported by the fact that only a few of these inequalities need be explicitly used to obtain an optimal solution to the problem.

### 2.1 Solving the MST problem by cutting planes

Consider the polyhedral region $R_0$ defined by:

$$x(E) = |V| - 1 \tag{2.5}$$

$$0 \leq x_e \leq 1 \quad e \in E. \tag{2.6}$$

Since $P_0 \subseteq R_0$,

$$\min\{\sum_{e \in E} c_e x_e : x_e \in R_0\} \tag{2.7}$$

gives an LP relaxation of (2.4). $R_0$ consists of constraint (2.1) plus SECs for subsets $S$ of $V$ with $|S| = 2$. It is therefore clear that, if an optimal solution to (2.7) implies a spanning tree of $G$, then this must also be an optimal solution to (2.4) as well. Whenever this is not the case, one or more inequalities (2.2) for $3 \leq |S| \leq |V| - 1$ are violated by that solution. One can then append violated inequalities to (2.7) and reoptimize the resulting program. Once again, if a spanning tree of $G$ is implied (by the resulting optimal solution) we are done. Otherwise, additional inequalities (2.2) for $3 \leq |S| \leq |V| - 1$ are violated at the new LP solution. As before, inequalities can be appended to the linear program which, in turn, would be reoptimized yet again. Proceeding in this way, in a finite number of steps, one would eventually solve the MST problem. Note here that the optimal simplex basis one has prior to introducing additional constraints is dual feasible for the newly defined problem. The dual simplex algorithm can thus be started from that initial (advanced) basis. We have used the XMP LP solver of Marsten [39] for the computational tests reported in this chapter.

A crucial step in the procedure discussed above is "given a point $\overline{x} \in \mathbb{R}^n$ find an inequality of type (2.2) that is violated by $\overline{x}$ or prove that no such inequality exists". This is known as the *separation problem* for subtour elimination constraints.

## 2.2 The separation problem for SECs

For an optimal solution $\{\overline{x}_e : e \in E\}$ to (2.7), the separation problem for SECs can be tackled by determining, in turn, for each $l \in V$,

$$\max\{ \sum_{e \in E(S)} \overline{x}_e - (|S| - 1) : \; S \subseteq V, \; l \in S\}. \tag{2.8}$$

Whenever this maximum is non-positive, no violated SEC, involving $l$, currently exists. Otherwise, $S$ is associated with the most violated SEC associated with vertex $l$. Let $z_i = 1$ if $i \in S$ and $z_i = 0$ otherwise. An optimal solution to the Boolean quadratic problem

$$\begin{aligned} &\text{maximize} && \textstyle\sum_{(i,j)\in E} \overline{x}_{(i,j)} z_i z_j - (\sum_{i \in V} z_i - 1) && (2.9)\\ &\text{subject to} && z_l = 1 && (2.10)\\ &&& z_i \in \{0,1\}, \quad i \in V && (2.11) \end{aligned}$$

implies a set $S \subseteq V$, $l \in S$, that solves (2.8).

### 2.2.1 *The separation of SECs as a maximum flow (minimum cut) problem*

Problem (2.9)–(2.11) can be reformulated as the problem of finding a maximum flow on a companion network and thus can be solved to optimality by LP (see Picard and Ratliff [44]). An algorithm for the separation of SECs, in this vein, due to Padberg and Wolsey [43], was used in our computational experiments (an earlier reference for the separation of SECs is the algorithm of Gomory and Hu [25]). For a description of the Padberg and Wolsey [43] algorithm, let $F = (\overline{V}, A)$ be the companion network. Network $F$ contains all vertices in $V$ associated with a non-zero $\overline{x}_e$, $e \in E$. Additionally, $F$ contains a source node and a sink node labelled, respectively, 0 and $n+1$. Arc set $A$ has, for each edge $e = (i,j)$, $e \in E$, with $\overline{x}_e > 0$, two arcs $[i,j]$ and $[j,i]$ of capacities $\xi_{ij} = \xi_{ji} = \overline{x}_e/2$. Denote by $\delta(i)$, $i \in V$, all the edges of $E$ with exactly one end point in $i$ and let $\xi_i = \sum_{j \in \delta(i)} \xi_{ij}$, for each $i \in \overline{V}$. The source node has arcs $[0,i]$, $i \in \overline{V}$ of capacity $\xi_{0i} = \max\{\xi_i - 1, 0\}$ associated with it. The sink node, on the other hand, has arcs $[i, n+1]$, $i \in \overline{V}$, of capacity $\xi_{i,n+1} = \max\{1 - \xi_i, 0\}$. For a given $l \in \overline{V}$, an optimal solution to (2.9)–(2.11) corresponds to a maximum flow (minimum cut) over $F = (\overline{V}, A)$ with $\xi_{0l}$ set to infinity. This problem can be solved to optimality in $O(|V|^3)$ time. Therefore separation for SECs can be carried out in $O(|V|^4)$ time (one maximum flow problem for every $l \in V$).

Since one can generate *duplicate* copies of the same violated subtour by following the procedure above, one can (see [43]), instead generate a set of violated subtours that contains *no duplicates.* This is done by solving $n-2$ maximum flow problems where, for problem $k$, $\xi_{0k} = \infty$, and if $k \geq 2$, $\xi_{v,n+1} = \infty$ for $v = 1, \cdots, k-1$. In this way, if $|S| \geq 3$, $S$ satisfies $k \in S$ and $\{1, \cdots, k-1\} \in V \backslash S$, for some $k = 1, \cdots, n-2$. Furthermore, every $(0, n+1)$-cut in $F$ having $|S| = 1$ has capacity 1 and any $(0, n+1)$-cut having $|S| = 2$ has capacity not less than 1. For the computational results presented in this chapter, we have used the code of Goldfarb and Grigoriadis [22] to solve maximum flow problems.

Repeatedly solving large-scale maximum flow problems tends to be fairly time consuming. As a result, an attempt to heuristically solve the separation problem for SECs is an attractive proposition. One such heuristic, proposed by Crowder and Padberg [10], has proved effective and involves a systematic shrinking of edges into vertices. In the example that follows we describe the use of a row (SEC) generation scheme to solve the MST problem.

## 2.3 Example

Let us consider a complete graph $G$ of 10 vertices in the Euclidean plane. Vertex coordinates were randomly chosen from the uniform distribution in the range [0, 100] and are respectively (1,10), (47, 87), (0, 23), (30, 75), (76, 86), (32, 22), (45, 68), (97, 74), (73, 70) and (66,69). For every edge $e = (i, j), \ \ e \in E, \ \ i, j \in V$, we associate a cost $c_e$ that equals the truncated Euclidean distance between the vertices.

### 2.3.1 *First iteration*

An optimal solution of value 155 is obtained from (2.7). Its support graph is shown in Fig. 2 and contains all the edges associated with non-zero variables in the solution. In all the support graphs for the simple example presented in this section the variable values in the solution to (2.7) were always integer (zero or one). For ease of reference we shall also index the sets $S$ in a sequential fashion.

The straightforward use of the separation procedure in Section 2.2.1 would result in a set $S_0 = \{1, 2, 4, 5, 7, 9, 10\}$ being returned in association with vertex 1. Nevertheless, it is clear from Fig. 2 that strictly speaking vertex 1 does not truly belong to a subtour. Indeed the only two obvious subtours in the support graph are implied by $S_1 = \{2, 4, 7\}$ and $S_2 = \{5, 9, 10\}$. Therefore, in principle, one has little to gain in working with $S_0$. Furthermore, it should be noticed that since vertex 1 has an edge cardinality of one (in the support graph) it could have been dropped from consideration without any loss of information. The same would apply to vertex 8 which also has edge cardinality one. This would result in a set $S_3 = \{2, 4, 5, 7, 9, 10\}$ being selected in association with vertex 2. For vertex 4, a set $S_4 = \{4, 5, 7, 9, 10\}$ would be the choice. Finally, for vertex 5, a set $S_5 = \{5, 9, 10\}$ would be the one returned by the separation procedure.

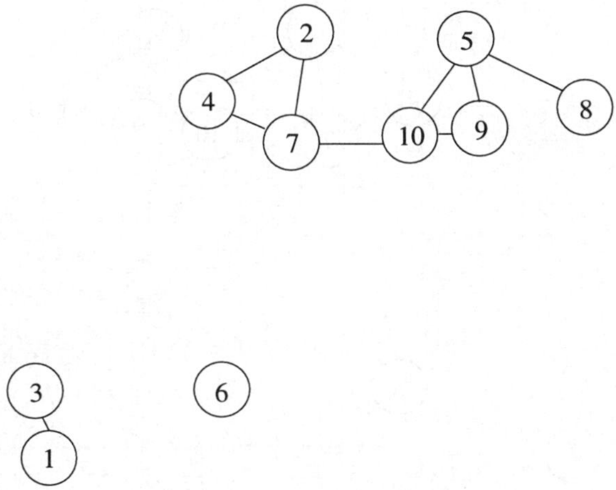


FIG. 2. Initial support graph

One may notice that $S_4$, although correctly implying a violated SEC, is not exactly adequate for our purposes. For such an SEC, the difference between left and right-hand sides is equal to one. Indeed, that equals the violation for the SEC implied by $S_5 \subset S_4$. Therefore, one has little to gain in using $S_4$ instead of simply using $S_5$. A clear message here is that, ideally, one should process the output of the separation routine in order to curb discrepancies like the ones described. Dropping from the support graph vertices with edge cardinality one (or less) was as far as we went in terms of processing the results.

### 2.3.2 *Second iteration*

We now strengthen $R_0$ (i.e. strengthen the LP relaxation) with the introduction of the SECs associated with $S_3$, $S_4$ and $S_5$. The resulting linear program has an optimal solution of value 170. The associated support graph is shown in Fig. 3. Notice that the SEC associated with $S_4$ became non-binding for the solution. It could therefore be dropped without any loss of information. That helps to keep the size of the linear program small and should, in general, be enforced.

Applying, once again, the separation procedure of Section 2.2.1 would return a set of vertices $S_6 = \{2, 4, 5, 7, 8, 9, 10\}$ in association with vertex 2. Notice that vertices 1, 3 and 6 are not taken into consideration for the present analysis since all of them have an edge degree (in the support graph) which is less than or equal to one. In addition, a closer examination of $S_6$ shows that it can be partitioned into two subsets, $\{2, 4, 7\}$ and $\{5, 8, 9, 10\}$ respectively, each of which implies a violated SEC. It would therefore be advantageous to replace the SEC associated with $S_6$ by those two SECs. Once again we did not process the output of the separation procedure in this respect. Therefore, we simply appended the SEC associated with $S_6$ to $R_0$. Proceeding with the separation, sets $S_7 = \{4, 5, 8, 9, 10\}$, $S_8 = \{5, 8, 9, 10\}$, and $S_9 = \{8, 9, 10\}$ are selected for vertices 4, 5 and 8, respectively. Notice, once again, that $S_7$ would be inadequate

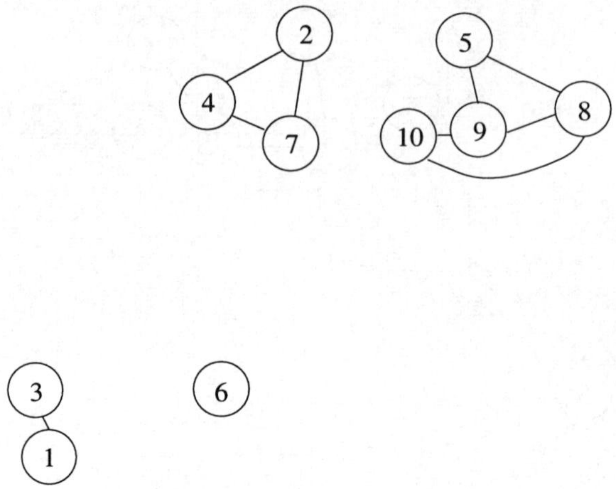

FIG. 3. Second support graph

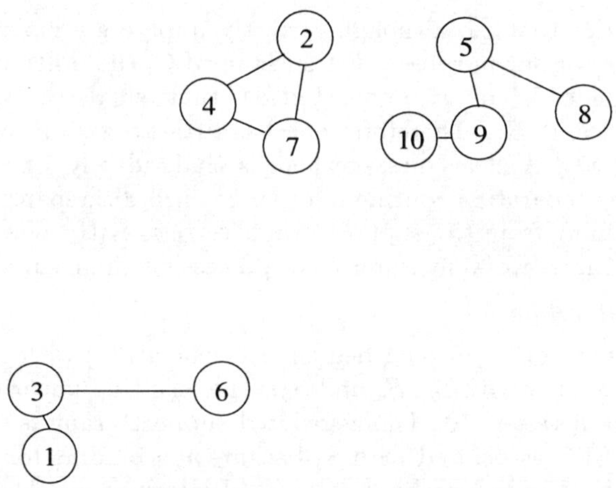

FIG. 4. Third support graph

to be used, in the same sense as $S_4$, before, was.

### 2.3.3 *Third iteration*

Now we strengthen the LP relaxation with the introduction of the SECs associated with $S_6$, $S_7$, $S_8$ and $S_9$. An optimal solution of value 180 is then obtained. The associated support graph is shown in Fig. 4. Subtours associated with $S_7$ and $S_9$ are non-binding for that solution and could therefore be dropped from the linear program. We have done so. The separation procedure would now return $S_{10} = \{2, 4, 7\}$ as the set of vertices associated with a violated SEC in the support graph. This is generated in association with vertex 2. This is indeed the correct action to be taken and no further remark applies here.

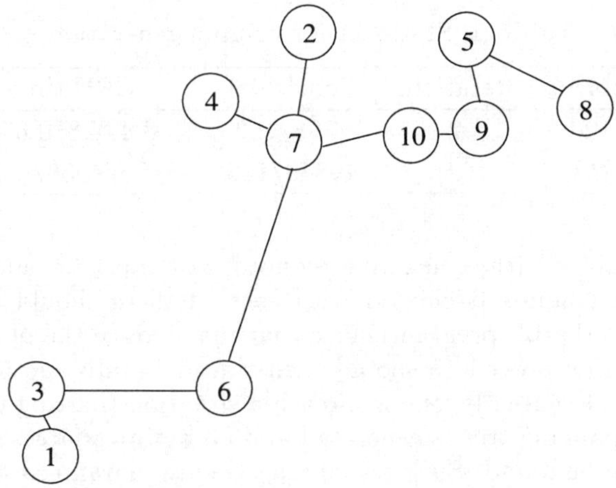


FIG. 5. Fourth support graph

### 2.3.4 *Fourth iteration*

Now we strengthen the LP relaxation with the introduction of the SEC associated with $S_{10}$. An optimal solution of value 195 is then obtained. The associated support graph is shown in Fig. 5. As can be appreciated from that figure, an MST is implied by the solution. This can be algorithmically checked by noticing that all non-zero variables have value one, all the vertices of $G$ appear in the support graph, and, finally, no violated SEC exists in the support graph.

## 2.4 The use of column generation

Assume one has made the unlikely choice of using LP to find MSTs. Then, for large-scale problems, it would be impracticable and inefficient to work with LP programs that explicitly contain all the variables in (2.4). Clearly, the number of non-zero variables in an optimal solution to (2.4), i.e. $|V| - 1$, is quite small. Furthermore, in general, it seems reasonable to assume that very few of the remaining variables would have much chance of assuming non-zero values in an optimal solution to one of the (intermediate) linear programs being generated. For those situations, one would, most likely, be better off resorting to *column generation*.

In column generation, only a small, conveniently chosen, subset of the set of all variables is initially used in (2.7). Such a subset is supposed to contain variables that imply a spanning tree of $G$. Once such a "relaxation" of the MST problem is defined one can proceed as if a proper relaxation of the problem is in place. As explained before, this would involve detecting violated SECs, appending them to the linear program and reoptimizing. Eventually, one would reach a stage where no violated inequalities can be found and, for this specific application, a spanning tree (that provides an upper bound on the original problem) is obtained. At this stage, one should compute reduced costs for those variables that were left out of the LP program. This is sometimes known as *pricing* the

**Table 1** Statistics for column generation

| Problem | Iterations | Constraints | CPU time |
|---|---|---|---|
| G120 | 2 | 398/153 | 8.4/4.8/0.075 |
| L318 | 2 | 49846/1503 | 1426/4956/0.496 |

variables. Variables with a negative reduced cost must be added to the "relaxation" (i.e. a column associated with each of them should be "generated" and appended to the LP program) since that may reduce the objective function value. The solution procedure should then continue, into another iteration, as described above. Eventually, one will reach a situation (barring numerical problems) where a spanning tree is generated and no more variables with a negative reduced cost can be found. Such a spanning tree is guaranteed to be an MST.

### 2.4.1 *Computational results*

As an example of the use of column generation to obtain MSTs, we have worked with two test problems defined over complete graphs. The first test problem (denoted by G120) has 120 vertices, which correspond to German cities, and has been generated by Grötschel [26]. The costs involved reflect distances between pairs of cities. The second test problem (denoted by L318) has 318 vertices. It has been generated using pseudo-Euclidean edge distances (costs). For column generation, our initial choice of variables was to select, for every vertex $i \in V$, the $m_c$ cheapest edges that contain the vertex. In addition, variables associated with the edges of the readily available Hamiltonian path (spanning tree) $\{(1,2),(2,3),\ldots,(n-1,n)\}$ were also selected. This guarantees the existence of a spanning tree among the variables (edges) selected. For the first test problem we fixed $m_c$ to 3 whilst for the second test problem we fixed it to 4. We have found it advantageous to append inequalities $\sum_{e\in\delta(i)} x_e \geq 1, i \in V$ to (2.7). Therefore our initial LP relaxation involved $|V|+1$ inequalities. Computational results are described in Table 1. The first entry in the table identifies the test problem. The second entry gives, for each problem, the number of column generation iterations before optimality was proven. The third entry gives, respectively, the total number of SECs that were generated during the solution procedure and the number of constraints in the linear program (i.e. the number of SECs plus $(|V|+1)$) at the time optimality was proven. The fourth entry gives, respectively, the CPU time to find an MST by using the combination of row and column generation, the CPU time to find an MST by simply using row generation and the CPU time taken by an implementation of the algorithm of Kruskal [31] to solve the problem. All codes involved were written in FORTRAN 77 and all the CPU times are in IBM RISC 6000/370 seconds.

A quick look at Table 1 confirms our assertion that it would be ludicrous to attempt to solve the MST problem by LP. Interestingly enough the table also shows, for problem L318, the advantage of the use of column generation as compared with dealing with all LP variables explicitly.

## 2.5 The early fixing of variables

In principle, one should always be capable of generating an MST from an optimal solution to relaxation (2.7). However, for the SPG, it can happen that one must resort to branching in order to reach optimality. That task can be made computationally cheaper if a reduction in problem size can be achieved prior to starting with branching. Let us generically assume that the problem under investigation is a minimization one (as for the SPG). Then, an LP relaxation bound gives a valid lower bound on the optimal solution value to the problem. Suppose also that an upper bound for the problem is known. The LP relaxation reduced costs could then be used, together with the lower and upper bounds, in an attempt to fix variables. It is simple to conclude that a variable can be eliminated if it has an LP relaxation reduced cost that exceeds the difference between the upper bound and the current lower bound. Tests based on this fact have been extensively used in branch and bound algorithms in general, and in branch and cut algorithms in particular. In the presence of column generation, a similar test for fixing variables can, indeed, be used much earlier than suggested in the literature. As stated previously we propose a scheme which allows for the early fixing of variables. We first present this scheme in general terms and then apply it to the MST.

### 2.5.1 *The early fixing of variables – general*

Let $C$ represent the set of constraints for the problem (a minimization problem). $C$ may include, apart from constraints in the original problem formulation, some additional valid inequalities. Let the variables be denoted by $X$. Denote by $Pr(C,X)$ the integer minimization problem involving $C$ and $X$ and let its optimal solution be $Z_{opt}$. Let $Z_{upper}$ be any upper bound (e.g. derived by some heuristic algorithm) on $Z_{opt}$.

Divide $X$ into two subsets $X_{in}$ and $X_{out}$ where $X = X_{in} \cup X_{out}$ and $X_{in} \cap X_{out} = \emptyset$. In column generation we solve the problem, but considering (at least initially) just the set $X_{in}$ of variables, i.e. we solve the LP relaxation of $Pr(C, X_{in})$. Accomplishing this by branch and cut entails finding a subset $C_{active}$ of $C$ that is sufficient for explicit consideration (i.e. the constraints $C \setminus C_{active}$ are satisfied implicitly). Let the solution to this LP relaxation be $X^*$ of value $Z^*$.

Without significant loss of generality assume that the set $X_{in}$ has been defined such there exists at least one integer feasible solution to the problem $Pr(C, X_{in})$. Then if $X^*$ is integer, $Z^*$ must be an upper bound on the optimal solution to the original problem $Pr(C,X)$, since $X^*$ explicitly satisfies the constraints $C_{active}$, implicitly satisfies all other constraints $(C \setminus C_{active})$, and is integer. In this case we can update $Z_{upper}$ using $Z_{upper}=\min[Z_{upper}, Z^*]$.

Take the original problem $Pr(C,X)$ and enlarge it to $Pr(C + C_{active}, X)$, i.e. duplicate the constraints $C_{active}$. This plainly leaves the optimal solution value $Z_{opt}$ unchanged. Suppose we were to attempt to obtain a Lagrangian lower bound on $Z_{opt}$ by relaxing the duplicated constraints, with as Lagrange multipliers, the dual variables associated with the constraints $C_{active}$ in the LP relaxation solution of $Pr(C_{active}, X_{in})$. We would produce a Lagrangian lower bound program $Pr(C, X)$ involving in its objective function costs for the variables $X$ plus a constant term. Because of the way we have defined our multipliers this constant term must be equal to $Z^*$. Moreover the Lagrangian costs for the variables $X$ will be equal to their LP reduced costs in the optimal solution of the LP relaxation of $Pr(C_{active}, X_{in})$. In particular these Lagrangian costs must be positive ($\geq 0$) for all variables in $X_{in}$ but can be negative ($< 0$) for variables in $X_{out}$, a negative reduced cost indicating that the variable should be introduced into the active set $X_{in}$ (i.e. priced into the active set in a column generation algorithm).

Hence we have that a valid (Lagrangian) lower bound on the optimal solution $Z_{opt}$ to the original problem is given by $Z^*$ plus the optimal solution to the problem $Pr(C, X)$ but involving LP reduced costs rather than the original costs. Let this lower bound be denoted by $Z_{lower}$.

Plainly solving $Pr(C, X)$ is hard, as it is the original problem we were trying to solve. However it may be that, from structural considerations (i.e. our knowledge of the particular problem being considered), we have a number of lower bounds available for $Pr(C, X)$. We now have that a valid lower bound $Z_{lower}$ on the optimal solution to the original problem is given by $Z^*$ plus *any* lower bound to the problem $Pr(C, X)$ but involving LP reduced costs rather than the original costs.

Having both an upper bound $Z_{upper}$ and a lower bound $Z_{lower}$ we can apply tests for variable fixing in a standard way, e.g. if the lower bound $Z_{lower}$ resulting when a particular variable is forced into the solution exceeds $Z_{upper}$ then that variable can never be in the solution.

### 2.5.2 *The early fixing of variables – MST*

Having set out the general scheme above the only particularization needed for the MST is the choice of an appropriate lower bound for the original problem ($Pr(C, X)$ above), in this case the MST. A simple lower bound for the MST is the sum of the $n - 1$ smallest edge costs. Hence a valid MST lower bound would be to add the $n - 1$ smallest reduced costs to the "relaxation" solution value ($Z^*$). As explained above this would clearly give a valid MST lower bound. In association, select a variable with the largest reduced cost, among those $n - 1$ variables chosen, say $x_k$. Then, finally, compute the difference between the LP reduced cost of a given variable and that of $x_k$. Whenever this value exceeds the difference between the MST upper and lower bounds, the variable could be fixed to zero, the underlying reasoning being that a spanning tree of $G$ with a minimum cost, which contains the edge associated with the variable in question, has a value that exceeds the known upper bound. Such a spanning tree is therefore clearly suboptimal.

### 2.5.3 *Computational results*

Table 2 gives computational results for the early fixing of variables for the problems of Table 1. The first entry in the table identifies the test problem. The

**Table 2** Early fixing of variables

| Problem | Upper bound | Lower bound | Optimal solution | Variables eliminated |
|---|---|---|---|---|
| G120 | 5220 | 5208.73 | 5211 | 95.14 % |
| L318 | 38000 | 32629.5 | 37832 | 97.82 % |

second entry gives the optimal solution value for the very first "relaxation" used (i.e. an MST upper bound, $Z^*$ in this case as the LP relaxation solution was naturally integer). The third entry gives the MST lower bound generated from the "relaxation" solution (i.e. $Z_{lower}$). The fourth entry gives the value of an MST. Finally, the fifth entry gives the percentage of variables fixed to zero as a result of the application of our test for the early fixing of variables.

### 2.6 Conclusions

We have described an LP-based approach to solve the MST problem. The motivation behind the study was to provide a smooth transition towards a more in-depth description of algorithmic aspects related to branch and cut algorithms (for NP-hard problems). MSTs are ideally suited for that role. Firstly, a complete non-redundant description of the MST polytope is available. Secondly, the MST problem is very closely related to the Steiner Problem in Graphs (SPG) which is an NP-hard problem. Therefore, by following what was outlined in this section, we are now going to use the SPG as a model in our study of branch and cut algorithms. As a result, a number of formulations for the SPG (one being a generalization of the MST formulation of this section) will be explored.

## 3 Formulations for the Steiner problem in graphs

For a finite undirected graph $G = (V, E)$, let $T \subseteq V$ be a set of terminal vertices to be connected together and let $N = V \setminus T$ be the set of non-terminal vertices. It is assumed that $T$ contains at least two elements, i.e. $T$ has cardinality $|T| \geq 2$. A Steiner tree is defined as a tree of $G$ that spans $T$ and the Steiner Tree Problem (STP) is to find a Steiner tree of least cost. Whenever costs $\{c_e : e \in E\}$ are all non-negative, one can restrict attention to Steiner trees where no *Steiner vertex* (i.e. a non-terminal vertex in a Steiner tree) appears as a leaf. In this case there is also no need to explicitly impose the condition that the solution be a tree. That is implied naturally by the non-negative costs involved. The STP is then commonly referred to as the SPG. For surveys on the STP see Hwang and Richards [30], Maculan [37] and Winter [48].

Following what was done before for MSTs, associate an incidence vector $x$ with every Steiner tree in $G$ such that $x_e = 1$ if edge $e \in E$ is in the Steiner tree and $x_e = 0$ otherwise. The Steiner tree polytope, $P_S$, is defined as the convex hull of incidence vectors of Steiner trees.

The STP is closely related to the MST problem. Indeed, whenever all vertices in $V$ are terminals, the STP reduces to that problem. For that case the STP can thus be solved in $O(|E| \log |V|)$ time. Another trivial case occurs when $|T| = 2$.

The problem then reduces to finding the shortest path between the two vertices in $T$ (see Gallo and Palotino [15] for a survey of shortest path algorithms). For that case the STP can be solved in $O(|V|^2)$ time. For the general case, however, the STP has been shown to be NP-hard even for planar graphs, bipartite graphs and grid graphs (see Garey and Johnson [16]). The STP (SPG) is therefore a very good candidate for branch and cut type algorithms.

As a first step in the design of a branch and cut algorithm for the SPG, we will analyse four different formulations of the problem.

### 3.1 The STP as an MST with side constraints

Assume that edge costs $\{c_e : e \in E\}$ can take any real value. Beasley [5] proposed an extended formulation of the STP that involves the introduction into $G$ of an artificial vertex together with some associated edges. Denote by $n+1$ the artificial vertex and let $r \in T$ be one (any) terminal vertex of $G$. For a set of vertices $N^+ = N \cup \{r\}$, introduce edges of cost 0 to link vertex $n+1$ with every vertex in $N^+$. The graph that results from this expansion of $G$ is denoted by $G' = (V', E')$ where $V' = V \cup \{n+1\}$ and $E' = E \cup \{(i, n+1) : i \in N^+\}$. Associate real-valued variables $x_e, e \in E$, with the edges of $G$ and $z_i, i \in N^+$, with those edges of $G'$ that have $n+1$ as an endpoint. Let $z(S), S \subseteq N^+$, denote the sum $\sum_{i \in S} z_i$, and consider the polyhedral region $R_1$ defined by

$$x_e + z_i \leq 1, \qquad i \in N,\ e \in \delta(i), \tag{3.1}$$

$$x(E) + z(N^+) = |V|, \tag{3.2}$$

$$x(E(S)) \leq |S| - 1, \qquad S \subseteq V',\ 2 \leq |S| \leq |V'| - 1, \tag{3.3}$$

$$0 \leq x_e \leq 1, \qquad e \in E, \tag{3.4}$$

$$0 \leq z_i \leq 1, \qquad i \in N^+. \tag{3.5}$$

A formulation of the STP (Beasley [5]) as a spanning tree of $G'$ subject to side constraints is

$$\min\{\sum_{e \in E} c_e x_e : (x, z) \in R_1 \cap \mathbb{Z}^{|E|+|N^+|}\}. \tag{3.6}$$

Consider a feasible solution to (3.6) and remove, from the corresponding spanning tree, edge $(r, n+1)$. Two subtrees will result. For the subtree associated with vertex $n+1$ all other vertices would have an edge degree of one, and the subtree would resemble a star centered on $n+1$. It is not difficult to check that the subtree associated with vertex $r$ is, in this case, a Steiner tree of $G$. The cost of the first subtree plus the cost of the edge which was removed before equals zero. The cost of a feasible solution to (3.6) is therefore equal to the cost of the Steiner tree, rooted at $r$, which is contained in the solution.

Notice that variables $z_i, i \in N$, control the use of non-terminal vertices as Steiner vertices. Whenever $z_i = 1, i \in N$, vertex $i$ will not be part of a Steiner tree. The reverse is true when $z_i = 0$. Notice, on the other hand, that $z_r$ will always be equal to one. Since $r$ is a terminal vertex, different from non-terminal vertices, $z_r = 1$ implies that $r$ is in the Steiner tree (as it obviously should be). We call (3.1) *exclusion inequalities* for the role they play in model (3.6). Strictly speaking, the use of an edge $(r, n+1)$ is unnecessary and is introduced so that the underlying problem being solved becomes that of finding a (particular) spanning tree for $G'$. This turns out to be quite helpful when solving the problem via Lagrangian relaxation, as in [5]. An alternative would be to eliminate that edge from $G'$ and impose the condition that the solution implies a spanning forest with exactly two connected components.

Take the LP relaxation of (3.6) given by

$$\min\{\sum_{e \in E} c_e x_e : (x, z) \in R_1\}. \tag{3.7}$$

It is clear that an optimal solution value for such a relaxation gives a valid lower bound for the SPG. Furthermore, (3.7) can be solved very much as described before for the MST. For instance, one could, at an initial stage, optimize (3.7) over $R_{11} \supseteq R_1$ instead of $R_1$, where $R_{11}$ is given by, say,

$$x(E) + z(N^+) = |V|, \tag{3.8}$$

$$0 \le x_e \le 1, \qquad e \in E, \tag{3.9}$$

$$0 \le z_i \le 1, \qquad i \in N^+. \tag{3.10}$$

Subsequently, one would strengthen the relaxation by appending to it violated inequalities in (3.1) and (3.3). The separation for exclusion inequalities is trivial and could be carried out by inspection. The separation for SECs has been described in Section 2.2.1 and can be carried out in $O(|V|^4)$ time. A branch and cut algorithm based on a strengthened version of (3.7) was studied in Beasley and Lucena [6].

Bearing in mind our definition of the Steiner tree polytope, (3.6) is an extended formulation of the STP. In order to be able to compare the different SPG formulations to be discussed in this section, we refer later to the projection of $R_1$ over the space of the $x$ variables. This projection is given by $R_{1x} = \{x : \ (x, z) \in R_1 \text{ for some } z\}$.

### 3.2 A formulation with cut-set inequalities for terminals

Assume that costs $\{c_e : e \in E\}$ are all non-negative and associate real-valued variables $x_e, e \in E$, with the edges of $G$. Choose one (any) terminal vertex to act as a root, say vertex $r$, and let $E(S, V \setminus S) \subseteq E$ denote the edges with one end point in $S$ and the other in $V \setminus S$. Consider the polyhedral region $R_2$ defined by

$$x(E(S, V \setminus S)) \ge 1, \qquad r \in V \setminus S, \ S \cap T \ne \emptyset, \tag{3.11}$$

$$0 \le x_e \le 1, \qquad e \in E, \tag{3.12}$$

where the inequalities (3.11) are known as Steiner cuts. A formulation for the SPG (see Aneja [1]) is given by, say,

$$\min\{\sum_{e\in E} c_e x_e : x \in R_2 \cap \mathbb{Z}^{|E|}\}, \tag{3.13}$$

and has an LP relaxation

$$\min\{\sum_{e\in E} c_e x_e : x \in R_2\}. \tag{3.14}$$

It has been shown by Goemans and Bertsimas ([21]) that the SPG lower bound given by LP relaxation (3.14) is within a factor $1/(2 - 2/|T|)$ of optimality.

In order to solve (3.14) one could follow the same basic solution approach outlined before. For instance, one could, at an initial stage, optimize (3.14) over $R_{21} \supseteq R_2$ instead of $R_2$, where $R_{21}$ is given by

$$x(\delta(i)) \geq 1, \qquad i \in T \setminus \{r\}, \tag{3.15}$$

$$0 \leq x_e \leq 1 \qquad e \in E. \tag{3.16}$$

Notice that (3.15) consists of all Steiner cuts for feasible sets $S$ of cardinality one. Subsequently, one could strengthen that relaxation by appending to it violated inequalities in (3.11).

3.2.1 *The separation of Steiner cuts*

The separation problem for Steiner cuts is to find minimum capacity cuts between root $r$ and vertices $i \in T \setminus \{r\}$. It can therefore be tackled by solving maximum flow problems in a companion network. In order to do so, consider a network $F = (\overline{V}, A)$ which contains all vertices $i \in V$ such that a non-zero $\overline{x}_e$, $e \in \delta(i)$, exists. Arc set $A$ has, for each edge $e = (i,j)$, $e \in E$, with $\overline{x}_e > 0$, two arcs $[i,j]$ and $[j,i]$ of capacities $\xi_{ij} = \xi_{ji} = \overline{x}_e/2$. No additional source and sink nodes are required here. Take vertex $r$ as a sink and then take in turn, as sources, the remaining vertices in $T$. Compute the maximum flow between the currently selected source and the sink. If the corresponding flow is less than one, a violated cut-set inequality in (3.11) is implied. The separation for inequalities (3.11) would therefore involve the solution of $|T| - 1$ maximum flow problems. The associated overall computational complexity is consequently $O(|V|^4)$.

## 3.3 A formulation with variables for vertices and edges

Assume that edge costs $\{c_e : e \in E\}$ can take any real value, and suppose one knows in advance which vertices are in an optimal Steiner tree. The STP would then reduce to the problem of finding an MST for the subgraph of $G$ induced by those vertices. This is the basic idea behind the STP formulation that follows.

Associate, as before, a real-valued variable $x_e$ with every edge $e \in E$ where $x_e = 1$ if edge $e \in E$ is in a Steiner tree and $x_e = 0$ otherwise. In addition, associate a real-valued variable $y_i$ with every vertex $i \in N$ where $y_i = 1$ if vertex $i$ is in a Steiner tree and $y_i = 0$ otherwise. In order to introduce another extended formulation of the STP, consider the polyhedral region $R_3$ defined by

$$x(E) = |T| + y(N) - 1, \tag{3.17}$$

$$x(E(S)) \leq |S \cap T| + y(S) - 1, \quad S \subseteq V, \; S \cap T \neq \emptyset, \tag{3.18}$$

$$x(E(S)) \leq y(S \setminus \{j\}), \quad j \in S, \; S \subseteq V, \; S \cap T = \emptyset, \tag{3.19}$$

$$0 \leq x_e \leq 1, \quad e \in E, \tag{3.20}$$

$$0 \leq y_i \leq 1, \quad i \in N. \tag{3.21}$$

Inequalities (3.18) and (3.19) are generalized SECs (GSECs) and can be seen as a lifting (see section II.2 in [40]) of SECs. An STP formulation (see Goemans [19], Lucena [33] and Margot *et al.* [38]), which can be seen as a generalization of the spanning tree polytope, is given by

$$\min\{\sum_{e \in E} c_e x_e : (x, y) \in R_3 \cap \mathbb{Z}^{|E|+|N|}\}. \tag{3.22}$$

An LP relaxation of (3.22) is

$$\min\{\sum_{e \in E} c_e x_e : (x, y) \in R_3\}. \tag{3.23}$$

One could solve (3.23) by following, yet again, the same basic solution approach outlined before. More specifically, one could, at an initial stage, optimize (3.23) over $R_{31} \supseteq R_3$ instead of $R_3$, where $R_{31}$ is given, for instance, by

$$x(E) = |T| + y(N) - 1, \tag{3.24}$$

$$0 \leq x_e \leq 1, \quad e \in E \tag{3.25}$$

$$0 \leq y_i \leq 1, \quad i \in N. \tag{3.26}$$

Subsequently one could strengthen this relaxation of (3.23) by appending to it violated inequalities in (3.18) and (3.19). The separation problem associated with inequalities (3.18) and (3.19) is solvable in $O(|V|^4)$ time. A procedure to do so is explained below.

### 3.3.1 *The separation of GSECs*

Let $\overline{x}_e, e \in E$, and $\overline{y}_i, i \in N$, denote the current LP relaxation solution and set $\overline{y}_i = 1$ for all $i \in T$. Let variables $z_i \in \{0,1\}, i \in V$, control the inclusion or not of vertex $i$ into a subset $S$ of $V$. The separation problem for GSECs, (3.18) and (3.19), is tackled by determining, for each $l \in V$, a subset $S \subseteq V, l \in S$, for which $\{\overline{x}(E(S)) - (\overline{y}(S) - \overline{y}_l)\}$ is maximum. Clearly, whenever this maximum is non-positive, no violated inequalities of types (3.18) or (3.19) (involving vertex $l$) currently exist. Otherwise, $S$ is associated with the most violated of these GSECs. For a given LP relaxation and a vertex $l \in V$, a subset $S \subseteq V, l \in S$, such as we just described, can be obtained by solving the following Boolean quadratic problem:

$$\begin{array}{lll} \text{maximize} & \sum_{(i,j)\in E} \overline{x}_{(i,j)} z_i z_j - (\sum_{i\in V} \overline{y}_i z_i - \overline{y}_l) & (3.27) \\ \text{subject to} & z_l = 1, & (3.28) \\ & z_i \in \{0,1\}, \quad i \in V. & (3.29) \end{array}$$

Notice that (3.27)–(3.29) is closely related to (2.9)–(2.11). Indeed, (2.9)–(2.11) corresponds to an instance of (3.27)–(3.29) where $\overline{y}_i = 1$ for every $i \in V$. Thus the separation problem for inequalities (3.18) and (3.19) can be seen as a generalization of the separation of SECs. Nevertheless, the problem can still be formulated as a maximum flow problem. In terms of the companion network $F = (\overline{V}, A)$, introduced before, the major difference here is the use of "$\overline{y}_i$" instead of "1" in the expression for arc capacities $\xi_{0i}$ and $\xi_{i,n+1}$. It is therefore immediate that the separation for GSECs can thus be carried out in $O(|V|^4)$ time. SPG lower bounds generated from the solution to (3.23) were studied in Lucena [36].

Since we are considering here, yet again, another extended formulation of the STP, in order to be able to compare it with the other SPG formulations in this section, let $R_{3x} = \{x : \ (x,y) \in R_3 \text{ for some } y\}$ be a projection of $R_3$ over the space of the $x$ variables.

3.3.2 *Relationship with the formulation of Section 3.1*

Variables $z_i, i \in N$, in (3.6), and $(1 - y_i), i \in N$, in (3.22), play similar roles. Whenever $z_i = 1$, vertex $i$ is not to be a Steiner vertex. The reverse is true when $z_i = 0$. The same situation applies to $(1 - y_i), i \in N$, in (3.22). Indeed, it is immediate to establish that after replacing $y_i, i \in N$, by $(1 - z_i)$ in (3.22) that formulation would clearly represent a strengthened version of (3.6). This results from the lifting of SECs into GSECs. Notice, in this context, that inequalities (3.1) are indeed GSECs for subsets $S$ where $|S| = 2$.

### 3.4 A bidirected formulation for the SPG

Assume that edge costs $\{c_e : e \in E\}$ can only take non-negative real values. Let us formulate the SPG as a Steiner Arborescence Problem (the directed version of the problem). To that effect introduce, in association with the undirected graph $G = (V, E)$, a directed graph $D = (V, A)$. Every edge $e = (i,j)$, $i,j \in V$, $e \in E$, gives rise to two arcs $[i,j]$ and $[j,i]$ in $D$ with costs $d_{ij} = d_{ji} = c_e$. Choose one (any) terminal vertex, say vertex $r$, to act as the root of the arborescence. Finally, associate a real-valued variable $w_{ij}$ with every arc $[i,j]$ in $D$ and consider the polyhedral region $R_4$ defined by

$$\sum_{[i,j]\in A(S,V\setminus S)} w_{ij} \geq 1, \quad \{r\} \in S, \ (V \setminus S) \cap T \neq \emptyset, \tag{3.30}$$

$$0 \leq w_{ij} \leq 1, \quad [i,j] \in A, \tag{3.31}$$

where $A(S, V \setminus S) = \{[i,j] \in A : \ i \in S, j \in V \setminus S\}$. Inequalities (3.30) are known as (directed) Steiner cuts. An extended (bidirected) formulation of the

SPG (see Chopra and Rao [7, 8]) is given by

$$\min\{\sum_{[i,j]\in A} d_{ij}w_{ij} : w \in R_4 \cap \mathbb{Z}^{|A|}\} \tag{3.32}$$

and has an LP relaxation

$$\min\{\sum_{[i,j]\in A} d_{ij}w_{ij} : w \in R_4\}. \tag{3.33}$$

One could solve (3.33) by following, yet again, the same basic solution approach outlined in previous subsections. For instance, one could, at an initial stage, optimize (3.33) over $R_{41} \supseteq R_4$ instead of $R_4$, where $R_{41}$ is given by, say,

$$\sum_{[i,j]\in\delta^-(i)} w_{ij} \geq 1, \quad i \in T \setminus \{r\}, \tag{3.34}$$

$$0 \leq w_{ij} \leq 1, \quad [i,j] \in A, \tag{3.35}$$

where $\delta^-(i)$ is the set of edges that have $i$ as a tail. Subsequently, one could strengthen that relaxation by appending to it violated inequalities in (3.30). The separation problem associated with inequalities (3.30) is solvable in $O(|V|^4)$ time. A procedure to do so is explained below.

### 3.4.1 *The separation of (directed) Steiner cuts*

Let $\{\overline{w}_{ij} : [i,j] \in A\}$ be an optimal solution to

$$\min\{\sum_{[i,j]\in A} d_{ij}w_{ij} : w \in R_{41}\}. \tag{3.36}$$

The separation problem for inequalities (3.30) can be formulated, yet again, as a maximum flow problem. To that effect consider a network $F = (\overline{V}, \overline{A})$ associated with $\{\overline{w}_{ij} > 0 : [i,j] \in A\}$. At this stage, take $\xi_{ij} = \overline{w}_{ij}$ as the capacity for arc $[i,j] \in \overline{A}$. In this formulation, vertex $r$ is to be the source and the remaining vertices in $T$ are, one at a time, to be the sinks. Compute the maximum flow between the source and the currently selected sink. If the corresponding flow is less than one, a violated Steiner cut is implied. The separation for inequalities (3.30) would thus involve the solution of $|T| - 1$ maximum flow problems. The overall computational complexity associated with separation is therefore $O(|V|^4)$. A branch and cut algorithm for the SPG, based on (3.33), was studied in Chopra *et al.* [9].

In order to be able to compare the extended formulation of this section with the ones previously described, define $R_{4wx} = \{(x,w) : w \in R_4 \text{ and } x_e = w_{ij} + w_{ji} \text{ for all } e = (i,j) \in E\}$. A projection of $R_{4wx}$ over the space of the $x$ variables is given by $R_{4x} = \{x : (x,w) \in R_{4wx} \text{ for some } w\}$.

### 3.5 Comparing the different formulations

Goemans and Myung [20] have shown that $R_{4x} = R_{3x}$ and that $R_{3x} \subseteq R_1$. We have also pointed out (Section 3.3.2) that $R_3 \subseteq R_2$ and that, consequently, $R_{3x} \subseteq R_{2x}$. Formulations (3.22) and (3.32) are therefore the two strongest formulations under consideration. The strength of an IP formulation is essentially measured in terms of the quality of the LP relaxation bounds it produces. From a practical viewpoint, our ability to efficiently generate (or approximate) those bounds is also critically important. For the MST problem, as we have seen before, we have the best of both worlds. Most important of all, the extreme points of (2.1)–(2.3) correspond exactly to incidence vectors of spanning trees in $G$. Therefore, linear programming suffices to find an optimal (integral) solution to that problem. For the SPG, however, the situation turns out to be radically different from that which applies for the MST.

#### 3.5.1 *Computational results*

We are going to computationally test the LP relaxations suggested for the formulations in this section (i.e. solve the LP relaxations through the successive generation of violated inequalities, as outlined above). Two sets of test problems with non-negative edge costs $\{c_e : e \in E\}$ are going to be used for this purpose. The first set consists of randomly generated problem instances where edge costs are also randomly generated (first ten problems in set C of Beasley [5]). The second test set consists of the first 15 problems, out of the test problem set of 46 rectilinear Steiner problems generated by Soukup and Chow [46] to closely resemble VLSI routing applications. For rectilinear Steiner problems, terminal vertices are positioned on the Euclidean plane and the distances involved follow the Manhattan metric. Such problems can be transformed into SPGs (see Hanan [29]) by drawing both a horizontal line and a vertical line through every terminal vertex. Then the intersection of two lines originating from different terminal vertices gives rise to a non-terminal vertex. The result is a grid graph, where, at most, $|V| = |T|^2$ vertices are involved. SPGs in this class are, in general, quite difficult to solve to optimality (because of the symmetry involved).

In Tables 3 and 4, the first entry identifies the test problem. The second entry gives, for each test problem, the initial number of vertices, edges and terminals involved. The next four entries are the lower bounds produced by each of the four relaxations presented in this section. The last entries are the optimal value and the CPU times (in IBM RISC 6000/370 seconds) required to produce each of the four lower bounds.

As one would have anticipated, relaxations (3.23) and (3.33) produce consistently better bounds than the other two relaxations. Furthermore, relaxations (3.23) and (3.33) have always produced optimal SPG solutions for the test problems considered. For the other two relaxations, (3.7) consistently produced better lower bounds than (3.14). In relation to (3.7), one should notice that, in spite of producing lower bounds of acceptable quality for the first set of test problems, duality gaps for rectilinear Steiner problems were, in some cases, quite large. Resorting to branch and cut in those situations may lead to unacceptably large

**Table 3** Sparse graph problems: C

| | | | | Lower bounds | | | | Opt | CPU time | | | |
|---|---|---|---|---|---|---|---|---|---|---|---|---|
| Prob | $\|V\|$ | $\|E\|$ | $\|T\|$ | (3.7) | (3.14) | (3.23) | (3.33) | value | | | | |
| C1 | 500 | 625 | 5 | 84.28 | 71 | 85 | 85 | 85 | 136 | 50 | 6.2 | 0.9 |
| C2 | | | 10 | 141.3 | 108.5 | 144 | 144 | 144 | 110 | 7 | 10.8 | 2.6 |
| C3 | | | 83 | 749.6 | 579.5 | 754 | 754 | 754 | 166 | 646 | 6.0 | 3.2 |
| C4 | | | 125 | 1079 | 854.75 | 1079 | 1079 | 1079 | 307 | 296 | 21.7 | 3.5 |
| C5 | | | 250 | 1576.5 | 1270.5 | 1579 | 1579 | 1579 | 1373 | 12 | 4.1 | 3.9 |
| C6 | | 1000 | 5 | 51.3 | 45.5 | 55 | 55 | 55 | 5956 | 26797 | 1590 | 11.9 |
| C7 | | | 10 | 102 | 83 | 102 | 102 | 102 | 1260 | 8634 | 68.6 | 17.8 |
| C8 | | | 83 | 509 | 379.25 | 509 | 509 | 509 | 2977 | 350 | 2068 | 66.5 |
| C9 | | | 125 | 705 | 507 | 707 | 707 | 707 | 6909 | 20429 | 3221 | 115.9 |
| C10 | | | 250 | 1093 | 815.5 | 1093 | 1093 | 1093 | 14929 | 145 | 81.4 | 22.5 |

**Table 4** Rectilinear Steiner problems

| | | | | Lower bounds | | | | Opt | CPU time | | | |
|---|---|---|---|---|---|---|---|---|---|---|---|---|
| Prob | $\|V\|$ | $\|E\|$ | $\|T\|$ | (3.7) | (3.14) | (3.23) | (3.33) | value | | | | |
| R1 | 15 | 22 | 5 | 177 | 159 | 187 | 187 | 187 | 0.39 | 0.15 | 0.08 | 0.09 |
| R2 | 12 | 17 | 6 | 164 | 112 | 164 | 164 | 164 | 0.10 | 0.05 | 0.35 | 0.29 |
| R3 | 28 | 45 | 7 | 218.1 | 154 | 236 | 236 | 236 | 0.71 | 0.08 | 0.68 | 0.22 |
| R4 | 64 | 112 | 8 | 169.5 | 155 | 254 | 254 | 254 | 8.80 | 10.0 | 39.7 | 8.44 |
| R5 | 12 | 17 | 6 | 226 | 153 | 226 | 226 | 226 | 0.12 | 0.06 | 0.17 | 0.12 |
| R6 | 24 | 38 | 12 | 239.6 | 154 | 242 | 242 | 242 | 0.29 | 0.10 | 0.19 | 0.15 |
| R7 | 30 | 49 | 12 | 245.5 | 164 | 248 | 248 | 248 | 0.37 | 0.10 | 0.31 | 0.19 |
| R8 | 24 | 37 | 12 | 230.6 | 151 | 236 | 236 | 236 | 0.24 | 0.09 | 0.21 | 0.16 |
| R9 | 15 | 22 | 7 | 164 | 127 | 164 | 164 | 164 | 0.13 | 0.07 | 0.01 | 0.01 |
| R10 | 36 | 60 | 6 | 140 | 107.2 | 177 | 177 | 177 | 1.30 | 0.48 | 2.46 | 0.52 |
| R11 | 30 | 49 | 6 | 139 | 121 | 144 | 144 | 144 | 0.50 | 0.55 | 0.53 | 0.25 |
| R12 | 27 | 42 | 9 | 171.6 | 110 | 180 | 180 | 180 | 0.57 | 0.08 | 0.36 | 0.19 |
| R13 | 42 | 71 | 9 | 130 | 95 | 150 | 150 | 150 | 2.13 | 0.44 | 3.21 | 0.84 |
| R14 | 36 | 60 | 12 | 180 | 110 | 260 | 260 | 260 | 0.79 | 0.54 | 0.88 | 0.48 |
| R15 | 100 | 180 | 14 | 114.8 | 92 | 148 | 148 | 148 | 32 | 8.2 | 49.7 | 15.8 |

CPU times. This would most certainly persist even if preprocessing and tests for fixing variables (see Section 4) were applied. Therefore, in order to develop a branch and cut algorithm for the SPG, the obvious path to follow is to concentrate on the two stronger formulations available. This decision is made simpler since the separation problems associated with the four competing LP relaxations have the same computational complexity.

The CPU times reported reflect, for relaxations (3.7), (3.14) and (3.23) in particular, the frequent occurrence of *tailing-off* (see [42] for details). No such behaviour was noticed for relaxation (3.33). Tailing-off occurs when, after repeatedly appending cutting planes to the LP program and reoptimizing, the LP relaxation solution value either fails to increase at all, or increases only very slightly. Among the three relaxations which experienced the problem, tailing-off was significantly less pronounced for relaxation (3.23), and is reflected in the small CPU times reported for that relaxation. In general, in a tree search procedure, tailing-off (once it is detected) is dealt with by branching. Another aspect worth mentioning is that the output of the separation routines for relaxations (3.7) and (3.23) ideally need some sort of postprocessing. In that respect, similar problems to those detected in association with the separation of SECs for the MST problem (see example in Section 2.3) were noticed. For the computational experiments performed throughout this chapter, however, we have done no such postprocessing. Therefore, CPU times reported for relaxations (3.7) and (3.23) could, most likely, be improved. For relaxation (3.33) we have followed the suggestion in [9] and "flipped" the support graph associated with LP relaxation solutions (i.e. changed the orientation of all arcs in the graph). Then, in addition to the separation described in Section 3.4.1, we also identified maximum cut sets (on the flipped graph) separating each vertex in $T \setminus \{1\}$ from vertex 1. Violated Steiner cuts, different from those obtained by the procedure in Section 3.4.1, may emerge as a result (see [9] for details).

## 4 Preprocessing and upper bounds for the SPG

Prior to attempting to solve to optimality a problem such as the SPG, the use of preprocessing (in order to reduce problem size) is highly recommended. In general, the tests involved are computationally cheap and can make a decisive impact on algorithmic performance. The basic idea of preprocessing can be extended further than the simple use of tests that exploit logical implications for the underlying problem being solved. For instance, Lagrangian relaxation can be used to derive lower bounds. This, coupled with heuristics that generate upper bounds, can result in efficient tests for fixing variables. Indeed one can go even further and repeatedly perform the two batteries of tests (i.e. logical and Lagrangian ones) for as long as they prove computationally successful. In this section we describe SPG preprocessing tests from the literature. We also use Lagrangian relaxation in conjunction with formulation (3.6) to generate SPG lower bounds. The Lagrangian relaxation solutions obtained are then used to guide a basic SPG heuristic.

### 4.1 Tests from the literature

A few tests that can be used for preprocessing are described below. Further tests can be found in Duin and Volgenant [12].

**Vertex degree tests** (see Winter [48], for instance)

- If $|\delta(i)| \leq 1$, $i \in V \setminus T$, then vertex $i$ can be eliminated.
- If $|\delta(i)| = 1$, $i \in T$, then the single edge associated with vertex $i$ can be shrunk into a vertex (since the edge must be part of an optimal

Steiner tree).

- If $|\delta(i)| = 2$, $i \in V \setminus T$, then one can form an edge linking the two vertices in $\delta(i)$ and eliminate $i$. The cost of this edge is to be set equal to the sum of the costs of the edges it replaces. If a duplication of edges result, the most expensive one should be discarded.

**Nearest vertex test** (Beasley [4])

For $i \in T$, let $i_1, i_2 \in V$ be, respectively, the nearest and the second nearest vertices to $i$. Then edge $(i, i_1)$ is guaranteed to be in an optimal Steiner tree if:

- $i_1 \in T$,
- $i_1 \in V \setminus T$ and $c_{(i,i_2)} \geq c_{(i,i_1)} + c_{(i_1,i_3)}$, where $i_3 \in T$.

**Minimum spanning tree test** (Balakrishnan and Patel [3])

If an edge $e \in E$ with end vertices $i_1, i_2 \in T$ is in a MST of $G$, then $e$ is guaranteed to be in an optimal Steiner tree.

**Through-vertex test** (Chopra *et al.* [9])

For a given vertex $i \in T$ and two vertices $i_1, i_2 \in V$ where $i_1, i_2 \neq i$, if $c_{(i_1,i_2)} > \max\{c_{(i,i_1)}, c_{(i,i_2)}\}$, then edge $(i_1, i_2)$ can be eliminated from $G$.

For sparse graphs, very often, $(i, i_1)$ or $(i, i_2)$, $i \in T$, may not exist. We have modified the through-vertex test given above to reflect this fact. As a result we have used costs $\gamma_{(i,i_1)}$ and $\gamma_{(i,i_2)}$, associated with the shortest paths from $i$ to $i_1$ and from $i$ to $i_2$, respectively, instead of $c_{(i,i_1)}$ and $c_{(i,i_2)}$. This should be restricted to only a small, a priori fixed, subset of the vertices in $T$ (because of the considerable computational effort involved). For the remaining vertices, the test should be conducted as explained before.

## 4.2 Lagrangian test and SPG upper bounds

Suppose one dualizes, in a Lagrangian fashion, inequalities (3.1) (in accordance with Beasley [5]). Let $\beta_{ei}, e \in \delta(i), i \in N$ be the corresponding multipliers. Then subgradient optimization (see Fisher [14], for instance) could be used to optimize the resulting unconstrained MST Lagrangian dual problem. Throughout subgradient optimization, we attempt to fix variables by using a valid SPG upper bound, say $v(ub)$; the current Lagrangian relaxation solution, $v(\beta)$; and reduced costs for variables in the Lagrangian problem (i.e. an MST for $G'$). Whenever the reduced cost of a variable exceeds $v(ub) - v(\beta)$ the variable (edge) can be fixed to zero (i.e. eliminated). This applies since, in that case, it can be guaranteed that the edge associated with the variable will not be in an optimal Steiner tree.

One can use the algorithm of Takahashi and Matsuyama [47], as suggested by Rayward-Smith and Clare [45], to generate feasible SPG solutions. This yields a three-step procedure. First, a shortest path tree is built between a given fixed terminal vertex and the remaining terminals. Then, an MST is computed for the subgraph of $G$ induced by the vertices in the shortest path tree. Finally, the MST is pruned of non-terminal vertices of degree one.

We use the upper bounding routine after performing the tests of Section 4.1 and whenever a new improved Lagrangian relaxation lower bound is obtained. For the latter case, we temporarily change to zero the original costs for edges that are part of the current Lagrangian relaxation solution (other edge costs remaining unaffected). As a result, edges with a modified cost of zero are made more attractive to be chosen in the shortest path tree part of the heuristic. At the second part of the procedure, we resort to the original edge costs, and compute the MST.

### 4.3 Computational results

The impact of preprocessing on the test problems of Tables 3 and 4 can be seen in Tables 5 and 6, respectively. The computational results shown there involved the repeated use of the two types of tests described. Tests were applied in sequence for as long as a percentage threshold of edges (fixed in this exercise to 10%) were eliminated between the start of a preprocessing round and the end of the Lagrangian test. In Tables 5 and 6, the first entry identifies the test problem. The second entry gives, for each test problem, the initial number of vertices, edges and terminals involved. The third entry indicates the reduction in problem size that results from the application of the tests, specifically we give there the number of vertices, edges and terminals which are left at the end of the procedure. The fourth entry gives the lower bounds generated and the sixth entry gives the upper bounds. The optimal solution values are given, as are the CPU times spent on the tests. An asterisk in entry one indicates that the problem has been solved to optimality by preprocessing. The CPU times are in IBM RISC 6000/370 seconds.

In order to evaluate the impact of preprocessing on the SPG algorithms of Section 3 we ran some computational experiments. We combined two of those algorithms with preprocessing and run-time tests for fixing variables (based on LP relaxation reduced costs and upper bounds). For the experiment we use the algorithms associated with models (3.23) and (3.33) respectively. In Table 7 we give the CPU times (in IBM RISC 6000/370 seconds) required by each algorithm to attain optimality for those test problems that were not solved to optimality by preprocessing.

We conclude, from the statistics presented, that the tests for fixing variables were very effective for the first set of test problems. The reverse was true for the second set. Upper bounds, however, were invariably sharp and were equal to the optimal solution values in all the cases considered.

## 5 Combining row and column generation

We have chosen (3.33) as a model for a description of the combined use of column and row generation to derive LP lower bounds for the SPG. As such we must select a (reasonably small) subset of the set of all variables in (3.33) and consider, for those variables selected, a polyhedral region $R_{42}$ (akin to $R_{41}$) to initiate the computations. Our initial choice of variables is guided by the upper bounding routine used at preprocessing (see Section 4). As a first step,

**Table 5** Sparse graph problems: C

| Prob | Initial status $\|V\|$ | $\|E\|$ | $\|T\|$ | Final status $\|V\|$ | $\|E\|$ | $\|T\|$ | Lower bounds | Upper bounds | Opt value | CPU time |
|---|---|---|---|---|---|---|---|---|---|---|
| C1* | 500 | 625 | 5 | - | - | - | 85 | 85 | 85 | 5.8 |
| C2 | | | 10 | 30 | 46 | 8 | 141.3 | 144 | 144 | 12.1 |
| C3* | | | 83 | - | - | - | 754 | 754 | 754 | 12.2 |
| C4* | | | 125 | - | - | - | 1079 | 1079 | 1079 | 5.2 |
| C5* | | | 250 | - | - | - | 1579 | 1579 | 1579 | 2.2 |
| C6* | | 1000 | 5 | - | - | - | 55 | 55 | 55 | 106 |
| C7* | | | 10 | - | - | - | 102 | 102 | 102 | 26 |
| C8* | | | 83 | - | - | - | 509 | 509 | 509 | 37 |
| C9* | | | 125 | - | - | - | 707 | 707 | 707 | 90 |
| C10* | | | 250 | - | - | - | 1093 | 1093 | 1093 | 16 |

**Table 6** Rectilinear Steiner problems

| Prob | Initial status $\|V\|$ | $\|E\|$ | $\|T\|$ | Final status $\|V\|$ | $\|E\|$ | $\|T\|$ | Lower bounds | Upper bounds | Opt value | CPU time |
|---|---|---|---|---|---|---|---|---|---|---|
| R1 | 15 | 22 | 5 | 9 | 14 | 4 | 151 | 187 | 187 | 0.59 |
| R2* | 12 | 17 | 6 | - | - | - | 164 | 164 | 164 | 0.02 |
| R3 | 28 | 45 | 7 | 20 | 35 | 7 | 221.3 | 236 | 236 | 1.53 |
| R4 | 64 | 112 | 8 | 60 | 108 | 8 | 158.3 | 254 | 254 | 3.53 |
| R5* | 12 | 17 | 6 | - | - | - | 226 | 226 | 226 | 0.02 |
| R6* | 24 | 38 | 12 | - | - | - | 242 | 242 | 242 | 0.60 |
| R7 | 30 | 49 | 12 | 7 | 11 | 3 | 246 | 248 | 248 | 1.48 |
| R8 | 24 | 37 | 12 | 18 | 31 | 10 | 230.7 | 236 | 236 | 0.61 |
| R9* | 15 | 22 | 7 | - | - | - | 164 | 164 | 164 | 0.01 |
| R10 | 36 | 60 | 6 | 30 | 53 | 6 | 103.7 | 177 | 177 | 1.26 |
| R11 | 30 | 49 | 6 | 26 | 45 | 6 | 111.9 | 144 | 144 | 0.96 |
| R12 | 27 | 42 | 9 | 23 | 38 | 9 | 164.2 | 180 | 180 | 0.80 |
| R13 | 42 | 71 | 9 | 37 | 65 | 9 | 129.8 | 150 | 150 | 1.75 |
| R14 | 36 | 60 | 12 | 35 | 59 | 11 | 110 | 260 | 260 | 1.35 |
| R15 | 100 | 180 | 14 | 96 | 176 | 14 | 114.7 | 148 | 148 | 6.68 |

**Table 7** CPU times: combined sparse and rectilinear test problems

| Relaxation | C2 | R1 | R3 | R4 | R7 | R8 | R10 | R11 | R12 | R13 | R14 | R15 |
|---|---|---|---|---|---|---|---|---|---|---|---|---|
| (3.23) | 24.5 | 1.1 | 3.6 | 41.6 | 3.1 | 1.4 | 3.8 | 2.3 | 1.9 | 6.1 | 3.4 | 67.4 |
| (3.33) | 32.0 | 1.2 | 3.4 | 14.1 | 2.9 | 1.8 | 3.1 | 2.2 | 1.7 | 4.2 | 3.2 | 28.3 |

one initially stores all edges associated with the SPG upper bounds obtained at preprocessing. Denote by $v(ub)$ the value of the best of those upper bounds. One then proceeds by selecting some of the edges (variables) that have been stored. The selection criteria is to only choose edges which appeared in feasible SPG solutions with a value of no more than 10% greater than $v(ub)$. Once the initial variables are chosen we follow very closely what was suggested before for the MST problem (see Section 2.4).

We introduce a bidirected version of the formulation of Beasley [5] (see Section 3.1). A relaxation of this bidirected formulation is to be used, together with LP relaxation reduced costs, to derive valid SPG lower bounds.

### 5.1 A bidirected version of Beasley's SPG formulation

Consider the expanded graph $G' = (V', E')$ of Section 3.1. From $G'$, construct a directed graph $D' = (V', A')$ by transforming edges of $G'$ into arcs of $D'$. Accordingly, every edge $e = (i, j)$ in $G'$ gives rise to two arcs in $D'$ of opposite orientation and costs $d_{ij} = d_{ji} = c_e$. Edges of the type $(i, n+1)$, $i \in N^+$, imply zero cost arcs in $D'$ of the type $[n+1, i]$ (each edge corresponding, in this case, to only one arc). A bidirected formulation of (3.6) models a minimum cost $(n+1)$-arborescence of $D'$ under some side constraints. Introducing real-valued variables $u_a$ for every arc $a \in A'$, these side constraints are, in this case, written as

$$u_{n+1,i} + u_a \leq 1, \qquad i \in N, \;\; a \in \delta^+(i), \tag{5.1}$$

where $\delta^+(i)$ is the subset of the arcs in $G'$ which have vertex $i$ as a tail.

For the purposes of this section, it is sufficient to notice that a minimum cost $(n+1)$-arborescence of $D'$ gives a valid lower bound for the original SPG. One should also notice that there is a one-to-one correspondence between variables $u_a, a \in A'$, which do not involve vertex $n+1$, and variables $w_a, a \in A$, in (3.33). Let us replace the cost of the former variables with the reduced costs of the latter ones. In doing so, the cost of the minimum $(n+1)$-arborescence of $D'$ (under the modified costs) can be added to the optimal value of the restricted linear program of type (3.33) (from which reduced costs were generated) to obtain a lower bound that is valid for (3.32).

### 5.2 Computational results

The combined row and column generation procedure of this section was tested for the problems shown in Tables 5 and 6. We report results for those problems that have not been solved to optimality by preprocessing. Table 8 gives a summary of the computational results obtained. The first entry in the table identifies the problem. The second entry gives the number of column generation iterations required to attain optimality. The third entry gives the total number of variables for each problem. The fourth entry gives the actual number of variables explicitly used in the LP program when optimality was proved. The fifth entry gives the number of variables that had already been eliminated at the beginning of the very

**Table 8** Test problems for row and column generation

| Prob | # Itrs | # Variables | # Variables used | # Variables eliminated | CPU time |
|---|---|---|---|---|---|
| C2 | 2 | 89 | 86 | 50 | 24.0 |
| R1 | 1 | 25 | 25 | - | 1.1 |
| R3 | 2 | 66 | 63 | 3 | 3.8 |
| R4 | 6 | 212 | 204 | 18 | 24.0 |
| R7 | 1 | 19 | 19 | - | 3.2 |
| R8 | 1 | 58 | 46 | - | 1.4 |
| R10 | 4 | 102 | 90 | 13 | 3.9 |
| R11 | 4 | 86 | 68 | 24 | 2.9 |
| R12 | 1 | 73 | 73 | - | 1.9 |
| R13 | 4 | 126 | 115 | 28 | 5.6 |
| R14 | 4 | 112 | 93 | 24 | 4.2 |
| R15 | 5 | 348 | 312 | 7 | 40.8 |

last column generation iteration (when more than one iteration is required to attain optimality). The final entry gives the CPU times (in IBM RISC 6000/370 seconds) required to prove optimality. Notice that CPU times in Table 8 relate to the preprocessing phase of the algorithm plus the row and column generation phase.

Column generation did not produce a significant impact on the test problems used. CPU times were not improved (see computational results in Section 4) nor were only a relatively small number of variables (columns) generated (as desired). Furthermore, the test for the early fixing of variables had only a modest impact. For this particular application the reasons for such a poor overall performance can be basically traced either to problem size (rectilinear problems) or to the effectiveness of the preprocessing tests (problem C2). One should expect, in general, column generation to perform better for larger SPGs.

# 6 Branching

As an example of the use of branching, within branch and cut, we give details of a modified (simpler) version of the algorithm in [6]. The resulting algorithm combines the formulation of Section 3.1 with the cut-set inequalities of Section 3.2. A binary, depth-first, branch and cut tree search procedure was implemented. The details of the algorithm, which does not use column generation, are given below.

## 6.1 Initial tree node

We first reduced the problem using the degree, nearest vertex and edge deletion reduction tests until no further reduction could be achieved. Notice that we use here a simplified version of the tests of Section 4.1. Furthermore no Lagrangian tests are applied. We then chose a vertex to act as the root vertex, i.e. that vertex $r \in T$ defined in Section 3.2. This was chosen to be the vertex of $T$ with the largest degree (ties broken arbitrarily).

#### 6.1.1 *Lower bounds, upper bounds and tests for fixing variables*

A relaxation of (3.6) given by

$$\min\{\sum_{e \in E} c_e x_e : (x, z) \in R_{11}\} \tag{6.1}$$

is solved to find an initial SPG lower bound. Denote, at any stage in the procedure, the current LP lower bound by $v(lp)$. Feasible solutions (upper bounds) were generated by using the upper bounding procedure of Section 4.2 (replacing Lagrangian relaxation solutions by LP relaxation solutions). An initial upper bound was obtained by applying the procedure to graph $G = (V, E)$ with edge costs $\{c_e : e \in E\}$. Subsequently, as new LP relaxations are derived (with the introduction of violated exclusion inequalities, SECs and cut-set inequalities) we modify the original edge costs and use the upper bounding routine again. Suppose the current LP relaxation solution is given by $\{\overline{x}_e, e \in E'\}$. Then edges in $E \subset E'$ have their costs modified to $\{(1 - \overline{x}_e)c_e : e \in E\}$ prior to applying the upper bounding procedure. Denote by $v(ub)$ the best upper bound that has been generated at a given point in time. Throughout the procedure LP reduced costs were combined with $v(ub)$ and $v(lp)$ to fix variables. Accordingly, if the current LP reduced cost of a given variable exceeds $(v(ub) - v(lp))$ then that variable can be fixed to zero.

Inequalities that have been added to $R_{11}$ and which eventually became non-binding (i.e. have a non-zero slack variable) were removed from the LP program. Note here that to avoid the possibility of cycling this removal was done at iterations at which the LP solution value changed (increased).

#### 6.1.2 *Termination*

The above procedure will terminate with a solution of value $v(lp)$ (associated with variables $\{\overline{x}_e : e \in E\}$) for which there are no violated exclusion inequalities, Steiner cuts or SECs. As such this solution is the solution of the LP relaxation of (3.7). It has either $\{\overline{x}_e : e \in E\}$ all integer (in which case the current LP solution is an optimal SPG solution); $\{\overline{x}_e : e \in E\}$ fractional and $v(lp) = v(ub)$ (i.e. $v(ub)$ is an optimal solution to SPG); or $\{\overline{x}_e : e \in E\}$ is fractional and $v(lp) < v(ub)$ (when branching is needed).

### 6.2 Other tree nodes

We branched in the search tree by setting edges into the solution, and adding violated inequalities as appropriate. In fractional branching we choose the (edge) variable $\overline{x}_e$ which had a fractional value in the current LP solution and for which $|\overline{x}_e - 0.5|$ was minimum and branched by setting that variable equal to one (in one branch of the tree) and equal to zero (in the other branch of the tree – a binary tree search). At each tree node we performed the same procedure described for the root node to obtain lower bounds. We can backtrack (i.e. climb back up the search tree) when $v(lp) \geq v(ub)$ at any tree node.

### 6.3 Computational results

We give, in Tables 9 and 10, computational results for the test problem described in Tables 3 and 4.

**Table 9** Sparse graph problems: C

| Prob | Final status $\|V\|$ | $\|E\|$ | $\|T\|$ | Lower bounds | Upper bounds | Opt value | # Tree nodes | CPU time |
|---|---|---|---|---|---|---|---|---|
| C1* | - | - | - | 85 | 85 | 85 | 1 | 5.8 |
| C2 | 30 | 46 | 8 | 141.3 | 144 | 144 | 3 | 12.1 |
| C3* | - | - | - | 754 | 754 | 754 | 1 | 12.2 |
| C4* | - | - | - | 1079 | 1079 | 1079 | 1 | 5.2 |
| C5* | - | - | - | 1579 | 1579 | 1579 | 1 | 2.2 |
| C6* | - | - | - | 55 | 55 | 55 | 1 | 106 |
| C7* | - | - | - | 102 | 102 | 102 | 1 | 26 |
| C8* | - | - | - | 509 | 509 | 509 | 1 | 37 |
| C9* | - | - | - | 707 | 707 | 707 | 1 | 90 |
| C10* | - | - | - | 1093 | 1093 | 1093 | 1 | 16 |

**Table 10** Rectilinear Steiner problems

| Prob | Final status $\|V\|$ | $\|E\|$ | $\|T\|$ | Lower bounds | Upper bounds | Opt value | # Tree nodes | CPU time |
|---|---|---|---|---|---|---|---|---|
| R1 | 9 | 25 | 4 | 182 | 187 | 187 | 3 | 0.09 |
| R2* | - | - | - | 164 | 164 | 164 | 1 | 0.01 |
| R3 | 20 | 62 | 7 | 221.5 | 242 | 236 | 19 | 1.13 |
| R4 | 60 | 213 | 8 | 175.5 | 258 | 254 | (500) | 263.0 |
| R5* | - | - | - | 226 | 226 | 226 | 1 | 0.02 |
| R6* | - | - | - | 242 | 242 | 242 | 1 | 0.02 |
| R7 | 18 | 48 | 10 | 247.5 | 254 | 248 | 5 | 0.31 |
| R8* | 18 | 31 | 10 | 236 | 236 | 239 | 1 | 0.07 |
| R9* | - | - | - | 164 | 164 | 164 | 1 | 0.01 |
| R10 | 30 | 102 | 6 | 154.75 | 184 | 177 | 29 | 6.01 |
| R11* | - | - | - | 144 | 144 | 144 | 1 | 0.37 |
| R12 | 23 | 67 | 9 | 171.7 | 180 | 180 | 31 | 2.11 |
| R13 | 33 | 110 | 8 | 132.6 | 150 | 150 | 213 | 35.0 |
| R14 | 17 | 34 | 8 | 260 | 260 | 260 | 1 | 0.02 |
| R15 | 96 | 341 | 14 | 121.16 | 149 | 148 | 107 | 504.2 |

Notice that the logical tests used in preprocessing here are weaker than those suggested in Section 4. Nevertheless, the tests for fixing variables (which are based on reduced costs) are at least as sharp as those described in Section 4. Whenever an entry for the number of tree nodes is given between brackets that implies the maximum number of tree nodes allowed was reached before optimality was proven. In this exercise we allowed a maximum of 500 tree nodes for each test problem.

## 7 Conclusions

We have described, at an introductory level, the use of branch and cut algorithms for the solution of integer programming problems. An attempt was made to present, in some detail, all the major steps involved in the approach. Computational results were presented to highlight the importance of each of those steps. In the process we did not concern ourselves so much with the efficiency of our computational implementation of algorithms. This is not to say that our implementations are inefficient. Nevertheless there is clear scope for them to be significantly improved provided we had assigned more time to this task. This brings us to a crucial point related to branch and cut algorithms: their computational implementation tends to be invariably time consuming for those involved! The reader who goes on from this chapter to experiment with branch and cut algorithms will discover this for themselves.

## Bibliography


1. Y.P. Aneja. An integer linear programming approach to the Steiner problem in graphs. *Networks*, 10:167–178,1980.
2. D. Applegate, R. Bixby, V. Chvátal and W. Cook. Finding cuts in the TSP. Technical report 95.05, DIMACS Center, Rutgers University, Piscataway, NJ 08855-1179, USA, 1995.
3. A. Balakrishnan and N.R. Patel. Problem reduction methods and a tree generation algorithm for the Steiner problem in graphs. *Networks*, 17:86–85, 1987.
4. J.E Beasley. An algorithm for the Steiner problem in graphs. *Networks*, 14:147–159, 1984.
5. J.E Beasley. An SST-based algorithm for the Steiner problem in graphs. *Networks*, 19:1–16, 1989.
6. J.E. Beasley and A. Lucena. A branch and cut algorithm for the Steiner problem in graphs. Technical report, The Management School, Imperial College, London SW7 2AZ, 1995.
7. S. Chopra and M.R. Rao. The Steiner tree problem I: formulations, compositions and extension of facets. *Mathematical Programming*, 64:209–230, 1994.
8. S. Chopra and M.R. Rao. The Steiner tree problem II: Properties and classes of facets. *Mathematical Programming*, 64:231–246, 1994.
9. S. Chopra, E.R. Gorres and M.R. Rao. Solving the Steiner tree problem on

a graph using branch and cut. *ORSA Journal on Computing*, 4:320–335, 1992.

10. H. Crowder and M. Padberg. Solving large-scale symmetric travelling salesman problems to optimality. *Management Science*, 26:495–509, 1980.
11. G.B Dantzig, D.R. Fulkerson and S.M. Johnson. Solution of a large-scale travelling salesman problem. *Operations Research*, 2:393–410, 1954.
12. C.W. Duin and A. Volgenant. Reduction tests for the Steiner problem in graphs. *Networks*, 19:549–567, 1989.
13. J. Edmonds. Matroids and the greedy algorithm. *Mathematical Programming*, 1:127–136, 1971.
14. M.L. Fisher. The Lagrangian method for solving integer problems. *Management Science*, 27:1–18, 1981.
15. G. Gallo and S. Pallotino. Shortest path methods: a unifying approach. *Mathematical Programming Study*, 14:127–136, 1986.
16. M.R. Garey and D.S. Johnson. *Computers and Intractability: a Guide to the Theory of NP-completeness.* W.H. Freeman and Co., San Francisco, 1979.
17. R.S. Garfinkel. Branch and bound methods for integer programming. In N. Christofides, A. Mingozzi, P. Toth, C. Sandi, (eds.), *Combinatorial Optimization.* John Wiley & Sons, Chichester, 1978.
18. R.S. Garfinkel and G.L. Nemhauser. *Integer Programming*, John Wiley & Sons, Chichester, 1972.
19. M.X. Goemans. The Steiner tree polytope and related polyhedra. *Mathematical Programming*, 63:157–182, 1994.
20. M.X. Goemans and Y-S. Myung. A catalog of Steiner tree formulations. Technical report, Department of Mathematics, MIT, Cambridge, MA 02139, 1991.
21. M.X Goemans and D. Bertsimas. Survivable networks, linear programming relaxations and the parsimonious property. *Mathematical Programming*, 60:145–166, 1993.
22. D. Goldfarb and M.D. Grigoriadis. A computational comparison of the Dinic and network flow simplex methods for maximum flow. *Annals of Operations Research*, 13:83–123, 1988.
23. R.E. Gomory. Outline of an algorithm for integer solutions to linear programs. *Bulletin of the American Mathematical Society*, 64:275–278, 1958.
24. R.E. Gomory. An algorithm for integer solutions to linear programs. In *Recent Advances in Mathematical Programming*, R. Graves and P. Wolfe, eds., McGraw-Hill,269–302, 1963.
25. R.E. Gomory and T.C. Hu. Multi-terminal network flows. *J. SIAM*, 9:551–570, 1961.
26. M. Grötschel. On the symmetric travelling salesman problem: solution of a 120-city problem. *Mathematical Programming Study*, 12:61–77, 1980.

27. M. Grötschel, M. Junger and G. Reinelt. A cutting plane algorithm for the linear ordering problem. *Operations Research*, 32:1195–1220, 1984.
28. M. Grötschel and M. Padberg. Polyhedral theory. In E.L. Lawler, J.K. Lenstra, A.H.G. Rinnooy Kan and D.B. Shmoys, (eds.), *The Travelling Salesman Problem*. John Wiley & Sons, Chichester, 1985.
29. M. Hanan. On Steiner's problem with rectilinear distances. *SIAM Journal on Applied Mathematics*, 14:255–265, 1966.
30. F.K. Hwang and D.S.Richards. Steiner tree problems. *Networks*, 22:55–89, 1992.
31. J.B. Kruskal. On the shortest spanning subtree of a graph and the travelling salesman problem. *Proceedings of the American Mathematical Society*, 7:48–50, 1956.
32. E.L. Lawler. *Combinatorial Optimization: Networks and Matroids*, Holt, Rinehart and Winston, 1976.
33. A. Lucena. Tight bounds for the Steiner problem in graphs. Talk given at the TIMS XXX - SOBRAPO XXIII Joint International Meeting, Rio de Janeiro, 1991.
34. A. Lucena. Steiner problem in graphs: Lagrangian relaxation and cutting-planes. *COAL Bulletin*, Mathematical Programming Society, 21:2–8, 1992.
35. A. Lucena. Steiner problem in graphs: Lagrangian relaxation and strong valid inequalities. Technical report, IRC for Process Systems Engineering, Imperial College, London SW7 2AZ, 1993.
36. A. Lucena. Tight bounds for the Steiner problem in graphs. Technical report, IRC for Process Systems Engineering, Imperial College, London SW7 2AZ, 1993.
37. N. Maculan. The Steiner problem in graphs. *Annals of Discrete Mathematics*, 31:185–222, 1987.
38. F. Margot, A. Prodon and T.M. Liebling. Tree polyhedron on 2-tree. *Mathematical Programming*, 63:183–192, 1994.
39. R. Marsten. The design of the XMP linear programming library. *ACM Transactions on Mathematical Software*, 7:481–497, 1981.
40. G.L. Nemhauser and L.E. Wolsey. *Integer and Combinatorial Optimization*, John Wiley & Sons, Chichester, 1988.
41. M. Padberg and S. Hong. On the symmetric travelling salesman problem: a computational study. *Mathematical Programming Study*, 12:78–107, 1980.
42. M. Padberg and G. Rinaldi. A branch-and-cut algorithm for the resolution of large-scale symmetric travelling salesman problems. *SIAM Review*, 33:60–100, 1991.
43. M. Padberg and L. Wolsey. Trees and cuts. *Annals of Discrete Mathematics*, 17:511–517, 1983.
44. J.C. Picard and H.D. Ratliff. A graph theoretic equivalence for integer programs. *Operations Research*, 21:261–269, 1973.

45. V.J. Rayward-Smith and A. Clare. On finding Steiner vertices. *Networks*, 16:283–294, 1986.

46. J. Soukup and W.F. Chow. Set of test problems for the minimum length connection networks. *ACM/SIGMAP Newsletter*, 15:48–51, 1973.

47. H. Takahashi and A. Matsuyama. An approximate solution for the Steiner problem in graphs. *Mathematica Japonica*, 6:573–577, 1980.

48. P. Winter. Steiner problems in networks: A survey. *Networks*, 17:129–167, 1987.

# 6

# Interior point algorithms for integer programming

**John E. Mitchell**
*Department of Mathematical Sciences,*
*Rensselaer Polytechnic Institute,*
*Troy, NY 12180-3590, USA*

## 1 Introduction

An algorithm for solving linear programming problems can be used as a subroutine in methods for more complicated problems. Such methods usually involve solving a sequence of related linear programming problems, where the solution to one linear program is close to the solution to the next, that is, it provides a *warm start* for the next linear program. The *branch and cut* method for solving integer programming problems is of this form: linear programming relaxations are solved until eventually the integer programming problem has been solved. Within the last ten years, interior point methods have become accepted as powerful tools for solving linear programming problems. It appears that interior point methods may well solve large linear programs substantially faster than the simplex method. A natural question, therefore, is whether interior point methods can be successfully used to solve integer programming problems. This requires the ability to exploit a warm start.

As discussed in Chapter 5 of this book, branch and cut methods for integer programming problems are a combination of cutting plane methods and branch and bound methods. In a cutting plane method for solving an integer programming problem, the linear programming relaxation of the integer program is solved. If the optimal solution to the linear program is feasible in the integer program, it also solves the integer program; otherwise, a constraint is added to the linear program which *separates* the optimal solution from the set of feasible solutions to the integer program, the new linear program is solved, and the process is repeated. In a branch and bound method for solving an integer programming problem, the first step is also to solve the linear programming relaxation. If the optimal solution is feasible in the integer program, it solves the integer program. Otherwise, the relaxation is split into two subproblems, usually by fixing a particular variable at zero or one. This is the start of a tree of subproblems. A subproblem is selected and the linear programming relaxation of that subproblem is solved. Four outcomes are possible: the linear programming relaxation is infeasible, in which case the integer subproblem is also infeasible, and the tree can be pruned at this node; the optimal solution to the linear programming relaxation is feasible in the integer subproblem, in which case it also solves the subproblem, and the tree can be pruned at this node; the optimal solution has a worse objective function value than a known integer solution to the original problem, in which case any solution to the integer subproblem is also worse than the known solution, and the tree can be pruned at this node; finally, none of these situations occur, in which case it is necessary to split the node into two further subproblems.

The difficulty with using an interior point method in a cutting plane algorithm is that the solution to one relaxation is usually not a good starting point for an interior point method, because it is close to the boundary of the feasible region. Thus, it is usually necessary to attempt to stop working on the current relaxation before it is completely solved to optimality: the earlier we are able to find good cutting planes, the better the initial solution to the next relaxation. Early termination obviously reduces the number of iterations spent solving the current relaxation; in addition, it reduces the number of iterations spent solving the next relaxation, because the initial point to the next relaxation is more centered. There are two potential disadvantages from trying to find cutting planes early: if the search for cutting planes is unsuccessful, we have wasted time; secondly, it may well be that superfluous constraints are added, with the result that the algorithm requires extra iterations and extra stages of adding cutting planes.

It is also necessary to be able to use a warm start when using an interior point method in a branch and bound algorithm; again, it is possible to use early termination to improve the algorithm. Usually, the only time it is necessary to solve the subproblem to optimality in the branch and bound tree is when the optimal solution to the linear programming relaxation is feasible in the integer subproblem.

We will concentrate on branch and cut algorithms for integer programming in the rest of this chapter, with the principal ideas being presented in Section 2. A cutting plane method can be regarded as a *column generation method* applied to the dual of the linear programming relaxation, because the addition of a cutting plane, or constraint, to the relaxation adds a column to the dual. There has been research on using interior point methods in column generation algorithms for various problems, and we will discuss some of this work later in the chapter, because of its relevance to branch and cut algorithms. For now, we give two examples. First, consider a linear programming problem with many more variables than constraints. If all the variables were included in the working set of columns, the matrix algebra would become impractical. Therefore, a pricing criterion is used to select a subset of the variables, and this subset is updated periodically – see Bixby *et al.* [3], Kaliski and Ye [20], and Mitchell [26], for example. Secondly, column generation methods are also useful for solving nonsmooth optimization problems. The constraints and the objective function are approximated by piecewise linear functions, and the approximation is improved as the algorithm proceeds. The problem solved at each stage is a linear programming problem, and refining the approximation corresponds to adding columns to the linear program. For more details, see, for example, the papers by Goffin and Vial and their coauthors [10, 12].

Another approach for solving integer programming problems using interior point methods has been presented by Karmarkar *et al.* [21]. We discuss this method in Section 3.2.

## 2 Interior point branch and cut algorithms

In this section, we discuss the use of interior point methods in cutting plane and branch and bound approaches to solving integer programming problems. The two approaches can be combined into a branch and cut algorithm; however, we intend to deal with them separately in order to make the basic ideas clear. Various observations about interior point algorithms will affect the design of efficient algorithms in both the cutting plane and branch and bound approaches. In particular, it is vital to use early termination in order to obtain good performance from these algorithms.

The standard form integer programming problem we consider has the form:

$$\begin{array}{lll} \text{minimize} & c^T x, & \\ \text{subject to} & Ax \geq b, & (IP) \\ & x \in \{0,1\}, & \end{array}$$

where $x$ and $c$ are $n$-vectors, $b$ is an $m$-vector and $A$ is an $m \times n$ matrix. Note that we restrict attention to problems where the integer variables are constrained to be binary. We define the polyhedron $Q$ to be the convex hull of feasible solutions to the problem $(IP)$. The problem $(IP)$ can be solved by solving the linear programming problem $\min\{c^T x : x \in Q\}$. (If this linear program has multiple optimal solutions, an interior point method may well return a point which is non-integral, but this can be rounded to give an integer optimal extreme point of $Q$ using the methods of, for example, Megiddo [24].) The difficulty with this approach, obviously, is that in general a closed form expression for $Q$ is not known. In a cutting plane method, we consider *linear programming relaxations* of $(IP)$, where the feasible region gradually becomes a better approximation to $Q$, at least in the neighbourhood of the optimal solution. The linear programming relaxation of $(IP)$ is

$$\begin{array}{lll} \text{minimize} & c^T x, & \\ \text{subject to} & Ax \geq b, & (LPP) \\ & 0 \leq x \leq e, & \end{array}$$

where $e$ denotes a vector of ones of the appropriate dimension. We denote the feasible region of this problem by $Q^{LPP} := \{x \in \mathbb{R}^n : Ax \geq b,\ 0 \leq x \leq e\}$. The dual of this problem is

$$\text{maximize} \qquad b^T y - e^T w,$$

$$\text{subject to} \quad A^T y - w \leq c, \quad (LPD)$$
$$y, w \geq 0,$$

where $y$ is an $m$-vector and $w$ is an $n$-vector. When necessary, we will let $s = Ax - b$ denote the slack variables for the primal problem $(LPP)$ and $z = c - A^T y + w$ denote the slack variables for the dual problem $(LPD)$. If $x$ denotes an $n$-vector then $X$ denotes the diagonal $n \times n$ matrix with $X_{ii} = x_i$. The diagonal matrices $Y$, $W$, $Z$, and $S$ are defined similarly. A positive iterate is considered *centered* if

$$XZe - \mu e = 0, \quad SYe - \mu e = 0, \quad (I - X)We - \mu e = 0 \tag{2.1}$$

for some appropriate scalar $\mu$; the iterate is *approximately centered* if the norm of each of these vectors is small. The scalar $\mu$ is called the *barrier parameter.*

## 2.1 Interior point cutting plane algorithms

Consider solving $(IP)$ using a cutting plane approach. The LP relaxation $(LPP)$ is solved, giving an optimal point $\hat{x}$, and an optimal dual solution $\hat{y}$, $\hat{z}$ for $(LPD)$. If $\hat{x}$ is feasible in $(IP)$ then it solves $(IP)$. Otherwise, we use a *separation routine* to find a cutting plane which separates $\hat{x}$ from the feasible region for $(IP)$. The separation routine depends on the problem being solved. For linear ordering problems, it is usually sufficient to examine the triangle inequalities (see Mitchell and Borchers [27]). Far more complicated separation routines have been developed for other problems, including the travelling salesman problem (see, for example, Padberg and Rinaldi [32]). Returning to the general case, we denote the cutting plane by

$$a_0^T x = b_0, \tag{2.2}$$

where $a_0$ is an $n$-vector and $b_0$ is a scalar. The point $\hat{x}$ satisfies

$$a_0^T \hat{x} < b_0 \tag{2.3}$$

and every feasible point $x$ for $(IP)$ satisfies

$$a_0^T x \geq b_0. \tag{2.4}$$

The constraint $a_0^T x \geq b_0$ is added to $(LPP)$, giving a new relaxation

$$\begin{array}{lll} \text{minimize} & c^T x, & \\ \text{subject to} & Ax \geq b, & (LPP0) \\ & a_0^T x \geq b_0, & \\ & 0 \leq x \leq e, & \end{array}$$

with a corresponding dual problem

$$\begin{array}{lll} \text{maximize} & b^T y + b_0 y_0 - e^T w, & \\ \text{subject to} & A^T y + a_0 y_0 - w \leq c, & (LPD0) \\ & y, y_0, w \geq 0, & \end{array}$$

where $y_0$ is a scalar. Notice that the current primal iterate $\hat{x}$ is not feasible in (*LPP0*), but $\hat{y}$, $\hat{w}$ is feasible in (*LPD0*), if we set $y_0 = 0$. A simplex-based cutting plane algorithm would reoptimize using the dual simplex algorithm. It is not so easy to restart an interior point method, because we no longer have an *interior* dual feasible point, that is, a point where the dual variables and the dual slacks are all strictly positive. This difficulty can be overcome by setting $y_0$ to a small positive value, although the dual iterate is then not feasible. An infeasible interior point method can be used to solve (*LPP0*) and (*LPD0*) – see, for example, Lustig *et al.* [22] or Zhang [39]. A more serious problem is that the primal and dual iterates are both extreme points (provided the LP relaxation is non-degenerate). Interior point methods tend to have poor performance when started from close to a nonoptimal extreme point. Typically, several iterations are spent moving towards the center of the feasible region before the algorithm moves towards the optimal point. One way to attempt to overcome this problem is to use *early termination*, that is, the problem $(LPP)$ is not solved to optimality, but we attempt to find cutting planes at a non-optimal point. This should ensure a better (i.e. more centered) starting iterate for problems (*LPP0*) and (*LPD0*).

A simple, conceptual interior point cutting plane algorithm could be written as

(1) **Initialize:** Pick initial $x$, $y$, $w$ and primal and dual slacks.
(2) **Approximately solve relaxation:** Solve the current relaxation to the desired degree of accuracy using an interior point algorithm. If the current iterate is a sufficiently accurate solution to the original problem $(IP)$, STOP.
(3) **Add cutting planes:** See if the current iterate violates any constraints. If not, tighten the desired degree of accuracy and return to Step (2); otherwise, add a subset of the violated constraints and go to Step (4).
(4) **Update the relaxation and restart:** Update the variables appropriately. Return to Step (2).

We will give a more formal algorithm later.

There are many different interior point methods and each one would be implemented in a slightly different manner in a cutting plane algorithm. Currently, the best algorithm for linear programming appears to be the primal-dual predictor–corrector barrier method (see Lustig *et al.* [22] and Mehrotra [25]), so, in this section, we consider modifying this algorithm for use in a cutting plane algorithm. Other interior point algorithms which maintain strictly positive primal and dual iterates can be modified in a similar manner; for a discussion of algorithms which only require either the primal or the dual iterate to be strictly positive, see Section 3.1.

### 2.1.1 *Early termination*

It is very desirable not to have to solve the current relaxation to optimality, but to attempt to add cutting planes early. Adding constraints early has two principal advantages:

- The initial iterates for the next subproblem are more centered, so fewer iterations are required to solve the next problem.
- Fewer iterations are required on the current problem.

The principal disadvantages of looking for constraints before solving to optimality are

- We may not be able to find any cutting planes, although we would be able to find cuts if we solved to optimality. Thus, time is wasted by looking early.
- Conversely, it may be that the optimal solution to $(LPP)$ solves $(IP)$, but the early iterates are not in the convex hull of feasible solutions, so the separation routines return cutting planes that are not necessary.

Attempting to find constraints early also has other, more subtle, advantages. First, the early iterates are closer than the optimal solution to the center of the feasible region for $(LPP)$, so it may be that the cuts which are found are *deeper* cuts, that is, they cut off more of the polyhedron $Q^{LPP}$. Secondly, it may also be that some of the cuts found by searching early would not have been found until the next relaxation if the relaxation had been solved to optimality. Consider, for example, the perfect matching problem.

**The perfect matching problem:** Given a graph $G = (V, E)$ with vertices $V$ and edges $E$, a *matching* is a subset $M$ of the edges such that no two edges in $M$ share an end vertex, and a *perfect matching* is a matching which contains exactly $|V|/2$ edges, where $|V|$ denotes the cardinality of $V$. Given a set of weights $w_e$ associated with each edge, the perfect matching problem is to find the perfect matching $M$ with smallest weight $w(M) := \sum_{e \in M} w_e$.

A complete polyhedral description for the perfect matching polytope was given by Edmonds [8]. He showed that the perfect matching problem is equivalent to the linear program

$$\begin{aligned}
\text{minimize} \quad & \sum_{e \in E} w_e x_e, \\
\text{subject to} \quad & \sum_{e \in \delta(v)} x_e = 1, && \forall v \in V, && (2.5) \\
& \sum_{e \in E(U)} x_e \le (|U| - 1)/2, && \forall U \subseteq V,\ |U| \text{odd}, && (2.6) \\
& x_e \ge 0, && \forall e \in E, && (2.7)
\end{aligned}$$

where $\delta(v)$ denotes the set of edges in $E$ which are incident to $v$ and $E(U)$ denotes the set of edges with both endpoints in the subset $U$ of $V$. In a cutting plane approach to this problem, the initial relaxation consists of the degree constraints (2.5) and the non-negativity constraints (2.7). The odd set constraints (2.6) are added as cutting planes. The odd set constraints prevent the existence of circuits with each edge $e$ in the circuit having value $x_e = 0.5$. For

FIG. 1. An illustration of the phenomenon of nested odd sets

details of a separation routine to find violated cutting planes, see Grötschel and Holland [13] and Mitchell and Todd [29]. Consider the graph shown in Fig. 1. Here, the edge weights $w_e$ are the Euclidean distances between vertices. An optimal perfect matching must contain one of the two edges $(v_4, v_6)$ and $(v_5, v_7)$. The optimal solution to the relaxation has $x_e = 0.5$ for the edges $(v_1, v_2)$, $(v_2, v_3)$, $(v_1, v_3)$, $(v_8, v_9)$, $(v_8, v_{10})$, and $(v_9, v_{10})$ and $x_e = 1$ for $(v_4, v_5)$ and $(v_6, v_7)$. The odd set constraints corresponding to the sets of vertices $\{v_1, v_2, v_3\}$ and $\{v_8, v_9, v_{10}\}$ are violated by this solution. After adding these constraints, the solution to the new relaxation has $x_e = 0.5$ for the edges $(v_1, v_2)$, $(v_2, v_5)$, $(v_4, v_5)$, $(v_3, v_4)$, $(v_1, v_3)$, $(v_6, v_7)$, $(v_7, v_9)$, $(v_9, v_{10})$, $(v_8, v_{10})$, and $(v_6, v_8)$. This violates the odd set constraints corresponding to the sets of vertices $\{v_1, v_2, v_3, v_4, v_5\}$ and $\{v_6, v_7, v_8, v_9, v_{10}\}$. However, if an interior point method is used and the search for violated constraints is conducted early, it may be possible to identify both of these sets of constraints at once, because the values of $x_e$ on the edges $(v_2, v_5)$ and $(v_3, v_4)$ tend to zero less quickly than the values on the edges $(v_5, v_7)$ and $(v_4, v_6)$. Thus, one fewer LP relaxation needs to be solved.

We have argued that it is necessary and beneficial to add cutting planes early when using an interior point cutting plane method. On the other hand, if we look for cutting planes when we are far from optimality, we may find unnecessary cuts. Thus, it is necessary to find a balance between looking for cuts too soon and waiting too long to look. This can be achieved by using a dynamically altered tolerance on the duality gap: we only look for cutting planes if the duality gap for the current relaxation falls below this tolerance. We are happy with a large tolerance if the current relaxation does not provide a good approximation to $(IP)$, because we do not want to spend a lot of time on poor relaxations. Conversely, once the relaxation becomes a good approximation, the tolerance should become small, because we want to be selective about adding additional cutting planes, and the relaxation is not likely to change dramatically, so the solution to the new relaxation should be close to the solution to the old relaxation. Additionally, the tolerance should be small when the relaxation provides a good approximation to $(IP)$ because we should make sure that the solution to the relaxation does not solve $(IP)$, in order to avoid adding unnecessary cutting planes. One way to modify the tolerance to achieve these aims is to increase it if we find a large number of violated constraints and decrease it if we find only a small number of (or no) violated constraints. This should have the effect of increasing the tolerance if the relaxation is a poor approximation to $(IP)$ and decreasing it if the relaxation is a good approximation to $(IP)$. For details on a particular way to modify this tolerance, see [27].

It is necessary to identify violated constraints early in an interior point cutting plane method in order for it to compete with a simplex cutting plane algorithm. It might be conjectured that a simplex cutting plane algorithm could be improved by using early identification. Such an approach is unlikely to be successful for several reasons. In the first place, the simplex algorithm moves from extreme point to extreme point, and thus it is likely to have relatively few fractional variables, and vital elements may be pivoted into the basis in the last few iterations. Complementary to this concern, it may be that the solution to the current relaxation solves the integer problem, but one of the vital basis elements only enters the basis on the last iteration, so the earlier iterates may violate constraints even if the optimal solution does not; it might then happen that unnecessary cutting planes are added to the relaxation. A third concern relates to the amount of time spent searching for violated cutting planes. Separation routines are generally far more expensive than simplex iterations, so it would be uneconomical to search every iteration; it may be possible to search every $K$ iterations for some number $K$.

### 2.1.2 *Restarting the algorithm and regaining primal feasibility*

Assume we have just added a cutting plane, so the relaxation now has the form of problem (*LPP0*), with corresponding dual problem (*LPD0*). We know a primal point $\hat{x}$ which satisfies $Ax \geq b$, but which does not satisfy $a_0^T x \geq b_0$. The current dual point $\hat{y}$, $\hat{w}$ is feasible in (*LPD0*) provided we set $y_0 = 0$. Since we used an interior point method to solve $(LPP)$ and since we terminated early, we have $\hat{x} > 0$, $\hat{y} > 0$, and the primal and dual slacks are strictly positive (except for the new primal constraint). We will use an infeasible interior point method to solve (*LPP0*), so we need an interior point and so we need positive values for $y_0$ and $s_0$. The simplest way to achieve this is to set these variables equal to some small positive amount $\epsilon$; it is also beneficial to reset any other variables which are smaller than $\epsilon$ to $\epsilon$. This works reasonably well in practice – the number of iterations required to solve the problem from this starting point is typically one third to one half of the number required to solve the relaxation from a cold start. Setting $y_0$ equal to $\epsilon$ makes the dual iterate infeasible; thus it is necessary to regain both primal and dual feasibility. This requires the careful setting of parameters to control the emphasis on regaining feasibility versus achieving optimality. Without care, the algorithm can drift far from optimality while it searches for feasibility. The predictor–corrector algorithm selects the barrier parameter $\mu$ automatically to achieve an appropriate balance between optimality and centering; if a different primal-dual method is used, it appears to be beneficial to take a pure centering step before starting to work towards optimality and feasibility.

The optimal solution $\bar{x}$ to (*LPP0*) will satisfy the added constraint at equality, if the constraint was a necessary cutting plane, so $\bar{y}_0$ will be strictly positive, since most interior point methods give solutions that satisfy strict complementary slackness (see Güler and Ye [16]). Thus, the algorithm must increase $y_0$ from $\epsilon$ to a non-trivial positive number, so it must move away from the face $y_0 = 0$.

Several papers (e.g. Anstreicher [1], Mizuno *et al.* [30], Zhang [39]) have discussed interior point algorithms for linear programming which move towards feasibility and complementary slackness simultaneously. A common feature of the analysis of these algorithms is the exploitation of the fact that they move towards feasibility at least as fast as they move towards complementary slackness. When restarting directly from the approximate solution to the previous relaxation $(LPP)$, the primal infeasibility is $s_0 + |b_0 - a_0^T \hat{x}|$. The total complementary slackness is $\hat{x}^T \hat{z} + \hat{s}^T \hat{y} + (e - \hat{x})^T \hat{w} + s_0 y_0$. To ensure that the ratio between complementary slackness and infeasibility remains reasonably large, it is necessary to set $s_0 y_0$ to a large value. This is going to result in an iterate which is far from centered, because it will be far from satisfying the condition of equation (2.1), with the 0th component of $SYe$ being far larger than the others.

An alternative approach that we have experimented with is to store and update a point $x^{FEAS}$ which is the relative interior of the polyhedron $Q$. When a cutting plane is added, the algorithm is restarted from this point, which is strictly feasible in (*LPP0*). The dual iterate is still infeasible, but only because $y_0$ must be chosen to be strictly positive. Thus, infeasibility is only a minor concern and it is possible to concentrate on regaining optimality. Initially, $x^{FEAS}$ can be set equal to a convex combination of feasible integer points. This also gives a good initial iterate for $(LPP)$. If the separation routines fail to find any cutting planes, the current iterate is in $Q$ (provided the separation routines are exact) so $x^{FEAS}$ can be updated to the current iterate. Even if the current iterate can be separated from $Q$, it is often possible to update $x^{FEAS}$ by moving towards the current iterate. Often, this extra work requires only one extra call to the separation routines to determine the feasibility of $x^{FEAS}$ exactly, and then a ratio test. For more details, see Mitchell and Borchers [27].

To the best of our knowledge, none of the interior point column generation algorithms proposed in the literature generate infeasible iterates. They either shift constraints so that the current iterate is still an interior point (e.g. [2, 26]), or they back up to a known feasible point (e.g. [17, 18, 20, 29]).

### 2.1.3 *Adding many constraints at once*

Usually, the separation routines will find several violated constraints. After adding these constraints, the problem has the form

$$\begin{array}{lll} \text{minimize} & c^T x, & \\ \text{subject to} & Ax \geq b, & (LPPA0) \\ & A_0^T x \geq b_0, & \\ & 0 \leq x \leq e, & \end{array}$$

with a corresponding dual problem

$$\text{maximize} \quad b^T y + b_0 y_0 - e^T w,$$

$$\begin{array}{llr} \text{subject to} & A^T y + A_0 y_0 - w \leq c, & (LPDA0) \\ & y, y_0, w \geq 0, & \end{array}$$

where $A_0$ is a matrix and $b_0$ and $y_0$ are vectors. Again, we can construct a dual feasible solution by taking $y_0 = 0$, and we set each component of $y_0$ to be a small positive number in order to obtain a strictly positive but slightly infeasible point. In the primal problem, the simplest way to restart is to set the slacks corresponding to the additional constraints to a small positive value, giving an infeasible iterate. Alternatively, as for the single constraint case, we can update to a point $x^{FEAS}$ in the relative interior of $Q$, if possible.

In many problems, the separation routines return a very large number of constraints. For example, the separation routines for a typical linear ordering problem often find thousands of violated constraints. It is inefficient to add all these constraints at once for several reasons. First, it will take many iterations to solve (*LPPA0*) because the linear programming relaxation has been dramatically modified, so the algorithm will take several iterations attempting to regain feasibility and recenter before trying to achieve optimality. Secondly, the large constraint matrix will make the linear algebra required at each of these iterations expensive. Thus, it is advisable to add only a subset of the constraints found by the separation routines. Perhaps the simplest way to decide between constraints is to see how badly they are violated by the current iterate – only constraints that are violated by more than some threshold are added. This is generally the method used in simplex-based cutting plane algorithms, although refinements have been used.

Consider for example the linear ordering problem, where the cuts correspond to directed cycles.

**The linear ordering problem:** Given a complete directed graph $G = (V, A)$ with weights $c_{ij}$ on the arcs, find a minimum weight *acyclic tournament*, that is, a subset of the arcs which contains exactly one of the arcs $(i, j)$ and $(j, i)$ for each pair $i$ and $j$, and which contains no directed cycles.

This problem has applications in economics and archaeology, as well as other areas. It can be formulated as

$$\begin{array}{llll} \text{minimize} & \sum_{(i,j)\in A} c_{ij}, & & \\ \text{subject to} & x_{ij} + x_{jk} + x_{ki} \leq 2, & \forall \{i, j, k\} \subset V, & (2.8) \\ & x_{ij} + x_{ji} = 1, & \forall \{i, j\} \subset V, & (2.9) \\ & x_{ij} \in \{0, 1\}, & \forall (i, j) \in A. & (2.10) \end{array}$$

Grötschel *et al.* [14] developed a simplex-based cutting plane algorithm where the initial relaxation consists of the simple bounds $0 \leq x_{ij} \leq 1$ and the equalities $x_{ij} + x_{ji} = 1$, and the 3-dicycle constraints (2.8) are added as cutting planes. They showed that these inequalities were sufficient to solve most of a set of real world problems. They experimented with two alternatives for deciding which constraints to add in their simplex-based cutting plane algorithm: firstly, the policy mentioned above of adding all constraints with violation greater than some threshold, and secondly, only adding a subset of the violated constraints with violation greater than some (smaller) threshold, where the cycles corresponding to two constraints in the subset are *arc-disjoint*. It should be noted that constraints added at different times may well share arcs. They found that implementations using the two policies required about the same amount of time, because the second policy resulted in the addition of fewer constraints at each relaxation but, by way of compensation, the solution of several more relaxations. For an interior point cutting plane algorithm, the benefit of adding arc-disjoint constraints is far more pronounced, because the projections required at each iteration can be calculated more efficiently due to the structure of the constraint matrix. This is because the columns of $A$ correspond to the arcs, so if two constraints share an arc then the inner product between the two rows is non-zero; thus, if two constraints correspond to arc-disjoint dicycles then the corresponding entry of $AA^T$ will be zero. It follows that if all the constraints are arc-disjoint dicycles then $AA^T$ is a diagonal matrix, so the Cholesky factor is also diagonal. Of course, constraints added at different times may share arcs, so there will be some fill-in in the Cholesky factor, but ensuring that each set of added constraints corresponds to arc-disjoint constraints does considerably reduce the fill-in.

It has also proved useful in other problems to consider the impact of a constraint on the non-zero structure of the constraint matrix; we discuss this in more detail later. It has been noted by Lustig *et al.* [23] that the runtime of an interior point method for linear programming is increased to a time comparable to that required by the simplex algorithm only when the Cholesky factors of the matrix product $AA^T$ are dense, i.e. they contain a large number of non-zeros. In [27], we first ordered the constraints by violation and then examined them in order to see if they were arc-disjoint, keeping only those that did not share any arcs with constraints with larger violations, and this improved the algorithm considerably.

There may be better measures of the "importance" of a constraint than the size of violation. One possibility was proposed by Atkinson and Vaidya [2]: it considers the violation divided by the size of the constraint in the norm defined by the Hessian of the barrier function. Larger values of this quantity correspond to constraints that will have a larger impact on the barrier function. Computational experiments are needed to determine the importance of this selection criterion.

#### 2.1.4 *Dropping constraints*

One way to improve the performance of a cutting plane algorithm is by judiciously dropping constraints that no longer appear to be important. This keeps the size of the relaxations reasonable, so less work is needed at each iteration to calculate the projections. A further benefit is that smaller linear programming problems require fewer interior point iterations. The simplest way to decide whether to drop a constraint is to check its current value: if the current iterate easily satisfies it, it is a candidate to be dropped.

Dropping a constraint changes the structure of the constraint matrix and the matrix $AA^T$, so it becomes necessary to find a new good ordering of the columns of $AA^T$ for performing the Cholesky factorization. For this reason, constraints were only dropped when other constraints were added in [27], with good results.

Other criteria can be used to measure the "importance" of a constraint. In particular, the measure proposed by Atkinson and Vaidya [2] can be used: a constraint is dropped if it is easily satisfied and it has little impact on the barrier function. Numerical experiments are necessary to determine the usefulness of this criterion.

#### 2.1.5 *Primal heuristics*

Primal heuristics, which generate integer solutions to $(IP)$ from fractional solutions to $(LPP)$, can be very useful in some circumstances. If the heuristics are cheap, they can be used at every iteration. However, it is usually more cost effective to call them only when the separation routines are also called.

One of the major advantages of using a cutting plane approach to solve an integer programming problem is that it provides a guarantee of optimality, and the current dual value provides a lower bound on the optimal value of the integer program. It is occasionally possible to stop the algorithm early by using the primal integer solution generated by the heuristics in conjunction with a good dual solution to $(LPD)$, especially if the entries of the objective function $c$ are integer.

If the interior point cutting plane method is converging to a fractional optimal point in the polyhedron $Q$, the primal heuristics can give an optimal solution to $(IP)$. This is obviously desirable in order to claim to have solved $(IP)$, and it may also prevent unnecessary calls to the separation routines.

The primal heuristics can also be used to update $x^{FEAS}$. Recall that $x^{FEAS}$ is supposed to be a good point in $Q$, so it can be updated by taking a convex combination of the old $x^{FEAS}$ and the good solution returned by the primal heuristics. Care should be taken to ensure that $x^{FEAS}$ does not get too close to a non-optimal integral point, because it would then not be a good restarting point, as argued above.

#### 2.1.6 *Multiple optimal solutions and degeneracy*

It is obvious that some integer programming problems have more than one optimal solution. In this case, the set of optimal solutions to the linear programming problem $\min\{c^T x : x \in Q\}$ is a face of $Q$ of dimension at least one. It is then likely that an interior point cutting plane algorithm will converge to a point in the interior of this optimal face, and thus it will not converge to an integer point. Perhaps the simplest way to find an integer optimal point is to use primal heuristics to convert the fractional point into a nearby integer point. Alternatively, methods have been developed in the linear programming context for converting from a point in the interior of the optimal face to an extreme point of that face – see, for example, Megiddo [24]. Thus, once it has been determined that we have found a point on the optimal face of $Q$, it should be possible to convert that point into an integer point.

If the process is converging to an optimal fractional point and if the primal heuristics do not produce an optimal integer point then a cutting plane algorithm may call the separation routines several times without success. This may produce a drag on performance of the algorithm, and this is the major cost involved with converging to a fractional optimal point. In such a situation, it may be worthwhile to invest more time in the primal heuristics in an attempt to find an optimal integer point. More computational testing is required to determine the importance of this issue.

In some cases, the fractional optimal solution may actually provide *more* information about the problem than an integer extreme point of the optimal face, because every integer optimal point has non-zero components only where the fractional point has non-zero components. For example, some of the linear ordering problems solved in [27] have multiple optimal solutions. In this case, the fractional optimal solutions provided by the cutting plane method correspond to partial orderings, and all optimal full orderings are implied by these fractional orderings.

Degeneracy of the linear programming relaxation is probably not a disadvantage for an interior point cutting plane algorithm, because interior point methods are not usually bothered by degeneracy. Thus, an interior point cutting plane method may possess a distinct advantage over a simplex cutting plane method if the relaxation is degenerate. If the problem has multiple optimal solutions then the final few relaxations are probably dual degenerate, so the simplex algorithm may slow down in the final few stages. Again, this is an issue that needs more computational investigation.

#### 2.1.7 *Fixing variables*

Simplex branch and cut algorithms use reduced costs to fix integral variables at zero or one as follows. Let $v_{UB}$ be the value of the best integer solution found so far, let $v_{LB}$ denote the optimal value of the current relaxation, and let $r_i$ denote the reduced cost for primal non-basic variable $x_i$. If $x_i$ has value zero in the optimal solution to the current relaxation, then any feasible primal solution to the current relaxation with $x_i = 1$ must have value at least $v_{LB} + r_i$. Thus, if $v_{LB} + r_i > v_{UB}$ then $x_i$ can be fixed at zero; no integral solution with $x_i = 1$ that is feasible in the current relaxation will be better than the best-known integer solution. Similarly, if $x_i = 1$ at the optimal solution to the current relaxation and if $v_{LB} - r_i > v_{UB}$ then $x_i$ can be fixed at one.

It is not necessary to have the optimal solution in order to fix variables, and it is possible to use the current dual solution to derive a criterion for fixing variables when using an interior point method. Any dual feasible solution provides a lower bound on the value of the relaxation. Let $\bar{x}$ be the current primal solution, let $\bar{y}$, $\bar{w}$ be the current dual solution, let $\bar{z}$ be the corresponding dual slack, and let $\bar{v}_{LB}$ be the dual value. Then $c = A^T\bar{y} - \bar{w} + \bar{z}$, so any primal feasible solution $x$

to the current relaxation $(LPP)$ will have value

$$\begin{aligned} c^T x &= (A^T \bar{y} - \bar{w} + \bar{z})^T x \\ &= \bar{y}^T A x + \bar{z}^T x - \bar{w}^T x \\ &\geq b^T \bar{y} + \bar{z}^T x - \bar{w}^T x \quad \text{(since } Ax \geq b \text{ and } \bar{y} \geq 0) \\ &= b^T \bar{y} - e^T \bar{w} + \bar{z}^T x + \bar{w}^T (e - x) \\ &= \bar{v}_{LB} + \bar{z}^T x + \bar{w}^T (e - x). \end{aligned}$$

Thus, any feasible solution to $(LPP)$ with $x_i = 1$ will have value at least $\bar{v}_{LB} + z_i$, so $x_i$ can be fixed at zero if $\bar{v}_{LB} + z_i > v_{UB}$. Similarly, $x_i$ can be fixed at one if $\bar{v}_{LB} + w_i > v_{UB}$. Note that we may be able to fix a variable at zero or one even if it is currently fractional. Fixing some variables using duality may have logical implications for the other variables. For example, if it is possible to fix one of the edges in a matching problem at one, then all other edges adjacent to the end points of that edge must have value zero. When variables are fixed, some constraints may become redundant, and these constraints should be dropped. If we again consider the matching problem with an edge fixed at one, the degree constraints (2.5) for the endpoints of that edge are now both redundant.

If it is possible to fix a large number of variables, the amount of linear algebra required at each iteration will be reduced; in addition, the number of iterations needed to solve the current relaxation will also probably be reduced. The drawback to fixing variables is that the current solution becomes infeasible; the restarting point $x^{FEAS}$ also has to be modified.

### 2.1.8 *The complete algorithm*

The discussion so far leads to a cutting plane algorithm which has the following form:

(1) **Initialize:** Set up the initial relaxation. Pick an initial iterate which strictly satisfies the required non-negativity bounds on all primal and dual variables (including slack variables). A standard method for initializing an interior point linear programming algorithm can be used, for example, the method given in Lustig *et al.* [23]. Set a tolerance $\tau_1$ for optimality. Set a tolerance $\tau_2 > \tau_1$ for searching for cutting planes.

(2) **Inner iteration:** Perform an iteration of the linear programming solver being used.

  (a) If the difference in value between the best integer solution and the best dual solution found so far is smaller than $\tau_1$, STOP with optimality.

  (b) If the duality gap between the current primal and dual values is smaller than $\tau_2$, go to Step 3.

  (c) If the duality gap between the current primal and dual values is smaller than $\tau_1$, the process has converged to a fractional point which is on the optimal face of $Q$, so use a basis refinement technique to obtain an optimal integer solution. Then STOP.

  Repeat this step.

(3) **Primal heuristics:** Use heuristics to obtain a good integer solution using the current fractional primal solution. If the heuristics return the best integer solution found so far, store this solution.

(4) **Look for cutting planes:** Use separation routines to find violated cutting planes. If no violated cutting planes are found, reduce $\tau_2$ and return to Step 2; otherwise, go to Step 5.

(5) **Add cutting planes:** Select and add a subset of the violated constraints to the relaxation. Increase $\tau_2$ if many violated constraints were found; decrease $\tau_2$ if only a few cuts were found.

(6) **Fix variables:** If the dual solution implies that variables must take certain values, fix those variables, if desired.

(7) **Drop cutting planes:** Drop any constraints which no longer appear to be important.

(8) **Modify current iterate:** Increase any small components to some value $\epsilon$. If $\mu$ is not selected by the predictor–corrector methodology, perform a centering step and update the centering parameter $\mu$, if appropriate. Return to Step 2.

This algorithm will converge to the optimal solution, provided the separation routines are guaranteed to find a violated cutting plane if one exists. If the separation routines may fail to find a violated constraint even if the current primal point is not in $Q$, then it may be necessary to use branch and bound to find an optimal solution to the integer program.

If it is possible to maintain and update a point $x^{FEAS}$ in the relative interior of $Q$, this should be done at Step 5. This point can then be used to update the primal solution in Step 8.

### 2.1.9 *The structure of the constraint matrix*

As mentioned above, interior point methods do not work so well if the Cholesky factors of the matrix $AA^T$ become dense. This should be taken into account when deciding which constraints to add. On a more fundamental level, it will also affect the choice of whether to use an interior point or cutting plane interior point method. If the structure of the constraint matrix is unsuitable, a simplex cutting plane algorithm will probably outperform an interior point one.

The paper by Mitchell and Todd [29] describes an interesting manifestation of the problem of decreased performance when the Cholesky factors become dense. In this paper, an interior point method was used to solve matching problems. Two different distributions were used to generate test problems. The first randomly placed points in a unit square. The odd set constraints that are generated tend to correspond to neighbouring points. This results in a matrix $AA^T$ that can be permuted to have small bandwidth, so the Cholesky factors are sparse. The second distribution randomly generated distances between vertices. In this case, the odd set constraints found by the algorithm are not nicely related to one another and the matrix $AA^T$ does not have a good structure, resulting in a lot of fill-in in the Cholesky factors. The upshot of this was that although problems generated using the first distribution required more iterations, they required considerably less time. When Grötschel and Holland [13] solved matching problems using a simplex-based cutting plane algorithm, they were able to solve problems generated using the second distribution considerably quicker than those generated using the first distribution, illustrating the fact that the structure of the constraint matrix is not so important with a simplex algorithm.

As pointed out by Saunders [35], Vanderbei [38] and others, there are alternative methods for finding projections than calculating Cholesky factors. It may happen that if these alternative methods are used then cutting plane interior point methods will not be so sensitive to fill-in of the matrix $AA^T$ and its Cholesky factors.

### 2.2 Interior point branch and bound methods

Branch and bound is a method of solving an integer programming problem $(IP)$ by solving a sequence of linear programming problems. The subproblems can be regarded as forming a tree, rooted at the linear programming relaxation $(LPP)$ of the integer programming problem. As we move down the tree, more and more integer variables are fixed in value. We provide a very brief description of the technique in order to highlight some aspects which prove important when using an interior point method. For a more detailed discussion of branch and bound and the options available, see, for example, Parker and Rardin [33].

When using branch and bound, one of four things can happen at each node of the tree. The subproblem could be unbounded; in an interior point method this can be detected by finding a ray in the dual problem. The subproblem could have optimal value worse than the value of a known integer feasible solution, so the node is fathomed by bounds; in an interior point method, this can usually be detected well before the subproblem is solved to optimality. The optimal solution could be an integer solution with value better than the best-known solution; in this case we need to solve the subproblem to optimality, but the node is then fathomed. The final possibility is that the optimal solution to the subproblem has optimal value smaller than the best-known solution, but the optimal solution is not feasible in the integer program; in this case, it is possible to use heuristics based upon the basis identification techniques described in El-Bakry *et al.* [9] to determine that one of the integer variables is tending to a fractional value, and therefore that we should branch early.

It should be noted that in only one case is it necessary to actually solve the relaxation to optimality, and in that case the node is fathomed. When we branch early, one constraint in $(LPD)$ is dropped, so the previous solution to $(LPD)$ is still feasible. One variable in $(LPP)$ is fixed, so infeasibilities are introduced into $(LPP)$. Despite this, it is still possible to solve the child node quickly [4, 5].

A branch and bound interior point algorithm has the following form:

(1) **Initialize:** Pick an initial relaxation. Choose an initial primal and dual iterate, using, for example, the method of Lustig *et al.* [23]. Set a tolerance $\tau$ for optimality. Initialize the branch and bound tree to contain only the

initial relaxation.

(2) **Pick a node:** Select a node of the branch and bound tree to solve next. Find an initial solution for this node. (We discuss a method for restarting at a node later.)

(3) **Perform an interior point iteration:** Perform one iteration of the interior point method for the current node.

   (a) Attempt to find a good integer solution by rounding the fractional solution in an appropriate manner. If this gives the best solution found so far, store it, and take its value as an upper bound on the optimal value of $(IP)$.

   (b) If the duality gap of the current relaxation is smaller than $\tau$ and if the primal value is better than the upper bound on the optimal value of $(IP)$ then

      i. If the primal solution is integral, fathom this node, update the upper bound on the optimal value of $(IP)$, and return to Step 2.

      ii. If the primal value is non-integral, go to Step 5.

   (c) If the dual value of the current node is greater than the best-known upper bound, prune this node.

   (d) If we can find a dual ray, showing that the primal problem is infeasible, prune this node.

   (e) If the current solution is dual feasible and if the relative primal infeasibility is smaller than some tolerance, go to Step 4.

   Repeat this step.

(4) **Check for non-integrality:** Check to see whether the solution to the current node appears to be fractional. If it appears that the solution to this node will be fractional, and if it appears unlikely that this node will be fathomed by bounds, go to Step 5; otherwise return to Step 3. (We discuss this step in more detail later.)

(5) **Branch:** Split the current node into two nodes. Pass the current primal and dual solutions to the child nodes as a warm start. Go to Step 2.

#### 2.2.1 *Terminating the current relaxation early*

The solution to a parent subproblem can be used as a warm start for a child subproblem. For this to be a good warm start for an interior point method, it should be somewhat centered, and therefore it is useful to terminate solution of the parent problem early. There is a risk associated with attempting to stop solution of the parent subproblem early: the parent may be split into two child subproblems, when it might have been possible to prune the parent if it had been solved to optimality. This could happen if the parent subproblem has worse objective function value than that of the best-known feasible solution to $(IP)$, or if it is infeasible, or if it has an integer optimal solution. Therefore, it is wise to include some safeguards to attempt to avoid this situation. Upper and lower bounds on the value of a subproblem are provided by the values of the current primal and dual solutions, respectively, and these can be used to regulate the decision to branch.

The paper by El-Bakry *et al.* [9] describes a good indicator for determining which variables are tending towards zero and which ones are tending towards non-zero values. In particular, if $x_i^k$ represents the $i$th component of the $k$th iterate then

$$\begin{aligned} \lim_{k\to\infty} x_i^{k+1}/x_i^k &= 1, \qquad \text{if } x_i^k \not\to 0, \\ \lim_{k\to\infty} x_i^{k+1}/x_i^k &= 0, \qquad \text{if } x_i^k \to 0. \end{aligned}$$

Similar observations apply to the primal slack variables, including those for the constraints $x \leq e$, and also to the dual variables. If a primal variable $x_i$ appears to be tending towards a non-zero value, and if this is confirmed by $(1-x_i)$ tending to a non-zero value and also the corresponding dual slack and $w_i$ tending towards zero, then it is reasonable to conclude that the optimal solution to the relaxation is fractional. It will follow that it will be necessary to split this subproblem into two children if the optimal value of the subproblem is smaller than that of the incumbent integer feasible solution to $(IP)$.

There are three tests used in [5] to prevent branching too early: the dual iterate must be feasible, the relative primal infeasibility must be no greater than 10%, and the dual objective function must not be increasing so quickly from iteration to iteration that it is likely that the node will be fathomed by bound within an iteration or two. Dual feasibility can usually be maintained throughout the branch and bound tree so the first criterion is basically just a technicality. Every time a variable is fixed, primal infeasibility is introduced; if the initial iterate for a subproblem is a good warm start, primal feasibility can be regained in a few iterations. Thus, the important criterion is the third one, regarding the increase in the dual value. This criterion prevents branching if the difference between the dual value and the value of the incumbent integer solution has been reduced by at least half in the last iteration, provided the current primal value is greater than the current integer solution if the current primal iterate is feasible.

### 2.2.2 *Warm starting at a node of the tree*

The exact method used for restarting at a node of the branch and bound tree depends upon the interior point algorithm used. In this section, we assume that the primal-dual barrier method is being employed (see, for example, Lustig *et al.* [22]). It should be noted that many of the observations we make will also be applicable if other interior point methods are used.

Assume a child problem will be created by fixing (at 0 or 1) a particular variable $x_0$ (say) in the parent problem

$$\begin{array}{lll} \text{minimize} & c^T x + c_0 x_0, & \\ \text{subject to} & Ax + a_0 x_0 \geq b, & (LP\ parent) \\ & 0 \leq x,\ x_0 \leq e, & \end{array}$$

where $A$ is an $m \times n$ matrix, $a_0$ and $b$ are $m$-vectors, $c$ and $x$ are $n$-vectors, and $c_0$ and $x_0$ are scalars. The child problem has the form

$$\begin{array}{lll} \text{minimize} & c^T x, & \\ \text{subject to} & Ax \geq \bar{b}, & (LP\ child) \\ & 0 \leq x \leq e, & \end{array}$$

where $\bar{b} = b$ if $x_0$ is fixed at zero, and $\bar{b} = b - a_0$ if $x_0$ is fixed at one. An approximate solution $x = x^*$, $x_0 = x_0^*$ to (*LP parent*) is known. Since we created this particular child problem, $x_0^*$ must be fractional, so $x^*$ is probably infeasible in (*LP child*). If we examine the dual problems

$$\begin{array}{lll} \text{maximize} & b^T y - e^T w - w_0, & \\ \text{subject to} & A^T y - w \leq c, & (LD\ parent) \\ & a_0^T y - w_0 \leq c_0, & \\ & y, w, w_0 \geq 0, & \end{array}$$

and

$$\begin{array}{lll} \text{minimize} & \bar{b}^T y - e^T w, & \\ \text{subject to} & A^T y - w \leq c, & (LD\ child) \\ & y, w \geq 0, & \end{array}$$

we notice that the approximate solution $y = y^*$, $w = w^*$ (and $w_0 = w_0^*$) to (*LD parent*) is still feasible in (*LD child*) – all that has changed is the objective function value. Therefore, it is possible to restart the algorithm using an infeasible interior point method. It may be that some of the components of $x^*$ may be very close to zero or one, and these components should be modified to give a slightly more interior point, with small components being increased and large components being decreased. Similarly, if some components of the dual variables $y^*$ and $w^*$ or the primal or dual slacks are smaller than some tolerance, they should be increased, at the cost of making the initial iterate for the subproblem slightly more infeasible. Thus, we can start the interior point method on the child problem if we have stored $x^*$, $y^*$ and $w^*$. It may be beneficial to use a pure centering step first before updating the barrier parameter $\mu$ in a standard manner.

It may be possible to start the solution of the child problem from an iterate for (*LP parent*) which was found before $x^*$. This earlier iterate would be further from optimality for (*LP parent*) than $x^*$, but it may be a better initial solution to (*LP child*) just because it is more centered, with the non-basic components being somewhat larger. Preliminary experiments show that this approach may hold some promise, but it needs considerably more investigation.

One cost of using the simplex algorithm in a branch and bound method is that it is necessary to perform an initial basis factorization for each child subproblem. The cost of this is clear when examining, for example, the performance of the branch and bound code for OSL [19] on solving integer programming problems: it often happens that the average time per iteration is about three times larger for subproblems than it is for the root node of the branch and bound tree. This extra time is almost completely attributable to the overhead required at each node to calculate a new basis factorization. A comparable slow down does not happen with interior point methods. One way to avoid this overhead would be to store the basis factorization of each parent node, but this usually requires too much storage and is hence impracticable. Of course, part of the reason that the slow down is so noticeable is that the simplex algorithm requires far fewer iterations to solve subproblems than to solve the root node, because the optimal solution to the parent node does provide a good simplex warm start for the child subproblem. At present, it does not seem possible to get a similar reduction with an interior point method, but the fact that the basis refactorization is so expensive means that it is not necessary to obtain as good a reduction in the number of iterations as enjoyed by the simplex algorithm.

#### 2.2.3 *Degeneracy*

The linear programming problems generated in a branch and bound tree can suffer from considerable degeneracy, which can greatly slow the simplex method. Degeneracy is generally not a problem for interior point methods, and at least one commercial package has installed a switch to change from simplex to an interior point method within the branch and bound tree if difficulties arise.

## 3 Extensions

In this section we describe some other aspects of interior point algorithms for integer programming problems. The algorithms discussed in Section 2 generally required that both the primal and the dual iterates be strictly feasible; in Section 3.1 we consider algorithms which only require that the dual iterates be strictly feasible. In Section 3.2 we discuss a different approach to solving integer programming problems using interior point methods; this method works by solving an equivalent non-convex quadratic programming problem. In Section 3.3 we briefly consider implementing branch and cut algorithms in parallel. Finally, in Section 3.4 we consider theoretical issues.

### 3.1 Algorithms which only require positive dual iterates

The primal-dual cutting plane algorithm described in Section 2.1 uses iterates which are interior in both the primal and the dual feasible region at each iteration. There has been some research on using interior point methods which only require an interior point in the dual problem; a solution to the primal problem is only used to provide an upper bound on the optimal value. These algorithms are generally adaptations of primal potential function methods.

The first experiments in using an interior point method in a cutting plane algorithm were in Mitchell and Todd [29]. They applied the primal potential reduction algorithm to the dual $(LPD)$ of the linear programming relaxation of $(IP)$. The advantage of this approach is that it is not necessary to have an iterate which is feasible in $(LPP)$. The primal solution is updated using the method of Todd and Burrell [36]. Mitchell and Todd found in their experiments on matching problems that it was best to search for cutting planes as soon as the primal solution was updated. When a constraint was added, they updated the solution to the new relaxation $(LPD0)$ by updating $y$ and $w$ and moving in a direction which is a projection of $a_0$ onto the row space of an appropriately rescaled version of the constraint matrix for $(LPD)$. They used heuristics to round the current solution to $(LPP)$ to a point feasible in $(IP)$. When the relaxation was updated, the best solution to $(IP)$ found so far was used as the new primal iterate. They were able to obtain promising results.

Goffin and Vial [10, 12] used a similar algorithm to solve non-smooth optimization problems using a column generation algorithm. They obtained results comparable with, or superior to, other algorithms for these problems.

### 3.2 Solving an equivalent quadratic programming problem

The integer programming feasibility problem can be stated as: find a point $x$ in $\mathbb{R}^n$ satisfying

$$Ax \geq b, \; x_i \in \{0,1\}, \; i = 1, \ldots, n, \qquad (IP\ feas)$$

where $A$ is an $m \times n$ matrix and $b$ is an $m$-vector. This can be solved by solving the non-convex quadratic programming problem

$$\begin{array}{lll} \text{minimize} & x^T(e-x) & \\ \text{subject to} & Ax \geq b & (QP) \\ & x \leq e & \\ & x \geq 0. & \end{array}$$

Note that for every binary point, the objective function $x^T(e-x)$ has value zero and for every fractional point this inner product has positive value. Kamath *et al.* [21] have investigated using an interior point method to solve this quadratic programming problem, and hence to solve the integer programming problem $(IP)$. Their algorithm attempts to minimize the potential function

$$(m+2n)\log(x^T(e-x)) - \sum_{i=1}^{m} \log s_i - \sum_{i=1}^{n} \log(x_i) - \sum_{i=1}^{n} \log(1-x_i),$$

where $s = Ax - b$. They show that $x^*$ is a global minimizer of this potential function if and only if it is a feasible solution to $(IP\ feas)$. The potential function may have local minimizers which are not global minimizers; therefore, the set of constraints is augmented to cut off local minimizers which are not feasible in $(IP\ feas)$, if necessary. They used their algorithm to solve satisfiability problems from inductive inference, obtaining promising results compared to an implementation of the Davis–Putnam procedure [6] (an implicit enumeration procedure which is similar to branch and bound).

### 3.3 Parallel implementations

Compared to the simplex algorithm, interior point methods typically require only a few iterations to solve a linear programming problem, with each iteration requiring a considerable amount of time. The linear algebra required at each iteration of an interior point method can be implemented in parallel, with the savings in time increasing as the number of processors increases. By contrast, the potential time saving from implementing the simplex algorithm in parallel are more limited.

This potential superiority of interior point methods for linear programming can be usefully exploited in a cutting plane algorithm. Furthermore, the separation routines can be run in parallel with the linear programming solver, making it possible to carefully choose when to add cutting planes. Separation routines typically require less time than the linear programming parts of an implementation, so it may be possible to implement more sophisticated separation routines if one or more processors are dedicated to them. A simplex cutting plane algorithm generally needs the optimal fractional solution to the relaxation before calling the separation routines, because the basic feasible solution can change dramatically in the last few iterations; thus, it is harder to run the separation routines and the linear programming solver in parallel when using the simplex algorithm.

Branch and bound can be implemented efficiently in parallel, even on machines with a large number of processors; see, for example, Eckstein [7]. Thus, the difficulty with parallelizing the simplex algorithm need not be a drawback for a branch and bound algorithm: once there are many nodes in the tree, the algorithm can work on them simultaneously.

### 3.4 Theoretical behaviour of column generation algorithms

Given an integer programming problem, a separation routine either confirms that a point is in the convex hull of the set of feasible integer points, or it provides a cutting plane which separates the point from the convex hull. If the separation routine runs in polynomial time in the size of the problem, then the ellipsoid algorithm can be used to solve the integer programming problem in polynomial time – see Grötschel *et al.* [15]. It is not necessary to drop any constraints when using this method. For the rest of this subsection, we assume that the separation routines require polynomial time.

To date, the only interior point algorithm which solves the integer program in polynomial time and which does not drop constraints is due to Vaidya [37]. This algorithm uses the *volumetric center*, so its analysis differs from that of more standard interior point methods.

Atkinson and Vaidya [2] developed a polynomial time cutting plane algorithm which used the analytic center. This algorithm drops constraints that become unimportant, and this is essential in their complexity analysis. Previous algorithms were often shown to be polynomial in the number of additional constraints, but without a proof that the number of added constraints is polynomial. Atkinson and Vaidya's algorithm finds a feasible point for a set of convex inequalities by finding an analytic center for a subset of the inequalities and using an oracle to test whether that point satisfies all the inequalities. If the oracle returns a violated inequality, a shifted linear constraint is added so that the analytic center remains feasible and close to the new analytic center.

Mitchell and Ramaswamy [28] developed a barrier function cutting plane algorithm using some of the ideas from [2]. They showed some links between the notion of a point being centered (see, for example, Roos and Vial [34]) and the criteria for a constraint to be added or dropped in [2]. Barrier function methods for linear programming have shown excellent computational performance and they can be constructed to have superlinear and quadratic convergence. It would thus appear desirable to employ these methods in a column generation algorithm.

Goffin *et al.* [11] presented a pseudopolynomial column generation algorithm which does not need to drop any columns. The number of iterations required to get the objective function value to within $\epsilon$ of optimality is polynomial in $\epsilon$, but this algorithm does not obtain a solution within $2^{-L}$ of optimality in time polynomial in $L$, where $L$ is the size of the data. Their paper used a result of Nesterov [31] which placed a bound on an important quantity in the analysis of column generation interior point methods.

## 4 Conclusions

We have concentrated on discussing the possibility of incorporating interior point methods into cutting plane and branch and bound algorithms for linear programming. In order to do this successfully, it is necessary to be able to use a warm start somewhat efficiently. The most important technique appears to be early termination: the current relaxation is only solved to within some tolerance of optimality before we attempt to refine the relaxation. Other techniques and potential pitfalls are discussed in Section 2.

Currently, interior point cutting plane methods do appear to be somewhat competitive with simplex cutting plane algorithms, at least for some problems. Interior point branch and bound algorithms still appear weaker than simplex-based algorithms, at least for the size of problem which can currently be solved. For linear programming, interior point methods start to outperform simplex for large problems, so a branch and bound interior point method would only be advantageous for large problems (thousands of variables and constraints). Pure integer problems of this size are currently difficult to solve. Thus, interior point branch and bound methods currently appear only to be useful for problems with a small number of integer variables, but a large number of continuous variables. As hardware improves, it will become possible to solve larger problems, and interior point branch and bound methods will become more attractive.

## Bibliography

1. K.M. Anstreicher. A combined phase I – phase II scaled potential algorithm for linear programming. *Mathematical Programming*, 52:429–439, 1991.

2. D.S. Atkinson and P.M. Vaidya. A cutting plane algorithm for convex programming that uses analytic centers. *Mathematical Programming*, 69:1–43, 1995.

3. R.E. Bixby, J.W. Gregory, I.J. Lustig, R.E. Marsten and D.F. Shanno. Very large-scale linear programming: A case study in combining interior point and simplex methods. *Operations Research*, 40:885–897, 1992.

4. B. Borchers. *Improved branch and bound algorithms for integer programming.* PhD thesis, Rensselaer Polytechnic Institute, Mathematical Sciences, Troy, NY, 1992.

5. B. Borchers and J.E. Mitchell. Using an interior point method in a branch and bound algorithm for integer programming. Technical Report 195, Mathematical Sciences, Rensselaer Polytechnic Institute, Troy, NY 12180, March 1991. Revised July 7, 1992.

6. M. Davis and H. Putnam. A computing procedure for quantification theory. *Journal of the ACM*, 7:201–215, 1960.

7. J. Eckstein. Parallel branch-and-bound algorithms for general mixed-integer programming on the CM–5. Technical Report TMC–257, Mathematical Sciences Research Group, Thinking Machines Corporation, 245 First Street, Cambridge MA 02142, September 1993.

8. J. Edmonds. Maximum matching and a polyhedron with 0, 1 vertices. *Journal of Research National Bureau of Standards*, 69B:125–130, 1965.

9. A.S. El–Bakry, R.A. Tapia and Y. Zhang. A study of indicators for identifying zero variables in interior-point methods. *Siam Review*, 36:45–72, 1994.

10. J.-L. Goffin, A. Haurie and J.P. Vial. Decomposition and non-differentiable optimization with the projective algorithm. *Management Science*, 38:284–302, 1992.

11. J.-L. Goffin, Z.-Q. Luo and Y. Ye. On the complexity of a column generation algorithm for convex or quasiconvex problems. In *Large Scale Optimization: The State of the Art.* Kluwer Academic Publishers, 1993.

12. J.L. Goffin and J.P. Vial. Cutting planes and column generation techniques with the projective algorithm. *Journal of Optimization Theory and Applications*, 65:409–429, 1990.

13. M. Grötschel and O. Holland. Solving matching problems with linear programming. *Mathematical Programming*, 33:243–259, 1985.

14. M. Grötschel, M. Jünger and G. Reinelt. A cutting plane algorithm for the linear ordering problem. *Operations Research*, 32:1195–1220, 1984.

15. M. Grötschel, L. Lovasz and A. Schrijver. *Geometric Algorithms and Combinatorial Optimization.* Springer-Verlag, Berlin, Germany, 1988.

16. O. Güler and Y. Ye. Convergence behaviour of interior point algorithms. *Mathematical Programming*, 60:215–228, 1993.

17. D. den Hertog. *Interior Point Approach to Linear, Quadratic and Convex*

*Programming, Algorithms and Complexity.* PhD thesis, Faculty of Mathematics and Informatics, TU Delft, NL–2628 BL Delft, The Netherlands, September 1992.

18. D. den Hertog, C. Roos and T. Terlaky. A build–up variant of the path-following method for LP. *Operations Research Letters*, 12:181–186, 1992.
19. IBM. *IBM Optimization Subroutine Library Guide and Reference*, August 1990. Publication number SC23–0519–1.
20. J.A. Kaliski and Y. Ye. A decomposition variant of the potential reduction algorithm for linear programming. *Management Science*, 39:757–776, 1993.
21. A.P. Kamath, N.K. Karmarkar, K.G. Ramakrishnan and M.G.C. Resende. A continuous approach to inductive inference. *Mathematical Programming*, 57:215–238, 1992.
22. I.J. Lustig, R.E. Marsten and D.F. Shanno. On implementing Mehrotra's predictor–corrector interior point method for linear programming. *SIAM Journal on Optimization*, 2:435–449, 1992.
23. I.J. Lustig, R.E. Marsten and D.F. Shanno. Interior point methods for linear programming: Computational state of the art. *ORSA Journal on Computing*, 6:1–14, 1994. See also the following commentaries and rejoinder.
24. N. Megiddo. On finding primal- and dual-optimal bases. *ORSA Journal on Computing*, 3:63–65, 1991.
25. S. Mehrotra and J. Sun. On the implementation of a (primal-dual) interior point method. *SIAM Journal on Optimization*, 2:575–601, 1992.
26. J.E. Mitchell. An interior point column generation method for linear programming using shifted barriers. *SIAM Journal on Optimization*, 4:423–440, 1994.
27. J.E. Mitchell and B. Borchers. Solving real-world linear ordering problems using a primal-dual interior point cutting plane method. *Annals of Operations Research*, 62:253-276, 1996.
28. J.E. Mitchell and S. Ramaswamy. An extension of Atkinson and Vaidya's algorithm that uses the central trajectory. Technical Report 37–93–387, DSES, Rensselaer Polytechnic Institute, Troy, NY 12180–3590, August 1993.
29. J.E. Mitchell and M.J. Todd. Solving combinatorial optimization problems using Karmarkar's algorithm. *Mathematical Programming*, 56:245–284, 1992.
30. S. Mizuno, M. Kojima and M.J. Todd. Infeasible-interior-point primal-dual potential-reduction algorithms for linear programming. *SIAM Journal on Optimization*, 5:52–67, 1995.
31. Y.E. Nesterov. Complexity estimates of some cutting plane methods based on the analytic barrier. *Mathematical Programming*, 69:149–176, 1995.
32. M.W. Padberg and G. Rinaldi. A branch-and-cut algorithm for the resolution of large-scale symmetric traveling salesman problems. *SIAM Review*,

33:60–100, 1991.

33. R.G. Parker and R.L. Rardin. *Discrete Optimization.* Academic Press, San Diego, CA 92101, 1988.
34. C. Roos and J.P. Vial. A polynomial method of approximate centers for linear programming. *Mathematical Programming*, 54:295–305, 1992.
35. M.A. Saunders. Major Cholesky would feel proud. *ORSA Journal on Computing*, 6:23–27, 1994.
36. M.J. Todd and B.P. Burrell. An extension of Karmarkar's algorithm for linear programming using dual variables. *Algorithmica*, 1:409–424, 1986.
37. P.M. Vaidya. A new algorithm for minimizing convex functions over convex sets. In *Proceedings of the 30th Annual IEEE Symposium on Foundations of Computer Science*, 338–343, Los Alamitos, CA, 1989. IEEE Computer Press.
38. R.J. Vanderbei. Interior-point methods: algorithms and formulations. *ORSA Journal on Computing*, 6:32–34, 1994.
39. Y.Zhang. On the convergence of a class of infeasible interior-point methods for the horizontal linear complementarity problem. *SIAM Journal on Optimization*, 4:208–227, 1994.

# 7

# Computational logic and integer programming

**H. Paul Williams and Sally C. Brailsford**

*Faculty of Mathematical Studies,*
*University of Southampton, Southampton SO17 1BJ, England*

## 1 Introduction

The enormous increase in the availability and speed of computers in the past few decades has led to a parallel increase in the number of applications. As with other advances in technology, as many problems are created as are solved. Although computers create a potential for solving many hitherto impossible problems the implementation is often much more difficult than might appear from cursory examination. However fast computer hardware becomes there will always be many problems whose computation takes a prohibitive amount of time. Many of these problems are of a "logical" nature. They arise in problems such as efficient computer circuit design, database retrieval systems, expert systems and artificial intelligence.

In order to explain the material in this chapter it is necessary to introduce some logical notation from the *propositional calculus* (also sometimes known as "sentential logic" or "boolean algebra").

We will use capital letters (sometimes suffixed) to represent *propositions* e.g. $A$, $B$, $X_1$, $X_2$, etc. These propositions may take two possible values, *true* (T) or *false* (F).

Such (atomic) propositions may be altered or combined by means of *connectives*.

| | | |
|---|---|---|
| e.g. | $\overline{A}$ | means "not $A$" |
| | $A \vee B$ | means "$A$ or $B$ (or both)" |
| | $A \cdot B$ | means "$A$ and $B$" |

The truth of the resulting *compound propositions* depend on the truth or falsity of the component *atomic propositions* in a way prescribed by *truth tables*. Those connectives which we will use are defined by the truth table below.

| $A$ | $B$ | $\overline{A}$ | $A \vee B$ | $A \cdot B$ | $A \leftrightarrow B$ | $A \rightarrow B$ |
|---|---|---|---|---|---|---|
| F | F | T | F | F | T | T |
| F | T | T | T | F | F | T |
| T | F | F | T | F | F | F |
| T | T | F | T | T | T | T |

" $\leftrightarrow$ " is referred to as "equivalent to" and " $\rightarrow$ " is referred to as "implies".

Atomic propositions, e.g. $A$, $X_1$, etc. or their negation, e.g. $\overline{A}$, $\overline{X}_1$, etc. are referred to as *literals.*

Any compound proposition will generally have many different *representations,* e.g. $(A \vee \overline{B}) \cdot C$ can be shown to always have the same truth value as $(A \cdot B \cdot C) \vee (\overline{B} \cdot C)$. In order to give comparable representations it is convenient to use one of two common *standard representations.* These are known as *disjunctive form* (DF) and *conjunctive form* (CF). Disjunctive form consists of a series of *conjunctions* (i.e. combined by " $\cdot$ ") of literals which are combined in a *disjunction* (i.e. by " $\vee$ "). For example, the following is a statement in DF:

$$(X_1 \cdot \overline{X}_2 \cdot X_3) \vee (\overline{X}_1 \cdot X_2 \cdot X_4) \vee (\overline{X}_3 \cdot X_5).$$

For convenience the brackets are often left out and the " $\cdot$ " connective assumed to be more binding than the " $\vee$ " connective. The component conjunctions, e.g. $X_1 \cdot \overline{X}_2 \cdot X_3$, are referred to as *conjunctive clauses.*

In contrast in CF a statement is written as a conjunction of disjunctive clauses, for example

$$(X_1 \vee X_2 \vee X_3) \cdot (\overline{X}_2 \vee X_3) \cdot (X_1 \vee \overline{X}_4 \vee \overline{X}_5)$$

is in CF where, e.g. $(X_1 \vee X_2 \vee X_3)$ is a disjunctive clause.

There are a number of standard equivalences between compound statements which can be used to convert statements into equivalent forms. The validity of these equivalences can readily be demonstrated by means of the truth table definitions above. Those equivalences which we will use are given below. We will use the symbol " $\equiv$ " to denote this (meta)equivalence, i.e. equivalence when referring *to* the system as opposed to sets of symbols *within* the system.

| | | | | |
|---|---|---|---|---|
| Reflexivity | | $\overline{\overline{A}}$ | $\equiv$ | $A$ |
| Symmetry | (i) | $A \vee B$ | $\equiv$ | $B \vee A$ |
| | (ii) | $A \cdot B$ | $\equiv$ | $B \cdot A$ |
| Distribution | (i) | $A \cdot (B \vee C)$ | $\equiv$ | $A \cdot B \vee A \cdot C$ |
| | (ii) | $A \vee B \cdot C$ | $\equiv$ | $(A \vee B) \cdot (A \vee C)$ |
| De Morgan's laws | (i) | $\overline{A \vee B}$ | $\equiv$ | $\overline{A} \cdot \overline{B}$ |
| | (ii) | $\overline{A \cdot B}$ | $\equiv$ | $\overline{A} \vee \overline{B}$ |
| Implies | | $A \rightarrow B$ | $\equiv$ | $\overline{A} \vee B$ |
| Equivalence | | $A \leftrightarrow B$ | $\equiv$ | $(A \rightarrow B) \cdot (B \rightarrow A)$ |

Although we have confined ourselves to the propositional calculus it should be mentioned that some applications of computational logic require use of the *predicate calculus.* This allows the use of *predicates* whose truth values depend on the interpretation of *arguments*, e.g. $F(a)$ might represent the predicate "Father of" whose truth depended on the substitution of a value for the argument "$a$". Once such interpretations have been made (a process known as *instantiation*) the statements are reduced to those of the propositional calculus. Although we make partial use of the predicate calculus in Section 2.4 its use is mainly beyond the scope of this chapter.

Further details and a rigorous coverage of the propositional and predicate calculus can be found in, e.g. Mendelson [18].

We will demonstrate some typical problems which arise in computational logic by means of small examples. To start with we will use English sentences but later revert to the more compact abstract notation of using capital letters for sentences. These problems are given here for illustrative purposes. Methods of solution are given in Sections 3, 4 and 5.

**Example 1.1** Valid deduction (taken from Mendelson [18])

Is the following deduction valid?

> *"If Jones did not meet Smith last night, then either Smith was the murderer or Jones is lying. If Smith was not the murderer, then Jones did not meet Smith last night and the murder took place after midnight. If the murder took place after midnight, then either Smith was the murderer or Jones is lying. Hence Smith was the murderer."*

It is convenient to rewrite this problem in symbolic form representing component sentences by capital letters as follows:

$X_1 \equiv$ "Jones did not meet Smith last night",
$X_2 \equiv$ "Smith was the murderer",
$X_3 \equiv$ "Jones is lying",
$X_4 \equiv$ "The murder took place after midnight".

The premises of the argument are

$$X_1 \rightarrow (X_2 \vee X_3), \tag{1.1}$$

$$\overline{X}_2 \rightarrow (X_1 \cdot X_4), \tag{1.2}$$

$$X_4 \rightarrow (X_2 \vee X_3). \tag{1.3}$$

The suggested conclusion is

$$X_2. \tag{1.4}$$

Whilst for a small example such as this the trying of all possible truth values for $X_1$, $X_2$, $X_3$ and $X_4$ by means of truth tables would test if the conclusion is valid such complete enumeration could be prohibitively demanding in space and time for larger problems. Methods of computational logic or integer programming (IP) could be used to better effect.

More complicated examples than these (often involving the predicate calculus) arise in database retrieval and expert system deduction.

**Example 1.2** Testing consistency (satisfiability)

Is the following set of statements consistent, i.e. is there a set of truth values for the atomic propositions that makes them all true?

$$\left.\begin{array}{c} X_{11} \vee X_{12} \\ X_{21} \vee X_{22} \\ X_{31} \vee X_{32} \\ \overline{X_{11} \cdot X_{21}} \\ \overline{X_{11} \cdot X_{31}} \\ \overline{X_{21} \cdot X_{31}} \\ \overline{X_{12} \cdot X_{22}} \\ \overline{X_{12} \cdot X_{32}} \\ \overline{X_{22} \cdot X_{32}} \end{array}\right\} \tag{1.5}$$

The satisfiability problem is of central importance in computational logic as many problems can be naturally reduced to this problem. For example, the problem of ascertaining if a deduction is valid can be made a satisfiability problem by testing if the negated conclusion is consistent with the premises, e.g. in Example 1.1, is $\overline{X}_2$ inconsistent with the three premises?

A variant of the satisfiability problem is the maximum satisfiability problem in which it is desired to find the maximum number of clauses which can be satisfied simultaneously.

**Example 1.3** Minimal representation

Write the following statement in CF using the minimum number of disjunctive clauses.

$$[(X_1 \vee X_2) \rightarrow (X_3 \vee X_4)] \cdot (X_3 \leftrightarrow X_1 \cdot X_4) \cdot (X_2 \rightarrow (\overline{X}_1 \cdot X_3) \cdot (X_4 \vee X_2 \cdot X_3). \tag{1.6}$$

This problem and variants (e.g. minimal representations in other standard forms) arise in computer circuit design. For example we might wish to minimize the number of components ("gates") in a logical device (see for example Williams [24]), or the number of inputs. Sometimes it is desired to minimize the number of literals as opposed to clauses.

Another problem, connected with this problem, is to make *all* valid deductions from a given statement. In order to make this problem precise the deductions must all be of a specified "minimum" form such as disjunctive clauses. This problem is in contrast to Example 1.1 where one specified deduction is stipulated, or Example 1.2 where the deduction of a contradiction indicates inconsistency.

# 2 Representation as integer programming models

## 2.1 Representations of logic connectives by integer programming variables and constraints

The truth values of atomic propositions in logic can be represented by the values of 0-1 integer variables. It is convenient to use lower case letters to represent such variables corresponding to the capital letters representing the propositions, e.g. $X_1$ is true when $x_1 = 1$ and false when $x_1 = 0$.

The standard connectives of the propositional calculus then correspond to equations or inequalities involving these variables, e.g.

$$\begin{array}{rll} \overline{X_1} & \text{corresponds to} & x_1 = 0, \\ X_1 \cdot X_2 & \text{corresponds to} & (x_1 = 1) \cdot (x_2 = 1) \text{ (or } x_1 + x_2 = 2), \\ X_1 \vee X_2 & \text{corresponds to} & x_1 + x_2 \geq 1, \\ X_1 \leftrightarrow X_2 & \text{corresponds to} & x_1 - x_2 = 0, \\ X_1 \rightarrow X_2 & \text{corresponds to} & x_1 - x_2 \leq 0. \end{array}$$

More complicated logical statements can also be so represented sometimes using more than one (in)equality and extra 0-1 variables. A systematic method of modelling logical statements in this way is given in Section 2.4. Often there is more than one representation. Which representation is best depends upon which algorithms are to be used. This aspect is again discussed in Section 5.

It is possible to model all three of the types of logic problem exemplified in Section 1 as IPs using these representations. This is done in Section 2.2. IP solution methods can then be applied to solving them.

Some of the classical applications of IP, arising generally in operational research are described in Williams [27]. Modelling of these applications and their extensions is also amenable to the use of logic as will be described in this section. In contrast some of the standard algorithmic methods of computational logic, which are described in Section 3, are also applicable to classical IP problems.

## 2.2 The formulation of IP models

An IP model involves maximizing or minimizing a linear expression subject to a set of linear constraints. *All* (the "pure" case) or *some* (the "mixed" case) of the variables are restricted to take integer values. If the variables are considered as representing the coordinates of points in multidimensional space then the constraints correspond to a region of this space. In some directions the coordinates are restricted to integer values. There is often more than one way of representing the constraints as (in)equalities with integrality conditions on some of the variables. The example below illustrates this for a pure IP problem.

**Example 2.1** Alternative representations of an IP

$$\begin{array}{lll} \text{maximize} & x_1 + x_2, & (2.1) \\ \text{subject to} & 2x_1 + 3x_2 \leq 5, & (2.2) \\ & x_1, x_2 \geq 0, & (2.3) \end{array}$$

$$x_1, x_2 \text{ integer}. \tag{2.4}$$

Figure 1 gives a geometrical representation of the feasible region. Clearly the feasible solutions are represented by the boldly marked points which are those integer points lying within the region $OAB$ specified by the inequalities (2.2), (2.3) and (2.4).

An alternate way of specifying constraint (2.2) is by the two constraints (2.5) and (2.6) below:

$$x_1 + x_2 \leq 2, \tag{2.5}$$

$$x_2 \leq 1. \tag{2.6}$$

This results in Fig. 2.

Many IP algorithms rely on first solving the *LP relaxation* (i.e. ignoring the integrality conditions on the variables). Different IP representations of the same problem may result in different LP relaxations. In the example above the feasible region for the LP relaxation of the first representation is $OAB$ in Fig. 1, whilst that for the second representation is $OPQR$ in Fig. 2. The second representation would often be regarded as more satisfactory since the solution to an LP (if using the simplex algorithm) yields a vertex solution. For the first representation this could be one of the *fractional* solutions at $A$ or $B$. However for the second representation it would always be an *integer* solution at $O$, $P$, $Q$ or $R$ obviating the need to pursue the algorithm further.

The boundaries of the feasible region of the LP relaxation in Fig. 2 are the "tightest" possible. Technically we have enclosed the feasible integer points by the smallest possible convex set. This is known as the *convex hull*. The boundaries $OP$, $PQ$, $QR$ and $OR$ are known as *facets* of this convex hull and the

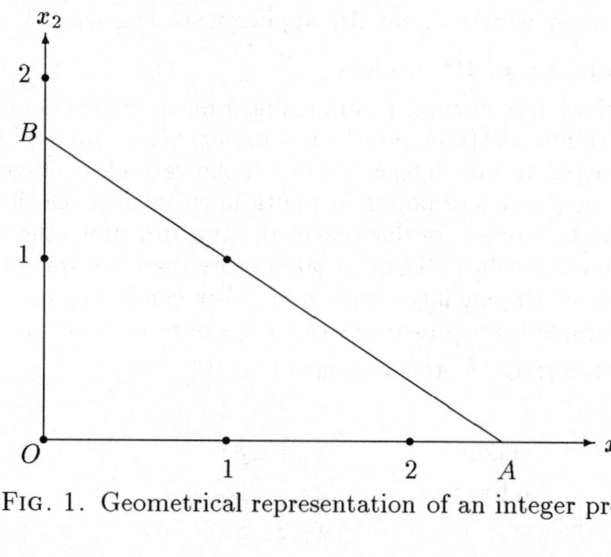


FIG. 1. Geometrical representation of an integer program

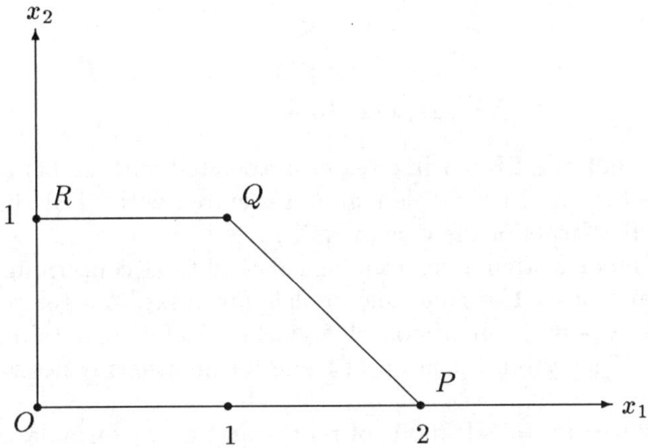


FIG. 2. Alternative geometrical representation of an integer program

corresponding inequalities (2.3), (2.5) and (2.6) known as *facet defining* inequalities. Hence there is theoretical (and often practical) advantage to be gained in seeking an IP representation in which the constraint (in)equalities define facets. Sometimes the number of facets proves astronomically large and impractical to either calculate or represent. This is particularly true of difficult combinatorial optimization problems such as the travelling salesman problem. However it is often possible to define *some* facets by (in)equalities or make a better approximation to the convex hull by constraints closer to the boundaries of the convex hull.

The use of logic can help in obtaining different representations although "minimal" representations of a problem in logic do not necessarily correspond to "minimal" representations in the geometrical sense outlined above. We illustrate the use of logic in a simple reformulation below.

**Example 2.2** Alternate logic representation leading to alternate IP representations

$$(X_1 \vee X_2) \rightarrow X_3. \tag{2.7}$$

If the statements $X_i$ are represented by "$x_i = 1$" then (2.7) leads naturally to the IP constraint:

$$x_1 + x_2 - 2x_3 \leq 0, \tag{2.8}$$

$$x_1, x_2, x_3 \in \{0, 1\}. \tag{2.9}$$

A systematic way of going from (2.7) to (2.8) is given in Section 2.4.

Using the logic manipulation rules, given in Section 1, (2.7) can also be written as

$$(X_1 \rightarrow X_3) \cdot (X_2 \rightarrow X_3). \tag{2.10}$$

This leads to the IP representation:

$$x_1 \quad - x_3 \leq 0, \tag{2.11}$$

$$x_2 - x_3 \leq 0, \tag{2.12}$$

$$x_1, x_2, x_3 \in \{0, 1\}. \tag{2.13}$$

It is easy to see that the LP feasible region associated with (2.11) and (2.12) is strictly contained in the LP feasible region associated with (2.8). In fact (2.11) and (2.12) describe facets of the convex hull.

Practical IP models often arise from logical conditions concerning constraints involving LP variables. The resulting models are mixed-integer programming (MIP) models. Again the methods of logic are useful in obtaining different representations. The systematization of these formulations is described in Section 2.3.

We further illustrate the IP mode of representation by formulating the computational logic examples of Section 1 as IP models.

**Example 2.3** Testing the validity of a deduction by an IP model

The premises (1.1), (1.2) and (1.3) of Example 1.1 give rise (in one possible representation) respectively to the constraints

$$-x_1 + x_2 + x_3 \geq 0, \tag{2.14}$$

$$x_1 + 2x_2 + x_4 \geq 2, \tag{2.15}$$

$$x_2 + x_3 - x_4 \geq 0, \tag{2.16}$$

$$x_1, x_2, x_3, x_4 \in \{0, 1\}. \tag{2.17}$$

The conclusion (1.4) corresponds to

$$x_2 \geq 1. \tag{2.18}$$

Hence the problem becomes one of ascertaining if (2.18) follows from (2.14)–(2.17). This is tested by solving the IP model

$$\begin{aligned} &\text{minimize} && x_2, && (2.19)\\ &\text{subject to} && (2.14) - (2.17). \end{aligned}$$

The LP relaxation of the above model is obtained by relaxing (2.17) to $0 \leq x_1, x_2, x_3, x_4 \leq 1$. Solving this relaxation yields the optimal solution

$$x_1 = 1, \ x_2 = 0, \ x_3 = x_4 = 1.$$

Since this solution is integer we have also obtained the integer optimum demonstrating that the conclusion (2.18) does not follow from the premises, i.e. Smith was not necessarily the murderer since we could have had the situation in which Jones did not meet Smith last night but Jones was lying and the murder still took place after midnight!

In fact we have managed to solve this model easily since we had a fairly good representation. It can be shown that (2.14) to (2.16) together with $0 \leq x_1, x_2, x_3, x_4 \leq 1$ do not give a convex hull representation. There are some vertices corresponding to fractional solutions. However with the objective function (2.19) we obtained an integer solution so demonstrating non-validity.

Even if a fractional solution is obtained as a result of solving the LP relaxation of a validity problem this will be sufficient to prove validity if the objective constraint is satisfied. The imposition of extra integrality conditions could not possibly then cause this constraint to be violated. If, however, the objective constraint is *not* satisfied by a fractional solution to the LP relaxation then this does not demonstrate non-validity since the integrality conditions might be sufficient to do this.

There is an important class of logic problems involving statements known as Horn clauses where the straightforward IP formulation has an LP relaxation whose fractional solution can be guaranteed to prove validity or non-validity by simply rounding the optimal objective value. Such problems are discussed in Section 4.

The reason for giving so much attention to using LP relaxations of IP representations is that LP models, unlike IP models, are generally easy to solve. Technically, for LP there is an algorithm for which the number of steps is never more than a polynomial function of the size of the data. For IP there is no known algorithm which is polynomial in this sense.

**Example 2.4** Testing consistency (satisfiability) by an IP model

The condition (1.5) can be modelled by the constraints

$$\left.\begin{aligned} x_{11} + x_{12} &\geq 1, \\ x_{21} + x_{22} &\geq 1, \\ x_{31} + x_{32} &\geq 1, \end{aligned}\right\} \tag{2.20}$$

$$\left.\begin{aligned} -x_{11} - x_{21} &\geq -1, \\ -x_{11} - x_{31} &\geq -1, \\ -x_{21} - x_{31} &\geq -1, \\ -x_{12} - x_{22} &\geq -1, \\ -x_{12} - x_{32} &\geq -1, \\ -x_{22} - x_{32} &\geq -1, \end{aligned}\right\} \tag{2.21}$$

$$x_{ij} \in \{0, 1\}, \quad \forall i, j. \tag{2.22}$$

We have deliberately written constraints (2.21) in the somewhat unnatural way as " $\geq$ " constraints (as opposed to " $\leq$ " constraints) for the purposes of the algorithms which will be discussed in Section 5.

In order to test consistency we can use standard IP methods. One such approach is to use Phase I of the simplex algorithm to test the LP relaxation (replacing (2.22) by $0 \le x_{ij} \le 1,\ \forall i,j$) for feasibility.

Solving this relaxation demonstrates the problem to have only one fractional solution where all variables take the value 1/2. Branch and bound (with any objective) would demonstrate integer infeasibility. In fact we could consider minimizing any of the expressions (2.20)–(2.21) subject to the other constraints in order to treat the problem as a verifiability problem, similarly to Example 2.2. Inconsistency would therefore be demonstrated by non-attainment of the corresponding right-hand side. Correspondingly, Example 2.2 could have been treated as a consistency problem by trying to find a feasible solution subject to constraints (2.14) to (2.18).

It is worth pointing out that the above problem has a clearly understood interpretation. If we wished to place three objects in two boxes then the statement $X_{ij}$ (i.e. $x_{ij} = 1$) could represent the decision to place object $i$ in box $j$. The constraints would then (i) ensure each object is placed in a box and (ii) ensure no more than one object goes in each box. Clearly this "pigeon-hole" problem has no solution when looked at semantically. Syntactically however it is not so straightforward to prove inconsistency. In fact the general pigeon-hole problem is a computationally difficult satisfiability problem (no polynomially bounded algorithm is known), although when each clause has at most two literals (as is the case here) we have a 2-satisfiability problem, which can always be solved in polynomial time.

### 2.3 Mixed-integer programming representability

In Section 2.2 it was pointed out that any compound logical statement in the propositional calculus can be expressed as IP constraints involving 0-1 variables. A systematic way of doing this is given in Section 2.4. It was also pointed out that there are generally a number of different IP representations.

We here give conditions under which a problem can be translated into a logical form in the propositional calculus where the atomic statements are LP constraints. This can then be translated into an MIP model. Such a problem is said to be "MIP representable". This concept is due to Meyer [19] and Jeroslow [14]. When a problem is MIP representable there is a theoretically realizable logical form which translates into a "sharp" IP representation, i.e. one in which the LP relaxation gives an integer solution for any objective.

So long as we have a *conjunction* of LP constraints (and a suitable objective) we have an LP model. IP becomes necessary when connectives other than " · " are also involved. Using the rules described in Section 1 it is always possible to convert the logical statements involving LP constraints into DF. We then have a disjunction of clauses each of which corresponds to a set of inequalities. The resultant expression may or may not be translatable into an MIP model.

**Example 2.5** Converting logical conditions on LP constraints into disjunctive form

$$(x, y \ge 0) \cdot (2x + 3y \le 4 \rightarrow x - y \ge 0). \tag{2.23}$$

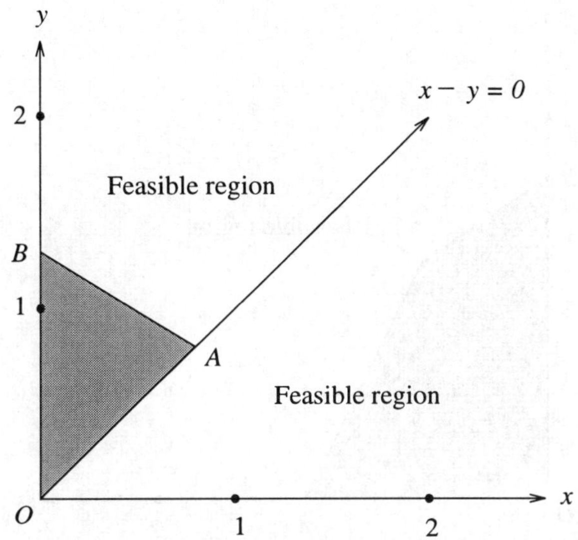


FIG. 3. A non-mixed-integer programming representable feasible region

We may assume $x, y \in \mathbb{R}$.

Using the rules given in Section 1 this translates into

$$x \geq 0 \cdot y \geq 0 \cdot (2x + 3y > 4 \vee x - y \geq 0),$$

giving

$$(x \geq 0 \cdot y \geq 0 \cdot 2x + 3y > 4) \vee (x \geq 0 \cdot y \geq 0 \cdot x - y \geq 0). \tag{2.24}$$

There are two reasons why (2.24) is not MIP representable. The first is easy to see but the second requires a further result. Firstly by negating the inequality $2x + 3y \leq 4$, in our first manipulation, we have produced the strict inequality $2x + 3y > 4$ which falls outside the constraints allowed in IP.

Such a result is probably not practically serious since (although this strictly alters the problem) it would probably be acceptable to replace the strict inequality by $2x + 3y \geq 4 + \varepsilon$ where $\varepsilon$ is a suitably small positive number.

The resultant feasible region associated with (2.24) is then shown in Fig. 3. It is the (non-convex) region of the positive orthant obtained by excluding $OAB$. Alternatively it can be regarded as the union of the two open (convex) regions bounded by (i) the $y$-axis, $AB$ and the line $x - y = 0$ and (ii) the $x$-axis and the line $x - y = 0$.

There is a condition for a disjunction of constraints to be MIP representable.

**Condition for MIP representability:** *The union of a finite set of convex polytopes is MIP representable if and only if they have the same recession directions.*

Recession directions are directions along which all points remain in the cor-

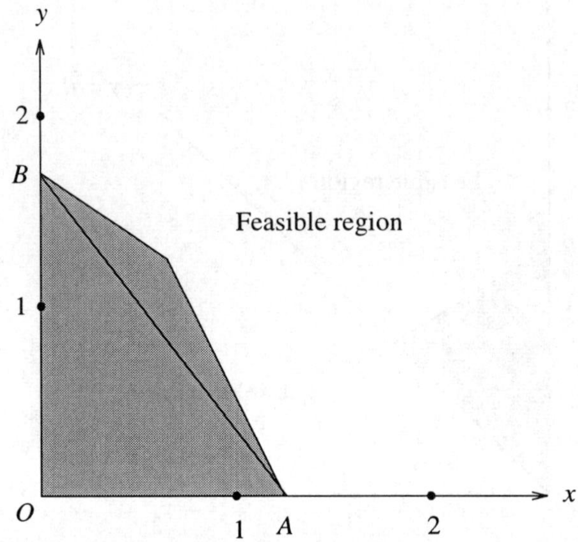


FIG. 4. A mixed-integer programming representable feasible region

responding polytope. For example, in Fig. 3 the $y$-axis is a recession direction of the first region corresponding to the first clause in the disjunction. The line $y = 2x$ is another recession direction. For the second region the $x$-axis and the line $y = x/3$ are examples of recession directions. Clearly these two regions do not have the same recession directions. In fact they only share one recession direction given by the line $x - y = 0$.

The above condition is proved in Jeroslow [14]. Clearly Example 2.5 breaks the condition and the logical condition (2.23) is not MIP representable.

Should all the regions corresponding to the clauses in a disjunction be closed (as is often the case) then no region has a recession direction. Vacuously they therefore have the same recession directions making the corresponding disjunction MIP representable.

**Example 2.6** An MIP representable disjunction

$$[(2x + 3y \geq 5) \vee (4x + 2y \geq 5)] \cdot x \geq 0 \cdot y \geq 0. \tag{2.25}$$

We assume $x, y \in \mathbb{R}$.

It is easy to see from Fig. 4 that this disjunction can be represented as the union of two polyhedra with the same recession directions (those of the positive quadrant).

By introducing a 0-1 variable $\delta$ whose possible values represent the two clauses of the disjunction we can model (2.25) as

$$2x + 3y + 5\delta \geq 5, \tag{2.26}$$

$$4x + 2y - 5\delta \geq 0, \tag{2.27}$$

$$x, y \geq 0, \ \ \delta \in \{0, 1\}. \tag{2.28}$$

For a disjunction with more than two clauses 0-1 variables could be introduced to correspond to each clause together with a constraint that they must sum to a value greater than, or equal to, one.

Whilst the above method of representation is always possible (so long as the polyhedra have the same recession directions) it does not always lead to the "best" representation. If, for example, we were to minimize the objective $x + y$ subject to the constraints (2.26)–(2.28) but to relax the condition $\delta \in \{0,1\}$ to $0 \leq \delta \leq 1$, then this LP relaxation would produce the solution $x = 5/6$, $y = 0$, $\delta = 2/3$ with an objective value of $5/6$.

If we were to obtain the facets of the convex hull associated with (2.25) we would (in this very simple case) produce

$$4x + 3y \geq 5, \tag{2.29}$$

$$x, y \geq 0. \tag{2.30}$$

This can easily be seen from Fig. 4 as (2.29) represents the facet $AB$.

Minimizing the objective $x + y$ subject to (2.29) and (2.30) produces the optimal solution $x = 5/4$, $y = 0$ with an objective value of $5/4$.

Balas [1] gives a characterization of all such "disjunctive cuts".

Such a convex hull representation is always theoretically possible for any MIP representable problem although in practice there may be an astronomical number of facets and no computationally practical way of generating them.

An alternative way of modelling disjunctions (and hence all MIP representable IPs) is possible by representing the problem in a higher-dimensional space. In order to do this it is first necessary to model the constraints in DF. This resultant IP representation is (slightly misleadingly) known as a *disjunctive formulation.* (In fact all IP formulations are in a sense disjunctive but the term has been reserved for this representation.) We illustrate this by the same example.

**Example 2.7** A disjunctive MIP representation

$$[(2x + 3y \geq 5) \vee (4x + 2y \geq 5)] \cdot x \geq 0 \cdot y \geq 0. \tag{2.31}$$

Putting (2.31) in DF gives

$$(2x + 3y \geq 5 \cdot x \geq 0 \cdot y \geq 0) \vee (4x + 2y \geq 5 \cdot x \geq 0 \cdot y \geq 0). \tag{2.32}$$

A disjunctive formulation then demands that each variable be "split" into components corresponding to each clause of the disjunction. In this example, since there are only two clauses, we split $x$ and $y$ into two components $x_1$, $x_2$ and $y_1$, $y_2$ respectively. Zero-one variables $\delta_1$ and $\delta_2$ correspond to each of the clauses, together with the constraints forcing each clause of the disjunction to hold with respect to the corresponding components when the corresponding value of $\delta_i$ is

1. The resulting formulation is

$$2x_1 + 3y_1 - 5\delta_1 \geq 0, \tag{2.33}$$
$$4x_2 + 2y_2 - 5\delta_2 \geq 0, \tag{2.34}$$
$$x - x_1 - x_2 = 0, \tag{2.35}$$
$$y - y_1 - y_2 = 0, \tag{2.36}$$
$$\delta_1 + \delta_2 = 1, \tag{2.37}$$
$$x_1, x_2, y_1, y_2 \geq 0,\ \delta_1, \delta_2 \in \{0, 1\}. \tag{2.38}$$

In order to see why this formulation "works", suppose for example that $\delta_1 = 1$; then $\delta_2 = 0$. (2.33) and (2.34) then give $2x_1 + 3y_1 \geq 5$, $4x_2 + 2y_2 \geq 0$. Since the two clauses in (2.32) have the same recession directions we have that $4x_2+2y_2 \geq 0$ together with $x_2, y_2 \geq 0$ implies $2x_2 + 3y_2 \geq 0$. Adding this inequality to $2x_1 + 3y_1 \geq 5$ gives $2x + 3y \geq 5$, i.e. the first clause of (2.32) is satisfied.

A remarkable and significant fact is that the formulation (2.33) to (2.38) represents the convex hull of (2.32) (in the space of $x_1, x_2, y_1, y_2, \delta_1, \delta_2$). Therefore if solved, with any objective, its LP relaxation will produce a solution to (2.32). For example minimizing $x + y$ can be verified to produce the LP solution $x_1 = y_1 = 0$, $x_2 = 5/4$, $y_2 = 0$, $x = 5/4$, $y = 0$, $\delta_1 = 0$, $\delta_2 = 1$ with an objective value of 5/4. This clearly satisfies (2.31).

The above method of formulating disjunctions (and hence IPs) is due to Jeroslow [14], who proves that the method results in a representation of the convex hull. Another proof based on LP duality is given in Williams [31]. The method demands that a problem be expressed in DF. Such a representation may be prohibitively large. However intermediate representations are usually improvements on "naive" formulations.

## 2.4 Systematic integer programming formulation

Although the representation of a logical problem in DF is always possible and sometimes theoretically desirable this representation in terms of the connectives " · ", " ∨ " and " ¯ " can be unduly restrictive. We often wish to model conditions such as "at least $r$ of these conditions must hold" or "at most $r$ of these conditions must hold". This is particularly true of the generalized covering type of constraints which arise in many IP formulations: see e.g. Granot and Hammer [9], Balas [2], Wolsey [34]. McKinnon and Williams [17] suggest representing the "at least $r$ " form by a "ge" predicate with the following meaning:

**ge**$(r : X_1, X_2, ..., X_n)$ means "greater than, or equal to, $r$ of $X_1, X_2, \ldots, X_n$ are true".

Clearly

$$\mathbf{ge}(1 : X_1, X_2, \ldots, X_n) \text{ means } X_1 \vee X_2 \vee \ldots \vee X_n$$

i.e. the " ∨ " connective on $n$ arguments, and

$$\mathbf{ge}(n : X_1, X_2, \ldots, X_n) \text{ means } X_1 \cdot X_2 \cdot \ldots \cdot X_n$$

i.e. the " · " connective on $n$ arguments.

Intermediate connectives between " $\vee$ " and " $\cdot$ " are more compactly represented. For example

$$\mathbf{ge}(2 : X_1, X_2, X_3)$$

is a more compact representation than that using the conventional " $\vee$ " connective as

$$X_1 \cdot X_2 \vee X_1 \cdot X_3 \vee X_2 \cdot X_3.$$

The "at most $r$" form is simply represented as

$$\mathbf{ge}(n - r : \overline{X}_1, \overline{X}_2, \ldots, \overline{X}_n).$$

In order to illustrate the use of the "**ge**" predicate we give an example.

**Example 2.8** A logical condition

$$[(f(x) \leq a) \vee (g(x) \geq b)] \longleftrightarrow [(s(x) \leq c) \vee (t(x) \geq d)] \tag{2.39}$$

Here $f$, $g$, $s$ and $t$ represent linear functions of the vector $x$.

We can successfully and systematically translate this into the following predicates from the "top down".

$$\begin{aligned}&\mathbf{ge}\,[2 : [(f(x) \leq a) \vee (g(x) \geq b)] \rightarrow [(s(x) \leq c) \vee (t(x) \geq d)]\,,\\ &\qquad [(s(x) \leq c) \vee (t(x) \geq d)] \rightarrow [(f(x) \leq a) \vee (g(x) \geq b)]]\,,\end{aligned}$$

$$\begin{aligned}&\mathbf{ge}\left[2 : \overline{[(f(x) \leq a) \vee (g(x) \geq b)]} \vee [(s(x) \leq c) \vee (t(x) \geq d)]\,,\right.\\ &\qquad \left.\overline{[(s(x) \leq c) \vee (t(x) \geq d)]} \vee [(f(x) \leq a) \vee (g(x) \geq b)]\right],\end{aligned}$$

$$\begin{aligned}&\mathbf{ge}\left[2 : \mathbf{ge}\left[1 : \overline{(f(x) \leq a) \vee (g(x) \geq b)}, s(x) \leq c, t(x) \geq d\right],\right.\\ &\qquad \left.\mathbf{ge}\left[1 : \overline{(s(x) \leq c) \vee (t(x) \geq d)}, f(x) \leq a, g(x) \geq b\right]\right],\end{aligned}$$

$$\begin{aligned}&\mathbf{ge}\,[2 : \mathbf{ge}\,[1 : \mathbf{ge}\,[2 : f(x) > a, g(x) > b]\,, s(x) \leq c, t(x) \geq d]\,,\\ &\qquad \mathbf{ge}\,[1 : \mathbf{ge}\,[2 : s(x) > c, t(x) < d]\,, f(x) \leq a, g(x) \geq b]]\,.\end{aligned} \tag{2.40}$$

The form of (2.40) demonstrates that (2.39) is not strictly MIP representable since at the "atomic level" we have strict inequalities of the form " $<$ " and " $>$ ". However, as in Example 2.5 it is possible to convert these inequalities to the " $\leq$ " and " $\geq$ " forms by introducing small corrections $\varepsilon$. We do not do this here, in order to preserve the "purity" of our derivation. Also, in order for (2.39) to be MIP representable the appropriate polytopes must have the same recession directions. For this illustrative example we will simply impose lower or upper bounds on the expressions $f$, $g$, $s$ and $t$, where needed, in order to close the polytopes. Neither will we give a disjunctive formulation for the "ge(1 : ...) ", i.e. " $\vee$ ", conditions above. In practice there would be virtue in so doing.

Zero-one variables can now be introduced from the "bottom up" so as to convert (2.40) into an IP form. Hence $\mathbf{ge}\,[2 : f(x) > a, g(x) < b]$ can be represented by

$$\left.\begin{array}{rcl} f(x) + (L_f - a)\delta_1 & > & L_f, \\ g(x) + (U_g - b)\delta_1 & < & U_g, \end{array}\right\} \tag{2.41}$$

$$\delta_1 = 1, \tag{2.42}$$

where $L_f$ and $U_g$ are lower and upper bounds respectively for $f(x)$ and $g(x)$.

Similarly, $\mathbf{ge}\,[2 : s(x) > c, t(x) < d]$ can be represented by

$$\left.\begin{array}{rcl} s(x) + (L_s - c)\delta_2 & > & L_s, \\ t(x) + (U_t - d)\delta_2 & < & U_t, \end{array}\right\} \tag{2.43}$$

$$\delta_2 = 1. \tag{2.44}$$

(2.40) can be represented by (2.41) and (2.43) together with

$$\begin{array}{r} \mathbf{ge}\,[2 : \mathbf{ge}\,[1 : \delta_1 = 1, s(x) \leq c, t(x) \geq d]\,, \\ \mathbf{ge}\,[1 : \delta_2 = 1, f(x) \leq a, g(x) \geq b]]\,. \end{array}$$

$\mathbf{ge}\,[1 : \delta_1 = 1, s(x) \leq c, t(x) \geq d]$ can be represented by

$$\left.\begin{array}{rcl} s(x) + (U_s - c)\delta_3 & \leq & U_s, \\ t(x) + (L_t - d)\delta_4 & \geq & L_t, \\ \delta_1 + \delta_3 + \delta_4 & \geq & 1. \end{array}\right\} \tag{2.45}$$

Similarly, $\mathbf{ge}\,[1 : \delta_2 = 1, f(x) \leq a, g(x) \geq b]$ can be represented by

$$\left.\begin{array}{rcl} f(x) + (U_f - a)\delta_5 & \leq & U_f, \\ g(x) + (L_g - b)\delta_6 & \geq & L_g, \\ \delta_2 + \delta_5 + \delta_6 & \geq & 1. \end{array}\right\} \tag{2.46}$$

Since the "top level" "**ge**" in (2.40) is of the form "**ge**$(2 : \ldots)$ ", i.e. " $\cdot$ ", we can simply state (2.41), (2.43), (2.45) and (2.46) as our IP formulation.

It is, in fact, possible to simplify this formulation somewhat. Relying on a result given in Williams [30], since $\delta_3$ will normally have a positive coefficient in the "opposite" inequalities of (2.45), it can be eliminated using Fourier–Motzkin elimination (see Section 5) by combining these inequalities. This is only possible because $\delta_3$ (an integer variable) has a unit coefficient in one of them. Similarly, $\delta_5$ can be eliminated from the inequalities in (2.46).

## 3 Solution methods in computational logic

### 3.1 Resolution and absorption

Resolution is a well-known technique which was shown by Robinson [22] to be a complete refutation method for the first-order predicate calculus. A *complete* method enables us to deduce every theorem which follows from a given set of formulae. In a *refutation* method, we prove that a goal formula $G$ follows logically from a set of formulae $\mathcal{S}$ by firstly negating $G$, adding it to $\mathcal{S}$, and showing that the resulting set of formulae ($\mathcal{S}$ plus not-$G$) is inconsistent. If we restrict ourselves to the propositional calculus, resolution is known as *ground resolution*.

Resolution works in the following way. Suppose we are given a set $\mathcal{S}$ of formulae, that is statements in the propositional calculus involving literals, their negations, and the logical connectives **and** and **or**. We want to prove that a formula $G$ follows logically from $\mathcal{S}$. Having negated $G$ and added it to $\mathcal{S}$, we convert the new set to a set $\mathcal{C}$ of clauses in conjunctive form. We then use resolution to attempt to show that $\mathcal{C}$ is contradictory, as follows:

(1) If the literals $X$ and $\overline{X}$ both appear in the same clause, this clause is a tautology and can be removed from $\mathcal{C}$.

(2) Take two clauses in which the same literal appears, negated in one but not in the other. For example, the clauses

$$X \vee X_2 \vee \overline{Y}$$

$$Y \vee Z_1 \vee Z_2$$

(3) Then if $Y$ is true $\overline{Y}$ must be false and it follows from the first clause that $X_1 \vee X_2$ must be true. Similarly, if $Y$ is false it follows from the second clause that $Z_1 \vee Z_2$ must be true. Thus, either $X_1 \vee X_2$ or $Z_1 \vee Z_2$ is true, and we have a disjunction

$$X_1 \vee X_2 \vee Z_1 \vee Z_2.$$

We have effectively "cancelled out" the literal $Y$. We say that we are *resolving* on the variable $Y$.

(4) This leaves the *resolvent* $X_1 \vee X_2 \vee Z_1 \vee Z_2$. The two original clauses are the parent clauses of this resolvent. Add this resolvent to the set $\mathcal{C}$.

(5) Repeat the above steps until the empty clause is obtained. This signals the fact that we have obtained a contradiction, and proves that $G$ follows logically from the set $\mathcal{S}$.

A variant of resolution in which one of the parent clauses must contain only a single literal is called *unit resolution*. Clearly, if a clause consists solely of the literal $X$ then $X$ must be true to satisfy this clause, and therefore any other clause in $\mathcal{C}$ containing $X$ will also always be true. Therefore we can remove all these clauses from $\mathcal{C}$ since we are seeking a contradiction. Moreover, we can also eliminate the literal $\overline{X}$ from any clauses in which it appears, since it will always be false. Thus we have reduced the size of the set we are seeking to prove inconsistent. Unit resolution is sometimes referred to as *chaining*.

Step 2 above involves choosing two clauses from $\mathcal{C}$ to be parents. If $\mathcal{C}$ contains a large number of clauses there are an enormous number of ways in which this can be done, and to make matters worse we are adding the resolvent clauses to

$\mathcal{C}$ at each stage. The choice of strategy for selecting clauses is important. The *breadth-first* approach is basically an exhaustive and systematic search which first resolves all combinations of the original clauses in $\mathcal{C}$ (known as the *base set*), then all their first-generation descendants plus the base set, and so on. The *set-of-support* approach always uses one of the descendants of not-$G$ as a parent. The *unit-preference* method seeks to use a unit clause as a parent wherever possible. *Linear input* always chooses at least one parent from the base set. This is illustrated in Section 4. The *ancestry-filtered* method is linear input plus the case when one parent is an ancestor of the other parent.

A clause is said to *absorb* another clause if the first clause implies the second, and thus the second is *redundant* (unnecessary). For example, consider the clauses $X_1 \vee X_2$ and $X_1 \vee X_2 \vee X_3$. The first clause implies the second, since if either $X_1$ or $X_2$ is true then clearly at least one of $X_1$, $X_2$ and $X_3$ must be true. Therefore the clause $X_1 \vee X_2 \vee X_3$ is redundant and can be eliminated. It can easily be seen that clause $A$ absorbs clause $B$ if and only if all the literals of $A$ are contained in $B$.

If resolution is applied to a set of disjunctive clauses, to generate extra implied clauses, interspersed with absorption wherever possible in order to remove redundant clauses, we reduce the system to a set of "minimal" disjunctive clauses. These clauses are minimal in the sense that they are implied by the original set of clauses, and each of these clauses on its own implies no longer clause. Such clauses are referred to as *prime implications.* There is an analogy, although not a correspondence, with facets in an IP representation. In a geometrical sense facets are the *strongest* inequalities one can obtain in an IP representation. Prime implications are the *strongest* clauses one can assert in a logical representation. Hooker [13] systematically classifies such "prime cuts" for a number of different classes of constraints (including those arising from disjunctive clauses). The full set of prime implications is logically equivalent to the original system. We illustrate the application of resolution to obtaining prime implications by referring to Example 1.3.

**Example 3.1** Obtaining all prime implications by resolution and absorption

The statement (1.6) in Example 1.3 can be written, using the logical manipulation rules of Section 1, as the set of disjunctive clauses

$$\overline{X}_1 \vee X_3 \vee X_4 \tag{3.1}$$

$$\overline{X}_2 \vee X_3 \vee X_4 \tag{3.2}$$

$$X_1 \vee \overline{X}_3 \tag{3.3}$$

$$\overline{X}_3 \vee X_4 \tag{3.4}$$

$$\overline{X}_1 \vee X_3 \vee \overline{X}_4 \tag{3.5}$$

$$\overline{X}_1 \vee \overline{X}_2 \tag{3.6}$$

$$\overline{X}_2 \vee X_3 \tag{3.7}$$

$$X_2 \vee X_4 \tag{3.8}$$

$$X_3 \vee X_4 \tag{3.9}$$

(3.1) and (3.2) are absorbed by (3.9) and are therefore removed. The resolvent of (3.4) and (3.9) is $X_4$. This absorbs (3.4), (3.8) and (3.9) which are removed. The resolvent of $X_4$ with (3.5) is $\overline{X}_1 \vee X_3$ which absorbs (3.5). The resolvent of (3.3) with (3.7) is $X_1 \vee \overline{X}_2$ which resolves with (3.6) to give $\overline{X}_2$ and is itself absorbed by $\overline{X}_2$. This absorbs (3.6) and (3.7). No other resolutions or absorptions are possible, giving the prime implications as

$$\overline{X}_2,\ X_4,\ \overline{X}_1 \vee X_3,\ X_1 \vee \overline{X}_3. \tag{3.10}$$

Having obtained all the prime implications if we wish (as in Example 1.3) to obtain a minimal set of disjunctive clauses of minimal length we have, in general, an additional problem. We have to obtain a *minimal subset* of the prime implications which "cover" all the clauses of the original problem. This type of problem is a well-known one arising in a number of contexts in operational research. It is known as a *set covering problem.* It can be formulated as an IP model by introducing 0-1 variables to represent the inclusion, or otherwise, of each prime implication.

In the above example it is easy to see that all the prime implications of (3.10) are needed, demonstrating that a minimal representation of (1.6) in CF is

$$\overline{X}_2 \cdot X_4 \cdot (\overline{X}_1 \vee X_3) \cdot (X_1 \vee \overline{X}_3).$$

If a set of statements is inconsistent then there will only be one prime implication, the empty clause representing a contradiction. Therefore in order further to illustrate resolution we apply it to demonstrating the inconsistency of Example 1.2, as an alternative to the IP approach of Example 2.4.

**Example 3.2** Demonstrating inconsistency (non-satisfiability) by resolution

The premises (1.5) can be written as disjunctive clauses

$$\begin{gathered} X_{11} \vee X_{12} \\ X_{21} \vee X_{22} \\ X_{31} \vee X_{32} \\ \overline{X}_{11} \vee \overline{X}_{21} \\ \overline{X}_{11} \vee \overline{X}_{31} \\ \overline{X}_{21} \vee \overline{X}_{31} \\ \overline{X}_{12} \vee \overline{X}_{22} \\ \overline{X}_{12} \vee \overline{X}_{32} \\ \overline{X}_{22} \vee \overline{X}_{32} \end{gathered}$$

A graphic way of illustrating the computational procedure is a *resolution tree.* Figure 5 illustrates this.

At each level of the tree two disjunctive clauses are resolved and the resolvent written directly below. It can be seen that ultimately $\overline{X}_{31}$ and $X_{31}$ are resolved to produce a contradiction, demonstrating the original premises to be inconsistent.

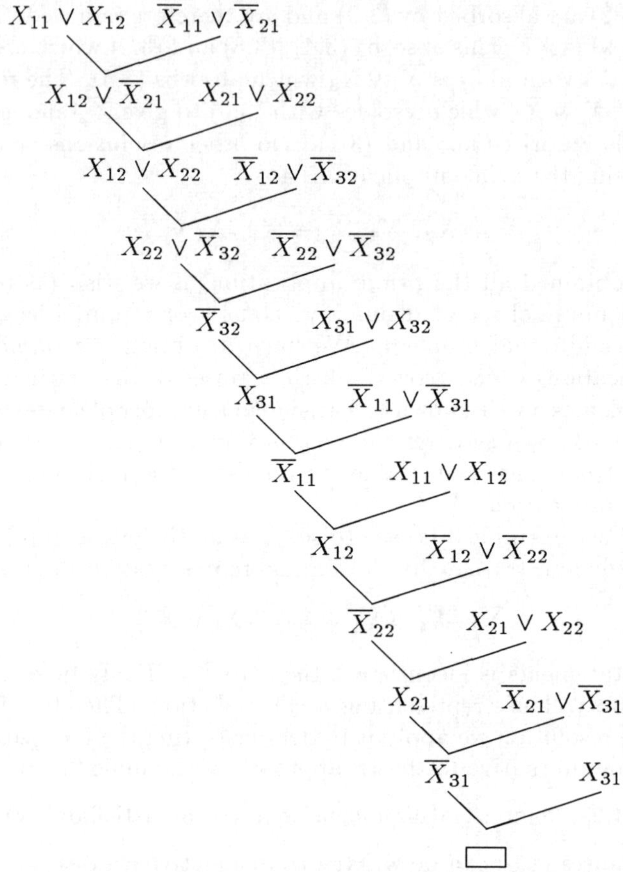


FIG. 5. A resolution tree

It must be emphasized again that resolution is a complete *refutation* procedure. If a set of statements is consistent (satisfiable) then other procedures are needed to find a satisfying set of truth values. Resolution is an inefficient method unless the problem has a special structure such as that discussed in Section 4. The above example demonstrates that even for a small problem resolution can take a large number of steps.

### 3.2 Resolution and cutting planes

It can be seen that the resolution process as described above is equivalent to the procedure for generating cutting planes in integer programming (see for example Williams [29], Hooker [12]). It can be shown that all such resolvents can be obtained by Chvátal's method [4]. For example, the two clauses

$$\begin{array}{ccccc} X & \vee & Y & \vee & Z, \\ X & & & \vee & \overline{Z} \end{array}$$

can be converted into IP constraints by defining variables $x$, $y$ and $z$ such that (for example) $x = 1$ if and only if $X$ is true, and $x = 0$ if and only if $X$ is false. We replace $X$ by $x$, $\overline{X}$ by $(1 - x)$, $\vee$ by $+$, and say that the resulting sum must be greater than one since at least one of the corresponding literals must be true. Thus the above two clauses are equivalent to the constraints

$$\begin{array}{ccccccc} x & + & y & + & z & \geq & 1\,, \\ x & & & - & z & \geq & 0\,. \end{array}$$

If we include the non-negativity constraint $y \geq 0$ and add these three constraints, we obtain

$$2x + 2y \geq 1.$$

Dividing through by 2 and rounding up the right-hand side, since all the variables are integer, we obtain

$$x + y \geq 1,$$

which is equivalent to the clause

$$X \vee Y,$$

which is the resolvent of the two original clauses. This extra constraint is known as a *cut*; it is satisfied by all feasible integer solutions of the original set of constraints, and may exclude some non-integer solutions if we are lucky. Hooker [12] gives conditions for clauses to be parents of *separating* resolution cuts, that is, cuts which lead to a strictly tighter LP relaxation. Since resolution is complete, resolution cuts are complete for IPs in clausal form.

### 3.3 Generalized resolution

The method of resolution described above applies to clauses which assert that at least one of their literals is true. A set of clauses will usually be converted to conjunctive form (CF) in order to use this method. Tseitin [23] gives a polynomial-time algorithm for converting a set of clauses to CF, and Blair, Jeroslow and Lowe [3] extended this to give a linear-time algorithm. However in the worst case resolution requires an exponential number of steps to obtain a refutation. For example, Haken [10] showed that this is the case for the pigeon-hole problem, which was described in Section 2 (Example 2.4).

However there is an extension of resolution known as *generalized resolution*, which applies to clauses in which we may require that more than one literal be true. Generalized resolution was first considered by Wilson [33] and was later developed more rigorously by Hooker [11]. For example, we may require that at least three of the literals in the clause

$$X_1 \vee X_2 \vee X_3 \vee X_4 \vee X_5$$

are true; in the notation of Section 2.3, we can write this as

$$\mathbf{ge}(3 : X_1, X_2, X_3, X_4, X_5).$$

This is equivalent to the constraint

$$x_1 + x_2 + x_3 + x_4 + x_5 \geq 3.$$

In general, a clause in which at least $\beta$ literals must be true is known as a $\beta$-clause: we say that the clause has degree $\beta$. Generalized resolution consists of repeated application of two types of resolution: *cancellation* resolution and *circulant* resolution. A cancellation resolvent of a $\beta_1$-clause $A$ and a $\beta_2$-clause $B$ is a clause obtained by reducing $A$ and $B$ to clauses of the same degree, by deleting some literals if necessary, and then applying ordinary resolution. For example, suppose we have a 2-clause

$$\mathbf{ge}(2 : X_1, \overline{X}_2, X_3, X_5) \tag{3.11}$$

and a 3-clause

$$\mathbf{ge}(3 : X_1, X_2, \overline{X}_3, X_4, X_5). \tag{3.12}$$

We can reduce (3.12) to a clause of degree 2 by deleting (say) the literal $X_1$. We are effectively saying that at least two of the literals in this new clause must be true. This then gives us a 2-clause

$$\mathbf{ge}(2 : X_2, \overline{X}_3, X_4, X_5). \tag{3.13}$$

We now resolve (3.11) and (3.13), which are both of degree 2, on (say) the literal $X_2$. This gives us a new 2-clause

$$\mathbf{ge}(2 : X_1, X_3, \overline{X}_3, X_4, X_5). \tag{3.14}$$

After removing the tautology $X_3 \vee \overline{X}_3$, this reduces to the 1-clause

$$\mathbf{ge}(1 : X_1, X_4, X_5). \tag{3.15}$$

(3.15) is a cancellation resolvent of (3.11) and (3.12); there are clearly others. It can be shown that cancellation resolvents can all be obtained as Chvátal cuts if the clauses are converted to constraints as described above.

Circulant resolution is more complex and is described rigorously by Hooker [11]. A set of clauses which exhibit a circulant pattern, such as (for example) the 1-clauses

$$\begin{array}{ccccc} X_1 & \vee & X_2 & & \\ X_1 & & & \vee & X_3 \\ & & X_2 & \vee & X_3 \end{array}$$

have as a circulant resolvent the 2-clause $X_1 \vee X_2 \vee X_3$. Circulant resolution also provides a method of generating Chvátal cuts: the above clause can easily be seen to be a Chvátal cut.

Before we describe the generalized resolution procedure, we first need to define the concept of *domination*. The definition of absorption given in Section 3.1 can be extended to $\beta$-clauses. We say that a $\beta_1$-clause $A$ *absorbs* a $\beta_2$-clause $B$ if $\beta_1 \geq \beta_2$ and $A$ can be obtained from $B$ by deleting zero or more literals from $B$.

We say that $B$ is a *reduction* of $A$ if $\beta_1 \geq \beta_2$ and $B$ can be obtained by deleting exactly $\beta_1 - \beta_2$ literals from $A$. For example, the 3-clause

$$\mathbf{ge}(3 : X_1, X_2, X_3, X_4)$$

absorbs the 2-clause

$$\mathbf{ge}(2 : X_1, X_2, X_3, X_4, X_5)$$

and the 2-clause

$$\mathbf{ge}(2 : X_1, X_2, X_3, X_4)$$

is a reduction of the 3-clause

$$\mathbf{ge}(3 : X_1, X_2, X_3, X_4, X_5).$$

The clause $A$ is said to *dominate* the clause $B$ if and only if some reduction of $A$ absorbs $B$.

In order to show that a set $\mathcal{S}$ of $\beta$-clauses (of various degrees) logically implies a $\beta$-clause $C$, we apply the generalized resolution procedure as follows.

(1) If some clause in $\mathcal{S}$ dominates $C$, stop: the original clauses imply $C$.

(2) If possible, find a pair of clauses in $\mathcal{S}$ with cancellation resolvent(s) which are not already dominated by clauses in $\mathcal{S}$. Delete from $\mathcal{S}$ all clauses dominated by these resolvents, and add the resolvents to $\mathcal{S}$.

(3) If possible, find a set of clauses in $\mathcal{S}$ with a circulant resolvent which is not already dominated by a clause in $\mathcal{S}$. Delete from $\mathcal{S}$ all clauses dominated by this resolvent, and add the resolvent to $\mathcal{S}$.

(4) If no clauses were added to $\mathcal{S}$ in steps 2 or 3, stop: the original clauses do not imply $C$. Otherwise, go to step 1.

Hooker [11] remarks that although generalized resolution cuts are complex to generate, they have strong separation conditions. Tseitin [23] extended resolution in another way by generating *extra variables.* He then showed that, for example, the pigeon-hole problem can be solved in polynomial time by this method.

### 3.4 The Davis–Putnam procedure

This is a well-known procedure for testing and finding the satisfiability of a set of clauses in the propositional calculus (Davis and Putnam [5]). Given a set $\mathcal{S}$ of clauses, we wish to know if there exists an assignment of truth values to the literals in these clauses such that $\mathcal{S}$ is satisfiable. The Davis–Putnam procedure assumes that $\mathcal{S}$ is already in CF, and involves successively replacing the set $\mathcal{S}$ (and its successors) by other sets: the rules given below are repeated until either

(1) all non-replaced sets are *explicitly contradictory*, i.e. contain unit clauses $X$ and $\overline{X}$ for some literal $X$, or

(2) the empty clause set is obtained.

If (1), $\mathcal{S}$ is unsatisfiable; if (2), $\mathcal{S}$ is satisfiable.

The following rules are applied successively for each set of clauses. Initially, we just have the original set $\mathcal{S}$.

(1) **One-literal clause rule:** If the set contains a one-literal clause $X$, delete from the set all clauses containing $X$ (including $X$ itself) and remove each occurrence of $\overline{X}$ from all the other clauses.

(2) **Affirmative-negative rule:** If $X$ appears in some clause but no occurrence of $\overline{X}$ is present in any other clause, delete all clauses containing $X$. Similarly, if $\overline{X}$ is present in some clause but $X$ does not appear in any clause, delete all clauses containing $\overline{X}$.

(3) **Splitting rule:** This was suggested by Loveland [16] as an extension to the original Davis–Putnam procedure, which uses resolution at this point. If Rules 1 and 2 do not apply, choose a literal $X$ such that $X$ and $\overline{X}$ both occur in the clause set, and replace the set by the following *two sets* of clauses:

   (a) the set with all clauses containing $X$ deleted and all occurrences of $\overline{X}$ removed;
   
   (b) the set with all clauses containing $\overline{X}$ deleted and all occurrences of $X$ removed.

For example, consider the clause set

$$\overline{X}_2 \vee X_3 \tag{3.16}$$
$$\overline{X}_1 \vee \overline{X}_3 \vee X_4 \tag{3.17}$$
$$X_2 \vee X_3 \vee \overline{X}_4 \tag{3.18}$$
$$X_1 \tag{3.19}$$

By Rule 1 we can eliminate the unit clause (3.19) and remove the literal $\overline{X}_1$ from (3.17). This leaves us with

$$\overline{X}_2 \vee X_3$$
$$\overline{X}_3 \vee X_4$$
$$X_2 \vee X_3 \vee \overline{X}_4$$

Rule 2 does not apply so we now apply Rule 3, for the sake of argument to the literal $X_3$. We thus obtain two sets of clauses:

**Set A.** Clauses containing $X_3$ deleted and the literal $\overline{X}_3$ removed:

$$X_4$$

**Set B.** Clauses containing $\overline{X}_3$ deleted and the literal $X_3$ removed:

$$\overline{X}_2$$
$$X_2 \vee \overline{X}_4$$

We now apply Rule 1 again to set A and delete the remaining unit clause: we are left with the empty clause set and therefore the original set is satisfiable.

The splitting rule closely parallels the use of the branch and bound approach for IP problems. It is effectively a branching procedure, where at each node one branch corresponds to fixing $x = 1$ and the other branch corresponds to fixing $x = 0$. Blair, Jeroslow and Lowe [3] develop techniques for solving satisfiability problems as integer programs using Loveland's [16] modification of the Davis–Putnam procedure, and Jeroslow and Wang [15] further extend this by using a variable-fixing heuristic to replace the LP in the branch and bound, which solves some satisfiability problems very quickly.

In order to demonstrate the Davis–Putnam procedure we apply it to Example 1.1 as an alternative to the IP approach of Example 2.3.

**Example 3.3** Deciding validity by the Davis–Putnam procedure

The premises (1.1)–(1.3) can be written as disjunctive clauses

$$\overline{X}_1 \vee X_2 \vee X_3 \tag{3.20}$$
$$X_1 \vee X_2 \tag{3.21}$$
$$X_2 \vee X_4 \tag{3.22}$$
$$X_2 \vee X_3 \vee \overline{X}_4 \tag{3.23}$$

In order to demonstrate validity we would have to show that these clauses, together with the negated conclusion

$$\overline{X}_2 \tag{3.24}$$

are not satisfiable.

We can apply Rule 1 above to delete $X_2$ from (3.20) to (3.23), leaving

$$\overline{X}_1 \vee X_3 \tag{3.25}$$
$$X_1 \tag{3.26}$$
$$X_4 \tag{3.27}$$
$$X_3 \vee \overline{X}_4 \tag{3.28}$$

This rule can twice be applied again to delete $\overline{X}_1$ from (3.25) and then $\overline{X}_4$ from (3.28), to leave the statement $X_3$ which is clearly satisfiable, so demonstrating the invalidity of the original argument. A satisfiable set of truth values is obtained by setting $X_3$ as true and $X_2$ as false. In order to satisfy clauses (3.21) and (3.22) we must have $X_1$ and $X_4$ true. Hence it is possible that Smith is not the murderer for the reasons already given in Example 2.3.

## 4 Horn clause systems

There is an important class of logical problems for which both the logical methods of resolution and the Davis–Putnam procedure, as well as the IP approach, take a particularly simple form. This is when all the disjunctive clauses are said to be "*Horn*". A Horn clause is a disjunctive clause in which at most one literal is non-negated, e.g.

$$X_1 \vee \overline{X}_2 \vee \overline{X}_3$$

is Horn, as is

$$\overline{X}_1 \vee \overline{X}_2.$$

Certain computational logic systems (e.g. PROLOG) demand that all the clauses in the input be Horn.

None of the logical problems considered so far has been Horn. However it is sometimes possible to convert a system into Horn clauses by negating some of the variables. Such systems are said to be "*essentially Horn*". The system exhibited in Examples 1.1, 2.3 and 3.3 is essentially Horn since it becomes Horn if we negate $X_2$ (and so regard $\overline{X}_2$ as one of the original atomic propositions). However the system exhibited in Examples 1.2, 2.4 and 3.2 cannot be made Horn.

If all the clauses in a set are Horn (or essentially Horn) then input resolution can be applied, i.e. one of the clauses in each resolution can always be one of the originals. Hooker [12], among others, gives a proof of this. We demonstrate it by a modification of Example 1.1 which involves appending an extra premise.

**Example 4.1** Applying input resolution to a Horn system

The premises are

$$\left.\begin{array}{llll} \overline{X}_1 & \vee\ X_2 & \vee\ X_3 & \\ X_1 & \vee\ X_2 & & \\ & \phantom{\vee\ }X_2 & & \vee\ X_4 \\ & \phantom{\vee\ }X_2 & \vee\ X_3 & \vee\ \overline{X}_4 \\ \overline{X}_1 & & \vee\ \overline{X}_3 & \vee\ \overline{X}_4 \end{array}\right\} \tag{4.1}$$

We will demonstrate that (with the extra premise) it is valid to deduce $X_2$. This system is essentially Horn since it becomes Horn if $X_2$ is negated.

The resolution tree is shown in Fig. 6. Notice that in each step of the resolution one of the original (*input*) clauses is used. This is in contrast to the tree shown in Fig. 5 where (for example) at one stage it was necessary to resolve the two non-input clauses $\overline{X}_{31}$ and $X_{31}$ for that non-Horn system. Input resolution allows for considerable computational economy since it is only necessary to maintain a list of the input clauses and the current clauses. The result is a polynomial algorithm.

It has already been pointed out in Section 3 that the resolution of two clauses has a counterpart in adding two inequalities and possibly rounding the right-hand side coefficient after dividing the coefficients by a common factor. If input resolution is used this rounding operation can be postponed to the end. We demonstrate the process by an analogous tree to that in Fig. 6, given in Fig. 7. Where necessary, the constraints $x_i \geq 0$ or $-x_i \geq -1$ have been added to give the coefficients a common factor.

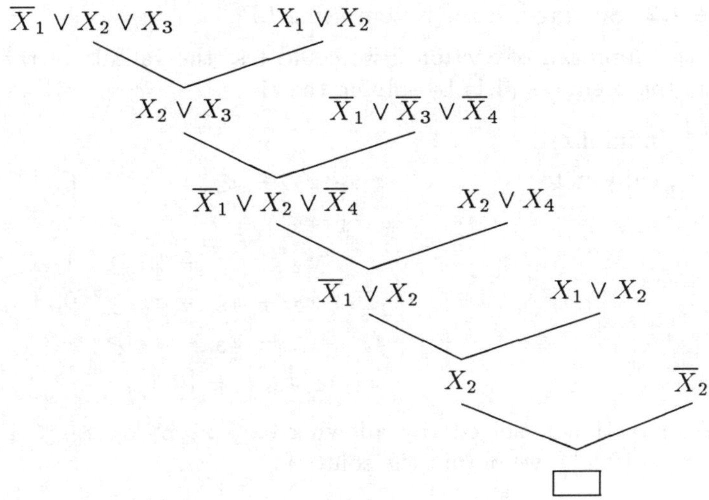


FIG. 6. An input resolution tree

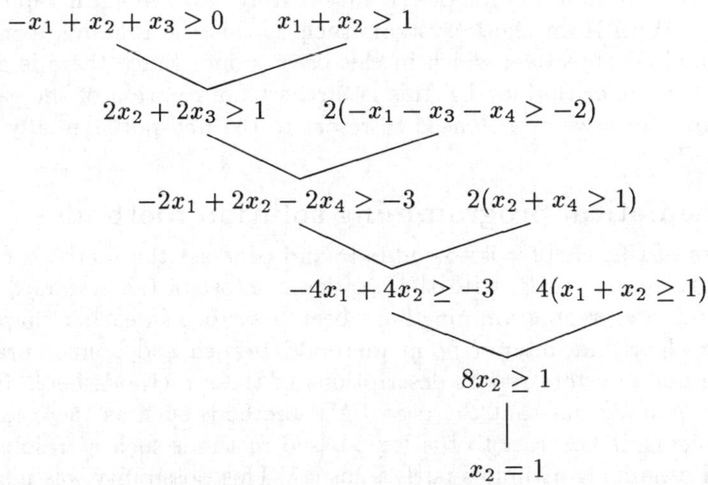


FIG. 7. Input resolution with inequalities

Instead of rounding after each resolution (e.g. after the first resolution we do not divide $2x_2 + 2x_3 \geq 1$ through by 2 and round up), we postpone the rounding to the end, using $8x_2 \geq 1$ to deduce $x_2 \geq 1/8$ and therefore $x_2 = 1$.

This property of Horn systems has an important implication for the IP approach to solving such a problem. In order to demonstrate this, we re-solve the example through an IP (LP) model.

**Example 4.2** Solving a Horn system using LP

Using the approach of Section 2 we could test the validity of the deduction of $X_2$ from the premises (4.1) by solving the IP

$$\begin{array}{llr} \text{minimize} & x_2, & \\ \text{subject to} & -x_1 + x_2 + x_3 & \geq 0, \\ & x_1 + x_2 & \geq 1, \\ & x_2 + x_4 & \geq 1, \\ & x_2 + x_3 - x_4 & \geq 0, \\ & -x_1 - x_3 - x_4 & \geq -2, \\ & x_1, x_2, x_3, x_4 \in \{0,1\}. & \end{array}$$

If the LP relaxation is solved (by allowing $0 \leq x_1, x_2, x_3, x_4 \leq 1$ instead of $x_1, x_2, x_3, x_4 \in \{0,1\}$), we obtain the solution

$$x_1 = \frac{3}{4},\ x_2 = \frac{1}{4},\ x_3 = \frac{1}{2},\ x_4 = \frac{3}{4}.$$

Since the objective function for the IP must always take integral values we can round it up. With Horn clause systems such as this the rounding corresponds to the optimal IP objective, which in this case is one. Since there is a polynomially bounded algorithm for LP this is further confirmation of the easiness of Horn systems, since we do not need to resort to the non-polynomially bounded methods of IP.

## 5 Mathematical programming solution methods

The purpose of this chapter is to compare and contrast the methods of mathematical programming (MP) with those of logic. Most of the standard methods for linear and integer programming have been described in earlier chapters, e.g. the simplex algorithm, interior point methods, branch and bound, branch and cut. We do not therefore repeat descriptions of these methods here. It should, however, be pointed out that the use of MP methods such as these for solving logical problems, in contrast to the logic-based methods such as resolution and the Davis–Putnam algorithm, is fairly unusual. This possibility was pointed out by Williams [26]. Blair, Jeroslow and Lowe [3], Hooker [12] and Williams [29], as well as others, also discuss this approach together with examples.

Some of these methods can be specialized to deal with the particular structures which arise in logic. Moreover, some of the methods of logic overlap with those of MP. For example, it has already been pointed out that the Davis–Putnam procedure with splitting (due to Loveland [16]) is a branch and bound method although it does not rely on LP relaxations. The cutting plane approach to solving IPs originated with Gomory [8]. It is now recognized that all cuts are special cases of Chvátal cuts [4]. The generality of Chvátal cuts demands that they must be specialized to deal with specific problems. As explained in Section 3, resolution when interpreted through inequalities can be regarded as one such specialization, which is particularly powerful for logic problems.

One approach to solving LP models which has not been discussed in other chapters is Fourier–Motzkin elimination. Although impractical as a general LP method it is particularly illustrative and valuable for some logic problems. The method is described in detail in Williams [28]. Essentially the method works by successively eliminating variables from an LP between all pairs of (like) constraints in which they have different signs. For logic problems involving only Horn clauses it has already been observed that the LP relaxation, together with one rounding, suffices to prove validity. We again use Example 4.2 to illustrate this.

**Example 5.1** Fourier–Motzkin elimination applied to an (essentially) Horn clause system

$$
\begin{array}{llr}
\text{maximize} & x_2, & (5.1)\\
\text{subject to} & -x_1 + x_2 + x_3 \geq 0, & (5.2)\\
 & x_1 + x_2 \geq 1, & (5.3)\\
 & x_2 + x_4 \geq 1, & (5.4)\\
 & x_2 + x_3 - x_4 \geq 0, & (5.5)\\
 & -x_1 - x_3 - x_4 \geq -2, & (5.6)\\
 & -x_1 \geq -1, & (5.7)\\
 & -x_2 \geq -1, & (5.8)\\
 & -x_3 \geq -1, & (5.9)\\
 & -x_4 \geq -1, & (5.10)\\
 & x_1 \geq 0, & (5.11)\\
 & x_2 \geq 0, & (5.12)\\
 & x_3 \geq 0, & (5.13)\\
 & x_4 \geq 0. & (5.14)
\end{array}
$$

Fourier–Motzkin elimination works by successively eliminating variables between *every* pair of inequalities in which they have opposite signs. Hence we eliminate $x_1$ between (5.2) and (5.3), (5.2) and (5.11), (5.6) and (5.3), (5.6) and (5.11), (5.7) and (5.3), and (5.7) and (5.11), which together with the inequalities not involving $x_1$ reduces the problem to

$$
\begin{array}{ll}
\text{maximize} & x_2,\\
\text{subject to} & 2x_2 + x_3 \geq 1,\\
 & x_2 + x_3 \geq 0,\\
 & x_2 - x_3 - x_4 \geq -1,\\
 & -x_3 - x_4 \geq -2,\\
 & x_2 \geq 0,
\end{array}
$$

$$\begin{aligned}
0 &\geq -1,\\
x_2 \qquad + x_4 &\geq 1,\\
x_2 + x_3 - x_4 &\geq -1,\\
-x_2 &\geq -1,\\
-x_3 &\geq -1,\\
-x_4 &\geq -1,\\
x_3 &\geq 0,\\
x_4 &\geq 0.
\end{aligned}$$

Ignoring the redundant inequality ($0 \geq -1$) and eliminating $x_3$ reduces the problem (after eliminating obviously redundant inequalities) to

$$\begin{aligned}
\text{maximize} \quad & x_2,\\
\text{subject to} \quad & 3x_2 - x_4 \geq 0,\\
& 2x_2 - x_4 \geq -1,\\
& 2x_2 - 2x_4 \geq -1,\\
& x_2 - 2x_4 \geq -2,\\
& x_2 - x_4 \geq -1,\\
& x_2 + x_4 \geq 1,\\
& x_2 \geq 0,\\
& -x_4 \geq -1,\\
& x_4 \geq 0.
\end{aligned}$$

$x_4$ is eliminated reducing the system (after eliminating redundant inequalities) to

$$\begin{aligned}
\text{maximize} \quad & x_2,\\
\text{subject to} \quad & 4x_2 \geq 1,
\end{aligned}$$

demonstrating $x_2 \geq 1/4$. Since $x_2$ is integer we can conclude that $x_2 = 1$ is the optimal objective value, and therefore $X_2$ must be true if the premises hold.

For integer programming Fourier–Motzkin elimination is, in general, inadequate since we have integrality conditions to observe as well as inequalities. Williams [25] gives a way of extending the method to IP models at great computational expense by the introduction of congruence relations and disjunctions. If, however, one of the coefficients of a variable to be eliminated is $\pm 1$ in each $(+, -)$ pair then Williams [30] observes it is still valid to apply the LP form of Fourier–Motzkin elimination. The simplification described at the end of Example 2.8 illustrates this. For Horn clause systems, such as that above, this will always be the case. By our eliminations we will not therefore have lost any of the "strength" of the original system. The only rounding takes place at the end.

Should we wish to find all the integer solutions corresponding to the optimal objective value obtained after the rounding we can backtrack.

In general, however, for non-Horn systems we need more sophisticated methods.

## 6 Further considerations

In this chapter we have demonstrated the close connections between computational logic and integer programming. We have not, however, answered definitively the question "Which methods are best?" It is very difficult to do this since the methods are still developing. MP methods have reached a high degree of sophistication in their implementation and use LP relaxations to good effect. They are, however, of very general applicability. Logic methods can be exploited to good effect on specialist problems. In practice what seems best is a blend of both approaches enhanced by the greater understanding which each brings to the other. Ultimately specialist, very efficient, methods will probably result.

We mention here some further considerations which lie beyond the scope of this chapter.

(1) A theoretical problem in the foundations of mathematics is finding "decidable fragments of arithmetic". Full arithmetic is not decidable, by the celebrated theorem of Gödel [7]. It is shown in Williams [28] that both LP and IP can be formalized in such decidable fragments and the methods used to demonstrate decidability constructively used to produce "logic algorithms" for LP and IP.

(2) The method of resolution applied to disjunctive clauses is predated by the equivalent methods of Quine [20] and [21]. Quine's methods apply to clauses in conjunctive form, but are equivalent in a logically "dual" sense. Williams [30] and Hooker [12] point this out. Quine proves completeness.

(3) An alternative approach to solving Horn systems has been developed by Gallo and Urbani [6] using directed hypergraphs. Each clause is then represented by a hyperedge for which the literals are the nodes, with the un-negated literal (if present) representing the "head" of the edge.

(4) This chapter has concentrated on the propositional calculus. Many computational logic problems are cast in terms of the predicate calculus. It is then necessary to "instantiate" the arguments within the predicates before the methods of the propositional calculus can be applied. Jeroslow [14] has however succeeded in applying the methods of IP to a limited class of problems within the predicate calculus.

## Bibliography

1. E. Balas. Intersection cuts from disjunctive constraints. Management Science Research Report 330, Graduate School of Industrial Administration, Carnegie Mellon University, 1974.

2. E. Balas. Facets of the knapsack polytope. *Mathematical Programming*, 8:146–164, 1975.

3. C.E. Blair, R.G. Jeroslow and J.K. Lowe. Some results and experiments in programming techniques for propositional logic. *Computers and Operations Research*, 13:633–645, 1988.
4. V. Chvátal. Edmonds polytopes and a hierarchy of combinatorial problems. *Discrete Mathematics*, 4:305–337, 1973.
5. M. Davis and H. Putnam. A computing procedure for quantification theory. *Journal of the ACM*, 7:201–215, 1960.
6. G. Gallo and G. Urbani. Algorithms for testing the satisfiability of propositional formulae. *Journal of Logic Programming*, 7:45–61, 1989.
7. K. Gödel. *On Undecidable Propositions of Formal Mathematical Systems.* Princeton University Press, Princeton, NY, 1934.
8. R.E. Gomory. Outline of an algorithm for integer solutions to linear programs. *Bulletin of the American Mathematical Society*, 64:275–278, 1958.
9. F. Granot and P.L. Hammer. On the use of boolean functions in 0-1 programming. *Methods of Operations Research*, 12:154–184, 1972.
10. A. Haken. The intractability of resolution. *Theory of Computer Science*, 39:297–308, 1985.
11. J.N. Hooker. Generalized resolution and cutting planes. *Annals of Operations Research*, 12:217–239, 1986.
12. J.N. Hooker. A quantitative approach to logical inference. *Decision Support Systems*, 4:45–69, 1988.
13. J.N. Hooker. Generalized resolution for 0-1 linear inequalities. *Annals of Mathematics and Artificial Intelligence*, 6:271–286, 1992.
14. R.G. Jeroslow. Logic based decision support. *Annals of Discrete Mathematics*, 40, 1989.
15. R.G. Jeroslow and J. Wang. Solving propositional satisfiability problems. *Annals of Mathematics and Artificial Intelligence*, 1:167–187, 1990.
16. D.W. Loveland. *Automated Theorem Proving: A Logical Basis.* North-Holland, Amsterdam, 1978.
17. K.I.M. McKinnon and H.P. Williams. Constructing integer programming models by the predicate calculus. *Annals of Operations Research*, 21:227–246, 1989.
18. E. Mendelson. *Introduction to Mathematical Logic.* Van Nostrand, 1964.
19. R.R. Meyer. Integer and mixed-integer programming models: general properties. *Journal of Optimization Theory and its Applications*, 16:191–206, 1975.
20. W.V. Quine. The problem of simplifying truth functions. *American Mathematical Monthly*, 59:521–531, 1952.
21. W.V. Quine. A way to simplify truth functions. *American Mathematical Monthly*, 62:627–631, 1955.
22. J.A. Robinson. A machine-oriented logic based on the resolution principle.

*Journal of the ACM*, 12:23–41, 1965.

23. G.S. Tseitin. On the complexity of derivation in the propositional calculus. In A.O. Slisenko, (ed.), *Structures in Constructive Mathematics and Mathematical Logic, part II*, 115–125. Plenum Press (translated from Russian), 1968.
24. H.P. Williams. The synthesis of logical nets consisting of NOR units. *The Computer Journal*, 11:173–176, 1968.
25. H.P. Williams. Fourier–Motzkin elimination extension to integer programming problems. *Journal of Combinatorial Theory*, 21:118–123, 1976.
26. H.P. Williams. Logical problems and integer programming. *Bulletin of the IMA*, 13:18–20, 1977.
27. H.P. Williams. *Model Building in Mathematical Programming.* Wiley, first edition, 1978.
28. H.P. Williams. Fourier's method of linear programming and its dual. *American Mathematical Monthly*, 93:681–695, 1986.
29. H.P. Williams. Linear and integer programming applied to the propositional calculus. *International Journal of Systems Research and Information Science*, 2:81–100, 1987.
30. H.P. Williams. The elimination of integer variables. *Journal of the Operational Research Society*, 43:387–393, 1992.
31. H.P. Williams. An alternative explanation of disjunctive formulations. *European Journal of Operational Research*, 72:200–203, 1994.
32. H.P. Williams. Logic applied to integer programming and integer programming applied to logic. *European Journal of Operational Research*, 81:605–616, 1995.
33. J.M. Wilson. A method of 0-1 integer programming using a boolean approach. In T.B. Boffey, (ed.), *Proceedings of CP77.* University of Liverpool, 1977.
34. L.A. Wolsey. Faces for a linear inequality in 0-1 variables. *Mathematical Programming*, 8:165–178, 1975.

# Index